T0184855

Matrix-Tensor Methods
in
Continuum Mechanics

Matrix-Tensor Methods
— in —
Continuum Mechanics

Second Edition

S. F. BORG

Professor Emeritus of Civil Engineering
Stevens Institute of Technology
Hoboken, NJ 07030, USA

World Scientific
Singapore • New Jersey • London • Hong Kong

Published by

World Scientific Publishing Co. Pte. Ltd.
P O Box 128, Farrer Road, Singapore 9128
USA office: 687 Hartwell Street, Teaneck, NJ 07666
UK office: 73 Lynton Mead, Totteridge, London N20 8DH

Library of Congress Cataloging-in-Publication data is available.

First printed in 1963 by D. Van Nostrand Company, Inc.

MATRIX TENSOR METHODS IN CONTINUUM MECHANICS

Copyright © 1990 by World Scientific Publishing Co. Pte. Ltd.

All rights reserved. This book, or parts thereof, may not be reproduced in any form or by any means, electronic or mechanical, including photocopying, recording or any information storage and retrieval system now known or to be invented, without written permission from the Publisher.

ISBN 981-02-0166-4
        981-02-0167-2 pbk

Printed in Singapore by JBW Printers & Binders Pte. Ltd.

IN MEMORY OF AUDREY

PREFACE TO THE FIRST EDITION

This book is based on a preprint edition, *An Introduction to Matrix-Tensor Methods in Theoretical and Applied Mechanics*, which was issued on a more or less interim basis and has been used by the author in at least six of his graduate courses. The present book represents a complete rewriting and bringing up to date of the earlier work, in the light of classroom experience. The purposes of the text are:

1. To introduce the engineer to the very important (and increasingly important) discipline in applied mathematics—tensor methods. Because the author's classroom experience has convinced him that the engineer can follow tensor theory most easily when it is presented in matrix form, this has been the method used in the text.

2. To show the fundamental unity of the different fields in continuum mechanics—with the unifying material formed by the matrix-tensor theory. Too often the student loses sight of the real connections between fields that we have artificially decompartmentalized. A truer understanding of the important and basic segment of engineering—mechanics of continua—can be obtained, the author feels, when the various portions of this field are presented as part of the complete fabric.

3. To present to the engineer modern engineering problems. For this reason, mathematical arguments have been kept to a minimum or avoided entirely where they would tend to add little to the physical understanding of the phenomenon being discussed. However, it should also be emphasized that the book is not to be thought of as nonmathematical. It requires of the student an understanding of differential and integral calculus, vector analysis, complex variable theory, mathematical analysis, and related topics usually considered to be the equipment of the graduate engineer or scientist. The fundamentals of matrix and tensor theory are covered in a form sufficient for the purposes of the text—and for additional advanced use as well.

In the first chapter the fundamentals of matrix algebra and calculus are presented, as well as a brief review of vector analysis and the introductory complex variable theory. This coverage, together with the current undergraduate mathematical training of engineers, should be sufficient preparation.

Chapter 2 presents the elements of tensor theory. The fundamental nature of the tensor is emphasized: the requirement that it behave in a certain manner under a transformation (rotation) of axes about the origin. The connection between the tensor and the matrix is brought out, and the groundwork in matrix-tensor analysis is laid.

Curvilinear coordinates, one of the most useful and important topics in applied mathematics to the engineer, is discussed in Chapter 3. The entire development, presented in matrix-tensor form, leads to expressions which permit one to put all of the equations of mathematical physics in any orthogonal curvilinear form whatever.

The remaining chapters indicate the applications of the theory to continuum mechanics—to fluids and to solids. Chapters 4 and 5 give the theory and some applications in the mathematical theory of elasticity. The essential tensors are derived, and their position in the theory is described in detail. Chapter 6 presents a discussion of matrix-tensor methods as they occur in structural engineering. Chapter 7 presents the application of matrix-tensor methods to plate and shell theory; Chapter 8 considers viscous flow phenomena, Chapter 9, plasticity. In all cases, the arguments and theory are presented from the matrix-tensor point of view, and the similarities (as well as essential differences) between the various fields are constantly brought out.

Chapter 10 presents a subject that is based squarely upon the matrix-tensor theory and that crosses all the fields considered in the text (and others). A form of dimensional analysis is described that is based upon tensoral invariance arguments, enabling one to give, without derivation, the qualitative form of many of the equations of mathematical physics and hence engineering.

A list of references to the standard works in the various fields considered, and to other special reports and books mentioned, is supplied. At the end of each chapter is a problems section.

In the author's graduate courses, he found it possible to complete essentially the entire book in a single three-hour-a-week semester. A graduate course in engineering mathematics was a prerequisite for this course. In his senior elective course the author was able to complete Chapters 1 through 4 in a three-hour-a-week semester. As a senior course, it should be possible to present the entire text in two semesters.

The author is indebted to Professor Francis Murnaghan whose inspiring lectures at Johns Hopkins University first introduced him to applied matrix-tensor methods. Professor Murnaghan's textbooks have been referred to liberally for basic source material. Several of the treatments presented are those given by Dr. Murnaghan in his lectures. However, in the interests of engineering simplification, the author has taken some liberties in the form of presentation. If there are errors in this material as given here, the fault lies with the author.

S. F. BORG
1962

PREFACE TO THE SECOND EDITION

The second edition generally follows the format and mode of presentation of the first one. Several typographical errors have been corrected and a number of new topics or extensions of the original material have been included in an Appendix following Chapter 10 at the end of the book.

The author wishes to express his sincere thanks to his publishers, World Scientific Publishing Co. for reprinting the text and for their continual help and encouragement in seeing the task through to completion.

S. F. BORG
March 3, 1990

CONTENTS

Matrix-Tensor Methods
in
Continuum Mechanics

Chapter 1

MATHEMATICAL PRELIMINARIES

1-1 Introduction. In this chapter a brief treatment of matrix algebra is presented. In addition a discussion is given, in abbreviated form, of vector analysis and complex variable theory. The presentation of the topics in this chapter is utilitarian in form and, insofar as the vector analysis and complex variable portions are concerned, it is more in the nature of a review and refresher of the introductory phases of these subjects. A knowledge of the material presented in this and the next chapter will give an adequate mathematical background for the later portions of the text.

1-2 Definition of a Matrix. A rectangular array of m rows and n columns of numbers or other quantities is called a *matrix*. We designate this matrix with a capital letter, as A, and show it in its expanded form as

$$A = \begin{pmatrix} a_{11} & a_{12} & a_{13} \\ a_{21} & a_{22} & a_{23} \end{pmatrix} \tag{1-1}$$

In the above expression, a_{ij} represents an *element* of the matrix. Note particularly that the subscripts of the elements carry a position significance. That is, the first subscript represents the row position of the element and the second subscript represents the column position.

A matrix is *not* a determinant.[1] As a reminder of this, the enclosing

---

[1] More precisely:
1. A determinant is a *quantity* associated with a *square* array of n^2 elements. Thus,

$$\begin{vmatrix} a_{11} & a_{12} \\ a_{21} & a_{22} \end{vmatrix}$$

is also given by the quantity $a_{11}a_{22} - a_{12}a_{21}$.
2. A matrix need not be square. It is simply a set of $m \times n$ elements in an ordered array. Thus

$$\begin{pmatrix} a_{11} & a_{12} \\ a_{21} & a_{22} \end{pmatrix}$$

does not imply any particular operation need be performed on the elements a_{ij}.

Footnote continued on page 2.

1

bars are shown curved as against the ordinary usage of straight bars for the determinant.

The number of rows in a matrix need not necessarily be the same as the number of columns. If the number of rows does equal the number of columns then the matrix is a *square matrix*.

A matrix which consists of elements in a single row is sometimes called a *row matrix*. If the elements are in a single column it is sometimes called a *column matrix*. No particular distinction, in general, need be made between the one and the other.

The elements of a matrix may or may not have any physical significance. For example, the elements may be pure numbers, as

$$(6 \quad -3.2 \quad 7\tfrac{1}{2}) \tag{1-2}$$

or the elements may be components of a velocity vector, as

$$V = \begin{pmatrix} u \\ v \\ w \end{pmatrix} \tag{1-3}$$

Indeed, they could even be colors, as

$$(red \quad blue \quad green)$$

or animals, as

$$\begin{pmatrix} cat \\ dog \\ hare \end{pmatrix} \tag{1-4}$$

or they could be mixtures of any or all of the above.

No significance must be attached to the use of a row for the numbers and colors and a column form for the velocity and animals in the above matrices.

The elements of a matrix may also be complex quantities, chemical symbols, equations, or, in fact, any quantity whatever.

The *zero, or null, matrix*, 0, has all elements equal to zero. Thus, we have

$$0 = (0 \ 0 \ 0) \tag{1-5}$$

or

$$0 = \begin{pmatrix} 0 & 0 \\ 0 & 0 \end{pmatrix} \tag{1-6}$$

Footnote continued from page 1.

3. We may, however, *define* the determinant of a square matrix. This is, by definition, the quantity obtained by treating the elements of the matrix as elements of a determinant—the position of the elements being the same in both the matrix and the determinant.

We shall also in the following sections indicate various operations (arithmetic, algebraic, and other) that may, by definition, be performed on and by matrices.

or any similar arrangement. Note: the zero matrix may be either a row or column or square matrix, or a general matrix of rectangular form.

The *unit matrix* E_n is an n-by-n-square matrix whose diagonal elements (top left to bottom right) equal unity and whose off-diagonal elements equal zero. That is,

$$\text{in } E_n, \quad \begin{cases} a_{ij} = 1 & \text{if } i = j \\ a_{ij} = 0 & \text{if } i \neq j \end{cases} \tag{1-7}$$

and, as an example,

$$E_3 = \begin{pmatrix} 1 & 0 & 0 \\ 0 & 1 & 0 \\ 0 & 0 & 1 \end{pmatrix} \tag{1-8}$$

A square matrix is *symmetrical* if

$$a_{ij} = a_{ji} \tag{1-9}$$

An example of a symmetrical matrix is the following:

$$\begin{pmatrix} x & e^t & z^2 \\ e^t & yz & 3t \\ z^2 & 3t & 1 \end{pmatrix} \tag{1-10}$$

A square matrix is *antisymmetric* or *skew-symmetric* if

$$a_{ij} = -a_{ji} \tag{1-11}$$

An example of a skew-symmetric matrix is

$$\begin{pmatrix} 0 & -3t \\ 3t & 0 \end{pmatrix} \tag{1-12}$$

Note that in a skew-symmetric matrix the main diagonal (upper left to lower right) elements must be zero, for only then will $a_{ij} = -a_{ji}$ be true for these elements.

The *transpose* of a matrix A is shown as $A^\star$ and is obtained by interchanging the rows and columns of A. Thus, if

$$A = \begin{pmatrix} z & xe^t & 2-y \\ 4 & 3xy & 0 \end{pmatrix} \tag{1-13}$$

then

$$A^\star = \begin{pmatrix} z & 4 \\ xe^t & 3xy \\ 2-y & 0 \end{pmatrix} \tag{1-14}$$

The foregoing represents the basic definitions or nomenclature in matrix theory.

1-3 Matrix Arithmetic, Algebra, and Calculus. Up to this point we have defined, in some detail, exactly what a matrix is and we have discussed some special matrices. If matrices are to be useful in engineering or physical applications, then they must behave in certain set ways when subjected to particular conditions. In our work in engineering and science we are primarily concerned with *quantitative* relations, and therefore we shall be most interested in the behavior of matrices in arithmetical and related mathematical operations. Matrices will be of use to us if, and only if, the theory of matrices can be developed along logical mathematical lines.

The simplest mathematical operations are those of arithmetic—equality, addition, subtraction, multiplication, and division. We discuss these first.

Two matrices A and B are *equal* only if each has the same number of rows and the same number of columns and if corresponding elements are equal. Thus, given

$$A = \begin{pmatrix} a_{11} & a_{12} & a_{13} \\ a_{21} & a_{22} & a_{23} \end{pmatrix} \tag{1-15}$$

$$B = \begin{pmatrix} b_{11} & b_{12} & b_{13} \\ b_{21} & b_{22} & b_{23} \end{pmatrix} \tag{1-16}$$

then if

$$a_{ij} = b_{ij} \tag{1-17}$$

it follows

$$A = B \tag{1-18}$$

Thus, the simple algebraic equations

$$\left. \begin{aligned} a &= p + 2u \\ b &= q + 7v \\ c &= r + 1.6w \\ d &= s + 17x \end{aligned} \right\} \tag{1-19}$$

may be given in matrix form as

$$\begin{pmatrix} a & b \\ c & d \end{pmatrix} = \begin{pmatrix} p + 2u & q + 7v \\ r + 1.6w & s + 17x \end{pmatrix} \tag{1-20}$$

We define the *sum* of two matrices A and B only if A and B have the same number of rows and of columns. The sum $A + B$ is then a matrix C of the same number of rows as A (and B) and the same number of columns as A (and B) and with

$$c_{ij} = a_{ij} + b_{ij} \qquad (1\text{-}21)$$

For example, the algebraic equations

$$\left.\begin{aligned}
c_{11} &= a_{11} + b_{11} \\
c_{12} &= a_{12} + b_{12} \\
c_{21} &= a_{21} + b_{21} \\
c_{22} &= a_{22} + b_{22}
\end{aligned}\right\} \qquad (1\text{-}22)$$

are equivalent[2] to

$$\begin{pmatrix} c_{11} & c_{12} \\ c_{21} & c_{22} \end{pmatrix} = \begin{pmatrix} a_{11} & a_{12} \\ a_{21} & a_{22} \end{pmatrix} + \begin{pmatrix} b_{11} & b_{12} \\ b_{21} & b_{22} \end{pmatrix} \qquad (1\text{-}23)$$

It may be shown (the student should verify this and the following statement for typical matrices) that the sum of two matrices is *commutative*. That is,

$$A + B = B + A \qquad (1\text{-}24)$$

Also, it may be shown that the addition of matrices is *associative*. That is,

$$(A + B) + C = A + (B + C) \qquad (1\text{-}25)$$

The *difference* of two matrices A and B is defined similarly. Thus,

$$C = A - B \qquad (1\text{-}26)$$

with

$$c_{ij} = a_{ij} - b_{ij} \qquad (1\text{-}27)$$

We may define multiplication of a matrix A by a scalar k as follows: the elements of kA are given by ka_{ij}, so that, for example, if

$$A = \begin{pmatrix} a_{11} & a_{12} \\ a_{21} & a_{22} \end{pmatrix} \qquad (1\text{-}28)$$

then

$$kA = \begin{pmatrix} ka_{11} & ka_{12} \\ ka_{21} & ka_{22} \end{pmatrix} \qquad (1\text{-}29)$$

---

[2] Alternatively, this may be expressed in the following essentially equivalent form:

$$\begin{pmatrix} c_{11} \\ c_{12} \\ c_{21} \\ c_{22} \end{pmatrix} = \begin{pmatrix} a_{11} \\ a_{12} \\ a_{21} \\ a_{22} \end{pmatrix} + \begin{pmatrix} b_{11} \\ b_{12} \\ b_{21} \\ b_{22} \end{pmatrix}$$

in which the positional significance of the subscripts has been waived.

Note that this is consistent with the usual notation

$$kA = \underbrace{A + A + \cdots + A}_{k \text{ times}} \tag{1-30}$$

An important property of square matrices which follows directly from the law of addition and subtraction is the following:

Any square matrix may be given as the sum of a symmetrical and antisymmetrical matrix. For, if A is a square matrix, then obviously

$$A = \frac{A + A^\star}{2} + \frac{A - A^\star}{2} \tag{1-31}$$

The first term on the right is a symmetrical matrix and the second term is an antisymmetrical matrix. This may be verified for a 2×2 matrix as follows:

If

$$A = \begin{pmatrix} a_{11} & a_{12} \\ a_{21} & a_{22} \end{pmatrix} \tag{1-32}$$

then

$$A^\star = \begin{pmatrix} a_{11} & a_{21} \\ a_{12} & a_{22} \end{pmatrix} \tag{1-33}$$

so that

$$\frac{A + A^\star}{2} = \begin{pmatrix} a_{11} & \dfrac{a_{12} + a_{21}}{2} \\ \dfrac{a_{21} + a_{12}}{2} & a_{22} \end{pmatrix} \tag{1-34}$$

and

$$\frac{A - A^\star}{2} = \begin{pmatrix} 0 & \dfrac{a_{12} - a_{21}}{2} \\ \dfrac{a_{21} - a_{12}}{2} & 0 \end{pmatrix} \tag{1-35}$$

A very important operation in matrix arithmetic is the *product* of two matrices. The previous operations are not too different from the more familiar ones of elementary arithmetic. The product operation, however, is quite different.

The *product* of two matrices is obtained as follows: given two matrices A and B such that the number of rows in B equals the number of columns in A, then the product AB is given by C, in which the element c_{ij} is obtained by multiplying each element of the ith row of A by the corresponding element of the jth column of B and adding. For example,

or[3]

$$\left.\begin{aligned}
C &= AB \\[6pt]
\begin{pmatrix} c_{11} \\ c_{21} \end{pmatrix} &= \begin{pmatrix} a_{11} & a_{12} & a_{13} \\ a_{21} & a_{22} & a_{23} \end{pmatrix} \begin{pmatrix} b_{11} \\ b_{21} \\ b_{31} \end{pmatrix} \\[6pt]
\begin{pmatrix} c_{11} \\ c_{21} \end{pmatrix} &= \begin{pmatrix} a_{11}b_{11} + a_{12}b_{21} + a_{13}b_{31} \\ a_{21}b_{11} + a_{22}b_{21} + a_{23}b_{31} \end{pmatrix} \\[6pt]
c_{11} &= a_{11}b_{11} + a_{12}b_{21} + a_{13}b_{31} \\
c_{21} &= a_{21}b_{11} + a_{22}b_{21} + a_{23}b_{31}
\end{aligned}\right\} \quad (1\text{-}36)$$

Note, in the above product, C is a 2×1 matrix. In general, if

$$C = AB \qquad (1\text{-}37)$$

and if k is the number of rows in A and l is the number of columns in B, then the matrix C will have k rows and l columns.

It will be obvious from the above example that, in general,

$$AB \neq BA \qquad (1\text{-}38)$$

that is, the position of a matrix in a matrix multiplication is *not* immaterial.

The student should also note that

$$CE = C \qquad (1\text{-}39)$$

where C is any matrix, E is a unit matrix of same number of rows as C has columns. For example,

$$\begin{pmatrix} c_{11} & c_{12} & c_{13} \\ c_{21} & c_{22} & c_{23} \end{pmatrix} \begin{pmatrix} 1 & 0 & 0 \\ 0 & 1 & 0 \\ 0 & 0 & 1 \end{pmatrix} = \begin{pmatrix} c_{11} & c_{12} & c_{13} \\ c_{21} & c_{22} & c_{23} \end{pmatrix} \qquad (1\text{-}40)$$

Thus, E plays the same role in matrix multiplication that unity plays in algebraic multiplication.

---

[3] Equation 1-36 indicates the usefulness of the given definition of matrix product. The first line of Equation 1-36 is a compact expression for the two linear equations shown in the last two lines of Equation 1-36. In general, systems of linear equations can be shown very compactly by utilizing the definition of matrix product.

As another example of matrix multiplication, the student should satisfy himself that the set of algebraic equations

$$\left.\begin{array}{l} a = 2ex + 3gy \\ b = 2ev + 3gw \\ c = -tx + s^2y \\ d = -tv + s^2w \end{array}\right\} \tag{1-41}$$

is equivalent to the matrix equation

$$\begin{pmatrix} a & b \\ c & d \end{pmatrix} = \begin{pmatrix} 2e & 3g \\ -t & s^2 \end{pmatrix}\begin{pmatrix} x & v \\ y & w \end{pmatrix} \tag{1-42}$$

It may be shown (the student should verify this and the following statement using simple matrices) that the product of matrices is *associative*. That is,

$$(AB)C = A(BC) \tag{1-43}$$

Also, the product of matrices is *distributive*. That is,

$$A(B + C) = AB + AC \tag{1-44}$$

The following expression is the statement of the very important *transpose product rule*:

$$(AC)^\star = C^\star A^\star \tag{1-45}$$

The student should verify this for a simple case.

Division of matrices is a non-unique process and therefore must remain undefined and not part of the algebra of matrices.[4] To illustrate what is meant by this, consider the product

$$AB = C \tag{1-46}$$

where

$$A = \begin{pmatrix} a_{11} & a_{12} & a_{13} \\ a_{21} & a_{22} & a_{23} \\ a_{31} & a_{32} & a_{33} \end{pmatrix} \tag{1-47}$$

$$B = \begin{pmatrix} 2 \\ 6 \\ 1 \end{pmatrix} \tag{1-48}$$

---

[4] Although division, as noted herein, is an undefined operation, division corresponding to the relation $AA^{-1} = E_n$ or $A^{-1} = E_n/A$ in which A^{-1} is the "inverse of A", is defined as shown on p. 10. When it exists, A^{-1} is analogous to the reciprocal of a number A in arithmetic.

and

$$C = \begin{pmatrix} 7 \\ 9 \\ 8 \end{pmatrix} \tag{1-49}$$

Now if division of matrices was permissible, it should be possible to obtain the unique matrix A from the following operation:

$$A = \begin{pmatrix} a_{11} & a_{12} & a_{13} \\ a_{21} & a_{22} & a_{23} \\ a_{31} & a_{32} & a_{33} \end{pmatrix} = \frac{\begin{pmatrix} 7 \\ 9 \\ 8 \end{pmatrix}}{\begin{pmatrix} 2 \\ 6 \\ 1 \end{pmatrix}} \tag{1-50}$$

But, the direct expansion of the product

$$\begin{pmatrix} a_{11} & a_{12} & a_{13} \\ a_{21} & a_{22} & a_{23} \\ a_{31} & a_{32} & a_{33} \end{pmatrix} \begin{pmatrix} 2 \\ 6 \\ 1 \end{pmatrix} = \begin{pmatrix} 7 \\ 9 \\ 8 \end{pmatrix} \tag{1-51}$$

gives

$$\left. \begin{array}{l} 2a_{11} + 6a_{12} + a_{13} = 7 \\ 2a_{21} + 6a_{22} + a_{23} = 9 \\ 2a_{31} + 6a_{32} + a_{33} = 8 \end{array} \right\} \tag{1-52}$$

and in these three equations we have *nine* unknowns, a_{ij}, any six of which may in general be arbitrarily chosen so that we still satisfy the equation. Thus, there are, in effect, an infinite number of matrices, A, which will satisfy the relation

$$A = \frac{\begin{pmatrix} 7 \\ 9 \\ 8 \end{pmatrix}}{\begin{pmatrix} 2 \\ 6 \\ 1 \end{pmatrix}} \tag{1-53}$$

and hence division of matrices is non-unique and therefore not permissible.

The *cofactor matrix* of any square matrix A, with n rows and columns (denoted by Co A) is the matrix obtained by replacing each element of A by its cofactor, the cofactor of a_{ij} being the product of the determinant of the matrix with $n-1$ rows and columns obtained by erasing the ith row and jth column of A, by $(-1)^{i+j}$. Thus, if A is given by

$$A = \begin{pmatrix} a_{11} & a_{12} \\ a_{21} & a_{22} \end{pmatrix} \qquad (1\text{-}54)$$

then

$$\text{Co } A = \begin{pmatrix} a_{22} & -a_{21} \\ -a_{12} & a_{11} \end{pmatrix} \qquad (1\text{-}55)$$

For *certain* square matrices A we can define the inverse of A, denoted by A^{-1} by the equation

$$A A^{-1} = E_n \qquad (1\text{-}56)$$

where E_n is the unit matrix and A^{-1} is given by[5]

$$A^{-1} = \frac{(\text{Co } A)^\star}{\text{determinant of } A} \qquad (1\text{-}57)$$

As a verification of this, suppose

$$A = \begin{pmatrix} 3 & 2 & 1 \\ 0 & -4 & 3 \\ 8 & 12 & -5 \end{pmatrix} \qquad (1\text{-}58)$$

Then

$$A^{-1} = \frac{1}{32} \begin{pmatrix} -16 & 22 & 10 \\ 24 & -23 & -9 \\ 32 & -20 & -12 \end{pmatrix} \qquad (1\text{-}59)$$

in which the first row, first column element, a_{11}^{-1} is obtained, for example, from

$$\frac{a_{11}^{-1}}{|A|} = \frac{\begin{vmatrix} -4 & 3 \\ 12 & -5 \end{vmatrix} (-1)^{(1+1)}}{|A|} = \frac{(20-36)(-1)^2}{32} = -\frac{16}{32} \qquad (1\text{-}60)$$

[5] This result is proven in Linear Equation Theory. It follows from an application of Cramer's Rule and properties of determinants to the solution of a set of simultaneous linear equations. It requires that determinant A is not equal to zero.

Then

$$AA^{-1} = \begin{pmatrix} 1 & 0 & 0 \\ 0 & 1 & 0 \\ 0 & 0 & 1 \end{pmatrix} \qquad (1\text{-}61)$$

as required.

Another way in which the inverse of a square matrix may be obtained (which is, in fact, the technique utilized in many electronic computer applications) is based upon the following theorem:

If A is a given square matrix, and if A^{-1} exists, then

$$[0]A \rightarrow E \qquad (1\text{-}62)$$

and

$$\big[[0]\big]E \rightarrow A^{-1} \qquad (1\text{-}63)$$

in which $[0]A \rightarrow E$ means "a series of matrix operations which transform A into E," and $\big[[0]\big]E \rightarrow A^{-1}$ means "these same matrix operations will transform E into A^{-1}." These matrix operations may be

1. The interchange of any two rows (or columns)[6]
2. Multiply any row or column by a non-zero constant.
3. Multiply any row or column by a non-zero constant and add to any other row or column.

An unsophisticated "proof" of the above is the following:

We have

$$[0]A \rightarrow E \qquad (1\text{-}64$$

multiply both sides by A^{-1}; then

$$[0]AA^{-1} \rightarrow EA^{-1} \qquad (1\text{-}65)$$

---

[6] The operations for inversion being discussed here represent, essentially, a series of matrix multiplications. Thus, the sequence of operations on rows corresponds to

$$C_1C_2C_3 \cdots C_nA = E$$

hence,

$$C_1C_2C_3 \cdots C_n = A^{-1}$$

(See Prob. 7 of Chapter 2.) Similarly, the sequence of operations on columns corresponds to

$$AD_1D_2D_3 \cdots D_n = E$$

hence,

$$D_1D_2D_3 \cdots D_n = A^{-1}$$

It is also possible to obtain A^{-1} by both post-multiplying and pre-multiplying. Thus, we can obtain E from

$$C_3C_2C_1AD_1D_2D_3 = E$$

so that

$$A^{-1} = D_1D_2D_3C_3C_2C_1$$

or

$$[[0]]E \to A^{-1}, \quad \text{Q.E.D.} \tag{1-66}$$

To illustrate, let us determine A^{-1} if

$$A = \begin{pmatrix} 3 & 2 & 1 \\ 0 & -4 & 3 \\ 8 & 12 & -5 \end{pmatrix} \tag{1-67}$$

Thus, transforming A and E concurrently,

$$A = \begin{pmatrix} 3 & 2 & 1 \\ 0 & -4 & 3 \\ 8 & 12 & -5 \end{pmatrix} \to E = \begin{pmatrix} 1 & 0 & 0 \\ 0 & 1 & 0 \\ 0 & 0 & 1 \end{pmatrix}$$

Interchanging R(3) and R(1)

$$= \begin{pmatrix} 8 & 12 & -5 \\ 0 & -4 & 3 \\ 3 & 2 & 1 \end{pmatrix} \quad \to \quad \begin{pmatrix} 0 & 0 & 1 \\ 0 & 1 & 0 \\ 1 & 0 & 0 \end{pmatrix}$$

Multiply R(2) by 3, add to R(1)

$$\begin{pmatrix} 8 & 0 & 4 \\ 0 & -4 & 3 \\ 3 & 2 & 1 \end{pmatrix} \quad \to \quad \begin{pmatrix} 0 & 3 & 1 \\ 0 & 1 & 0 \\ 1 & 0 & 0 \end{pmatrix}$$

Multiply R(1) by $\tfrac{1}{8}$

$$\begin{pmatrix} 1 & 0 & \tfrac{1}{2} \\ 0 & -4 & 3 \\ 3 & 2 & 1 \end{pmatrix} \quad \to \quad \begin{pmatrix} 0 & \tfrac{3}{8} & \tfrac{1}{8} \\ 0 & 1 & 0 \\ 1 & 0 & 0 \end{pmatrix} \tag{1-68}$$

Multiply R(2) by $-\tfrac{1}{4}$

$$\begin{pmatrix} 1 & 0 & \tfrac{1}{2} \\ 0 & 1 & -\tfrac{3}{4} \\ 3 & 2 & 1 \end{pmatrix} \quad \to \quad \begin{pmatrix} 0 & \tfrac{3}{8} & \tfrac{1}{8} \\ 0 & -\tfrac{1}{4} & 0 \\ 1 & 0 & 0 \end{pmatrix}$$

Multiply R(1) by -3, add to R(3)

$$\begin{pmatrix} 1 & 0 & \tfrac{1}{2} \\ 0 & 1 & -\tfrac{3}{4} \\ 0 & 2 & -\tfrac{1}{2} \end{pmatrix} \quad \to \quad \begin{pmatrix} 0 & \tfrac{3}{8} & \tfrac{1}{8} \\ 0 & -\tfrac{1}{4} & 0 \\ 1 & -\tfrac{9}{8} & -\tfrac{3}{8} \end{pmatrix}$$

Multiply R(2) by -2, add to R(3)

$$\begin{pmatrix} 1 & 0 & \frac{1}{2} \\ 0 & 1 & -\frac{3}{4} \\ 0 & 0 & 1 \end{pmatrix} \quad \rightarrow \quad \begin{pmatrix} 0 & \frac{3}{8} & \frac{1}{8} \\ 0 & -\frac{1}{4} & 0 \\ 1 & -\frac{5}{8} & -\frac{3}{8} \end{pmatrix}$$

Multiply R(3) by $\frac{3}{4}$, add to R(2)

$$\begin{pmatrix} 1 & 0 & \frac{1}{2} \\ 0 & 1 & 0 \\ 0 & 0 & 1 \end{pmatrix} \quad \rightarrow \quad \begin{pmatrix} 0 & \frac{3}{8} & \frac{1}{8} \\ \frac{3}{4} & -\frac{23}{32} & -\frac{9}{32} \\ 1 & -\frac{5}{8} & -\frac{3}{8} \end{pmatrix}$$

Multiply R(3) by $-\frac{1}{2}$, add to R(1)

$$\underbrace{\begin{pmatrix} 1 & 0 & 0 \\ 0 & 1 & 0 \\ 0 & 0 & 1 \end{pmatrix}}_{E_3} \quad \rightarrow \quad \underbrace{\begin{pmatrix} -\frac{1}{2} & \frac{11}{16} & \frac{5}{16} \\ \frac{3}{4} & -\frac{23}{32} & -\frac{9}{32} \\ 1 & -\frac{5}{8} & -\frac{3}{8} \end{pmatrix}}_{A^{-1}}$$

$\left.\begin{array}{c} \\ \\ \\ \\ \\ \\ \\ \\ \\ \\ \end{array}\right\}$ (1-68)
continued

We can then factor A^{-1}, to obtain

$$A^{-1} = \frac{1}{32}\begin{pmatrix} -16 & 22 & 10 \\ 24 & -23 & -9 \\ 32 & -20 & -12 \end{pmatrix} \tag{1-69}$$

which checks the value obtained before, see Eq. (1-59).

The *derivative of a matrix* with respect to a variable, say x, is obtained by differentiating each element separately. Thus

$$\frac{d}{dx}\begin{pmatrix} x & x^2 & 2 \\ e^x & 3x^3 & 0 \end{pmatrix} = \begin{pmatrix} 1 & 2x & 0 \\ e^x & 9x^2 & 0 \end{pmatrix} \tag{1-70}$$

The student should verify by considering simple matrices that

$$\frac{\partial(AB)}{\partial x} = \frac{\partial A}{\partial x}B + A\frac{\partial B}{\partial x} \tag{1-71}$$

Note that the order of terms in the product must be maintained.

The *integral of a matrix* with respect to a variable, say x, is obtained by integrating each element separately. Thus

$$\int \begin{pmatrix} x & x^2 & 2 \\ e^x & 3x^3 & 0 \end{pmatrix} dx = \begin{pmatrix} \frac{x^2}{2} & \frac{x^3}{3} & 2x \\ e^x & \frac{3x^4}{4} & c \end{pmatrix} + \begin{pmatrix} c_1 & c_2 & c_3 \\ c_4 & c_5 & 0 \end{pmatrix} \tag{1-72}$$

The above represents the elements of a matrix arithmetic and algebra and calculus sufficient for our purposes. It may also be noted that the matrix operations described herein are of interest, not only in connection with tensor analysis, but also because they occur in some electronic computer operations, in which an essential step is the formulation of the problem (generally in finite-difference notation) in a matrix form. See Refs. (1) and (2) for additional information on matrices.

1-4 Introduction to Vector Analysis. The vector is a quantity of fundamental importance in physics, engineering and mathematics. The usual definition of a vector as a "quantity having direction and magnitude" loses meaning when we extend our ideas to more than three dimensions (as must frequently be done in certain fields of physics). In the next chapter, we will define the vector more precisely from the mathematical and physical point of view. This later definition of a vector will follow naturally and simply from our notions of matrices and tensors. Indeed, we will show that the vector is one of a family which includes other equally important engineering and physical quantities. For the present, and as an introduction to later work in this book, we review briefly the algebra and calculus of vectors.

In the three-dimensional space a vector has three components in the x, y and z directions indicated as follows:

$$\bar{A} = a_1 i + a_2 j + a_3 k \tag{1-73}$$

in which i, j, and k are unit vectors in the x, y, and z directions, respectively, and a_1, a_2, a_3 are the numerical quantities which represent the components of the vector in these directions.

Note that in the above expression, the "directional" property is accounted for by the unit vectors i, j, and k and the "magnitude" property is introduced by means of the terms, a_1, a_2, and a_3.

However, we may represent the vector equally as well by using a matrix form,

$$A = (a_1 \quad a_2 \quad a_3) \tag{1-74}$$

in which we *define* the three components of the vector A in the x, y, and z directions as a_1, a_2, and a_3. An equivalent and equally acceptable notation is

$$A = \begin{pmatrix} a_1 \\ a_2 \\ a_3 \end{pmatrix} \tag{1-75}$$

In these representations, the "magnitude" property is accounted for directly by the terms, and the directional property is accounted for by means of the position of these terms in the matrix—the first representing the x direction, the second the y direction, and the third the z direction.

We emphasize that the definitions given above are arbitrary but, as will be seen shortly, they lead to consistent mathematical operations.

It must also be noted that there are other possible and equally acceptable ways for representing vectors.[7] However, the above are sufficient for our purposes.

Just as in the case of the matrix, in developing the arithmetic or algebra of vectors, we must define addition, subtraction, multiplication, and division. While the definitions may be arbitrary, they will be of interest to engineers and physicists only if they form a consistent mathematical system and if they have a rational physical significance. The student should note particularly that this last requirement is satisfied in all cases.

We say that the sum of two vectors $\vec{A}$ and $\vec{B}$ is the sum of the components in each direction. Thus (using both notations given above), if

$$\left.\begin{aligned} \vec{A} &= a_1\vec{i} + a_2\vec{j} + a_3\vec{k} \\ A &= (a_1 \quad a_2 \quad a_3) \end{aligned}\right\} \tag{1-76}$$

and

$$\left.\begin{aligned} \vec{B} &= b_1\vec{i} + b_2\vec{j} + b_3\vec{k} \\ B &= (b_1 \quad b_2 \quad b_3) \end{aligned}\right\} \tag{1-77}$$

then

$$\left.\begin{aligned} \vec{A} + \vec{B} &= (a_1 + b_1)\vec{i} + (a_2 + b_2)\vec{j} + (a_3 + b_3)\vec{k} \\ A + B &= (a_1 + b_1 \quad a_2 + b_2 \quad a_3 + b_3) \end{aligned}\right\} \tag{1-78}$$

We see that this second form is consistent with our definition of matrix addition.

[7] In fact, vector analysis as described herein and as it exists today represents a later development of the "quaternion." The quaternion, invented by Hamilton about 1840–50 and extensively developed by him and also by Tait, is a quantity consisting of four terms, three of which correspond to the vector components considered here and the fourth a scalar. Although many very useful theorems and relations are obtained in quaternion theory, and although Hamilton and Tait thought this theory would find extensive applications in mathematical physics, for various reasons the "vector analysis" (and tensor analysis) have been found much more useful and have largely replaced, in applied work, the quaternion. Incidentally, the term *tensor* was used by Hamilton to define the positive square root of the sum of the squares of the four elements of the quaternion. Needless to say, this definition has nothing whatever to do with the tensor which we are considering in this book.

An interesting outline of the controversy which arose between quaternions and vectors (and between the proponents of each) will be found in *The Scientific Papers of J. Willard Gibbs*, Vol. II, republished by Dover Publications, 1961.

The difference of two vectors is defined similarly.   Thus

$$\bar{A} - \bar{B} = (a_1 - b_1)\bar{i} + (a_2 - b_2)\bar{j} + (a_3 - b_3)\bar{k} \atop A - B = (a_1 - b_1 \quad a_2 - b_2 \quad a_3 - b_3)\Bigg\}$$

(1-79)

Also, the student should note that this definition of vectors leads to the parallelogram law of vector addition (see Fig. 1.1).   In this figure,

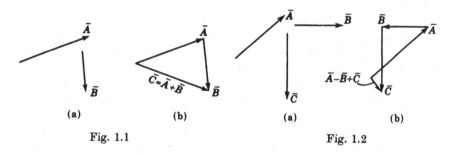

(a)   (b)   (a)   (b)

Fig. 1.1   Fig. 1.2

to obtain $\bar{A} + \bar{B}$, we lay off vector $\bar{A}$ to scale insofar as magnitude is concerned and in the given direction.   An arrow on the end indicates the direction in which the vector acts.   (Note: the negative of a vector is one which acts in the opposite direction, i.e. has an arrow on the other end.)   Then, place the tail of vector $\bar{B}$ on the head (arrow end) of vector A and draw vector $\bar{B}$ to the same scale as $\bar{A}$ and in its shown direction.   To find $\bar{A} + \bar{B}$ simply connect the tail of $\bar{A}$ to the head of $\bar{B}$. (This is the same as completing the parallelogram of $\bar{A}$ and $\bar{B}$, taking $\bar{C}$ as the diagonal of this parallelogram.)

It may be shown (the student should verify this and the following statement) that the addition of vectors is commutative.   That is,

$$\bar{A} + \bar{B} = \bar{B} + \bar{A}$$

(1-80)

Also, it may be shown that the addition of vectors is associative That is,

$$(\bar{A} + \bar{B}) + \bar{C} = \bar{A} + (\bar{B} + \bar{C})$$

(1-81)

This in turn leads to the following procedure for determining the sum of more than two vectors.   For example, in Fig. 1.2 are shown vectors $\bar{A}$, $\bar{B}$, and $\bar{C}$.   To determine $\bar{A} - \bar{B} + \bar{C}$, first lay off vector $\bar{A}$, to the assumed scale in the given direction.   At the head (arrow) of $\bar{A}$ construct $-\bar{B}$, i.e., the vector in the opposite direction of $\bar{B}$, to the given scale, placing the tail of $-\bar{B}$ at the head of $\bar{A}$.   Now place the tail of $\bar{C}$ at the head of $-\bar{B}$.   This corresponds to adding $\bar{C}$ to $\bar{A} - \bar{B}$, and finally, the quantity $\bar{A} - \bar{B} + \bar{C}$ is given, in value and direction by the scaled vector going from the tail of $\bar{A}$ to the head of $\bar{C}$.

In discussing the products of vectors, we come upon a more complex situation based upon different possible definitions of vector products. First, we know that if we have a vector $\bar{A}$, then a vector $\alpha\bar{A}$ which is α times as large as $\bar{A}$ must have components which are α times as large as the components of $\bar{A}$. That is,

$$\underbrace{\bar{A} + \cdots \bar{A} + \bar{A}}_{\alpha \text{ times}} = \alpha\bar{A} \tag{1-82}$$

or

$$\alpha\bar{A} = (\alpha a_1)\bar{i} + (\alpha a_2)\bar{j} + (\alpha a_3)\bar{k} \tag{1-83}$$

and this defines the product of a vector and a constant. This relation also holds if α is any constant, not necessarily an integer.

Insofar as the product of two vectors is concerned, it is known from physics and engineering that at least two fundamentally different types of products are possible:

1. *A vector times a vector may equal a scalar.*[8] For example, force times a distance equals work. Force is a vector quantity, distance is a vector quantity, and work is a scalar quantity.

Another much more significant scalar is the length (or squared length) of a vector.[9] That is, we must define a product of a vector by itself such that the result is a scalar, this being the squared length of the vector. Let us first denote this by

$$\bar{A}_s\bar{A} = |A^2| = a_1{}^2 + a_2{}^2 + a_3{}^2 \tag{1-84}$$

in which the right-hand sign value is known to be true from elementary geometry and where the s denotes scalar product. This is rather cumbersome, however. We can economize by using a "dot" notation, as

$$\bar{A} \cdot \bar{A} = a_1{}^2 + a_2{}^2 + a_3{}^2 \tag{1-85}$$

where the left-hand side reads "$\bar{A}$ dot $\bar{A}$" and this, by definition, must represent the right-hand side, which is the known value of the squared length of the vector. Also, the dot has meaning only when it appears between two vectors, unit vectors or other vectors. Now let us multiply

[8] Although we have not, up to this point, defined a "scalar" (this will be done in the next chapter), it will be sufficient for our present purposes to say that a scalar is a quantity (such as a pure number, or the pressure at a point, or the temperature at a point, etc.) that can be given unambiguously without the necessity of specifying an accompanying coordinate system. On the other hand, a vector quantity is not fully specified unless it is connected with a particular system of axes.

[9] Obviously the *length* of a vector (say 10 in. or 100 lb or 60 ft/sec) does not depend upon the coordinate system but is an absolute quantity independent of the frame of reference—hence a scalar.

the full forms of $\bar{A}$ as on the left-hand side. This gives

$$(a_1 i + a_2 j + a_3 k) \cdot (a_1 i + a_2 j + a_3 k) \tag{1-86}$$

which equals

$$a_1 a_1 i \cdot i + a_1 a_2 i \cdot j + a_1 a_3 i \cdot k + a_2 a_1 j \cdot i + a_2 a_2 j \cdot j$$
$$+ a_2 a_3 j \cdot k + a_3 a_1 k \cdot i + a_3 a_2 k \cdot j + a_3 a_3 k \cdot k \tag{1-87}$$

and which must also equal

$$a_1{}^2 + a_2{}^2 + a_3{}^2 \tag{1-88}$$

In order that the last two equations be equal, we must define

$$\left. \begin{array}{l} i \cdot i = j \cdot j = k \cdot k = 1 \\ i \cdot j = j \cdot k = k \cdot i = j \cdot i = k \cdot j = i \cdot k = 0 \end{array} \right\} \tag{1-89}$$

Equation 1-89 is the key to the definition of scalar product. From it, it follows that the scalar product of vectors is commutative, or

$$\bar{A} \cdot \bar{B} = \bar{B} \cdot \bar{A} = a_1 b_1 + a_2 b_2 + a_3 b_3 \tag{1-90}$$

and in matrix notation this is equal to

$$AB^\star = (a_1 \; a_2 \; a_3) \begin{pmatrix} b_1 \\ b_2 \\ b_3 \end{pmatrix} = BA^\star \tag{1-91}$$

In addition, it may be shown that the distributive law of multiplication holds for scalar products (the student should verify this). That is,

$$(\bar{A} + \bar{B}) \cdot \bar{C} = \bar{A} \cdot \bar{C} + \bar{B} \cdot \bar{C} \tag{1-92}$$

Also (see Prob. 7 at the end of this chapter) the scalar product is numerically equal to (1), the numerical value of $\bar{A}$ times the component of $\bar{B}$ in the direction $\bar{A}$, and (2) the numerical value of $\bar{B}$ times the component of $\bar{A}$ in the direction of $\bar{B}$.[10]

2. *A vector times a vector may result in a vector.* That is, we must define

$$\bar{A} \vee \bar{B} = \text{a vector} \tag{1-93}$$

where the v now denotes a "vector product." This may be expressed in the more usual form $\bar{A} \times \bar{B}$, read "$\bar{A}$ cross $\bar{B}$" and the cross has meaning only when it appears between two vectors, unit vectors or otherwise.

Two simple examples of cross-products are (1) force (a vector) times distance (a vector) equals moment (a vector) and (2), much more

[10] In other words, $\bar{A} \cdot \bar{B} = |A||B| \cos \theta$, in which θ is the angle between $\bar{A}$ and $\bar{B}$.

significant from a basic point of view (and we now are restricting our discussion to the three-dimensional physical space), distance times distance equals area, where area is a vector quantity, its direction defined by a normal to its plane.

The definition of area vector requires that,

(a) if $\bar{A}$ is a vector $a_1 i$ and
(b) if $\bar{B}$ is a vector $b_2 j$ then

$$\bar{A} \times \bar{B} = \bar{C} \tag{1-94}$$

where $\bar{C}$ must be numerically equal to $a_1 b_2$ and must have a direction normal to $\bar{A}$ and $\bar{B}$. This last statement requires some further clarification. There are two possible directions normal to $\bar{A}$ and $\bar{B}$, each of which is opposite the other. We must now define the direction that we want, and (again in accordance with usual American practice) we use the "right-hand rule." That is, the vector $\bar{C}$ has the direction of the thumb of the right hand cupped so that the fingers go in the direction from $\bar{A}$ to $\bar{B}$. This means that

$$\bar{A} \times \bar{B} = -\bar{B} \times \bar{A} \tag{1-95}$$

so that the commutative law of multiplication does *not* hold for vector products.

If we use the value of $\bar{A}$ and $\bar{B}$ given above, we must have

$$a_1 b_2 i \times j = a_1 b_2 k \tag{1-96}$$

This means that the unit vectors, i, j, and k (and hence the x, y, and z axes) are related by the right-hand convention defined above.

The above reasoning, applied successively to the other sets of unit vectors in pairs, leads to the following definition of vector product:

$$\left.\begin{array}{l} i \times j = -j \times i = k \\ j \times k = -k \times j = i \\ k \times i = -i \times k = j \\ i \times i = 0 \\ j \times j = 0 \\ k \times k = 0 \end{array}\right\} \tag{1-97}$$

and from this it follows that the vector product of two general vectors $\bar{A}$ and $\bar{B}$ is given by the determinant which follows:[11]

---

[11] As an alternate development we could simply have *defined* $\bar{A} \times \bar{B}$ as a vector $\bar{V}$ whose length is the product of the lengths of $\bar{A}$ and $\bar{B}$ and the sine of the angle between them, and whose direction is perpendicular to the plane of $\bar{A}$ and $\bar{B}$ and so sensed that a right-hand screw turned from $\bar{A}$ toward $\bar{B}$ through the smaller of the angles between these vectors would move in the direction of $\bar{V}$.

$$\bar{A} \times \bar{B} = \begin{vmatrix} \bar{i} & \bar{j} & \bar{k} \\ a_1 & a_2 & a_3 \\ b_1 & b_2 & b_3 \end{vmatrix} \tag{1-98}$$

$$= (a_2 b_3 - a_3 b_2)\bar{i} + (a_3 b_1 - a_1 b_3)\bar{j} + (a_1 b_2 - b_1 a_2)\bar{k} \tag{1-99}$$

The scalar and vector products defined above represent the elementary products of vector analysis. There is, however, another type of product which is extremely important and which will be used in all later sections of this book, a "dyadic product" introduced by the American mathematician–physicist, Willard Gibbs; see Ref. (9). This product is obtained by multiplying two vectors $\bar{A}$ and $\bar{B}$ in the usual way, considering each as a simple sum of terms, thus,

or
$$\left.\begin{aligned} \bar{A}\bar{B} &= (a_1\bar{i} + a_2\bar{j} + a_3\bar{k})(b_1\bar{i} + b_2\bar{j} + b_3\bar{k}) \\ \bar{A}\bar{B} &= a_1 b_1 \bar{i}\bar{i} + a_1 b_2 \bar{i}\bar{j} + a_1 b_3 \bar{i}\bar{k} \\ &\quad + a_2 b_1 \bar{j}\bar{i} + a_2 b_2 \bar{j}\bar{j} + a_2 b_3 \bar{j}\bar{k} \\ &\quad + a_3 b_1 \bar{k}\bar{i} + a_3 b_2 \bar{k}\bar{j} + a_3 b_3 \bar{k}\bar{k} \end{aligned}\right\} \tag{1-100}$$

In the above, the products of the unit vectors may be used to signify position in an array. That is, we can define the dyadic as

$$\bar{A}\bar{B} = \begin{pmatrix} a_1 b_1 & a_1 b_2 & a_1 b_3 \\ a_2 b_1 & a_2 b_2 & a_2 b_3 \\ a_3 b_1 & a_3 b_2 & a_3 b_3 \end{pmatrix} \tag{1-101}$$

and we see that the first unit vector of any term represents row position and the second unit vector represents column position. For example, $\bar{j}\bar{k}$ is the term in the jth row (2nd row) and kth column (3rd column). Also, it is apparent that, in terms of matrix products,

$$\bar{A}\bar{B} = A \star B = \begin{pmatrix} a_1 \\ a_2 \\ a_3 \end{pmatrix} (b_1 \quad b_2 \quad b_3) \tag{1-102}$$

The physical significance of this type of vector product will be discussed more fully in Chapter 2.

Division of vectors is not defined. The following discussion will indicate the reason for this.

Consider the vector $\bar{V}$. Now,

$$\bar{V} \cdot \bar{V} = v^2 \tag{1-103}$$

where v^2 is the scalar, squared length of $\bar{V}$.

If division of vectors was defined, then it would be necessary that

$$\frac{v^2}{\widetilde{V}} = \vec{V}, \quad \text{unambiguously} \qquad (1\text{-}104)$$

where $\sim$ represents some vector form of division. In other words, if one were asked to give the answer to

$$\frac{v^2}{\widetilde{V}} \qquad (1\text{-}105)$$

it would seem reasonable to expect one and only one answer, namely, $\vec{V}$. But,

$$v^2 = \vec{V} \cdot \vec{V} = \vec{V} \cdot (\vec{V} + \overline{W}) \qquad (1\text{-}106)$$

where $\overline{W}$ is any vector perpendicular to $\vec{V}$. Thus, it can be seen that the answer to $\frac{v^2}{\widetilde{V}}$ may have an infinity of values. This cannot be permitted in any mathematical system to be used in physics and engineering. We must therefore rule out division of vectors as undefined. A similar argument may be applied to the vector product $\vec{A} \times \vec{B} = \vec{C}$.

A full discussion of the differential and integral calculus of vectors is beyond the scope of this book. However, brief mention will be made of some of the more important terms and relations. Of first importance is the "∇ vector," that is, the vector defined by

$$\nabla = \frac{\partial}{\partial x}\vec{i} + \frac{\partial}{\partial y}\vec{j} + \frac{\partial}{\partial z}\vec{k} \qquad (1\text{-}107)$$

This vector has meaning only when it operates on either a scalar or a vector. Thus, some possible forms of simple combinations into which this vector enters are

(a) $\nabla \cdot \vec{V}$, called divergence $\vec{V}$ or div $\vec{V}$,
(b) $\nabla \times \vec{V}$, called curl $\vec{V}$,
(c) ∇w, called gradient w, w is a scalar function,
(d) $\nabla \cdot \nabla w = \nabla^2 w$, called Laplacian w.

A physical interpretation of these terms, which occur in all the fields of mathematical physics (and which will appear in later sections of this book), may be of interest. We shall indicate fluid mechanics applications.

(a) The divergence $\vec{V}$, $\nabla \cdot \vec{V}$ or div $\vec{V}$, in which $\vec{V}$ is the fluid velocity vector having components, u, v, w in the x, y, z directions, in fluid mechanics, for example, is a measure of the rate of loss of fluid per unit volume. Hence if, for example, the fluid is incompressible, there can be neither a gain nor a loss, and so we obtain the so-called *continuity* equation (which is essentially a statement of conservation of mass),

$$\nabla \cdot \vec{V} = 0 \tag{1-108}$$

(b) the expression curl $\vec{V}$ or $\nabla \times \vec{V}$ also occurs in fluid mechanics, in which field it represents the rotation or vorticity in the fluid. A major class of fluids is that for which the rotation is zero, i.e., the flow is *irrotational*—and thus, for this fluid

$$\nabla \times \vec{V} = 0 \tag{1-109}$$

(c) the gradient of ϕ in which ϕ is a scalar is an expression which also occurs in fluid mechanics. ϕ is called the "velocity potential" and

$$\nabla \phi = \frac{\partial \phi}{\partial x} \vec{i} + \frac{\partial \phi}{\partial y} \vec{j} + \frac{\partial \phi}{\partial z} \vec{k} \tag{1-110}$$

In this expression, the derivative of ϕ in any direction represents the fluid velocity in that direction. Thus

$$\left. \begin{aligned} \frac{\partial \phi}{\partial x} &= u \\[1em] \frac{\partial \phi}{\partial y} &= v \\[1em] \frac{\partial \phi}{\partial z} &= w \end{aligned} \right\} \tag{1-111}$$

and

A velocity potential for a particular fluid exists when the fluid flow is irrotational since (as the student should verify) this implies

$$\nabla \times \vec{V} = \nabla \times (\nabla \phi) = 0 \tag{1-112}$$

and we see that $\vec{V}$ can be given by

$$\vec{V} = \nabla \phi \tag{1-113}$$

(d) The Laplacian ϕ, or $\nabla^2 \phi$ or $\partial^2 \phi / \partial x^2 + \partial^2 \phi / \partial y^2 + \partial^2 \phi / \partial z^2$ occurs in fluids as a result of the combination (a) and (c) given above. Thus, if a fluid is incompressible and the flow is irrotational we have

$$\nabla \cdot \vec{V} = 0 \tag{1-114}$$

and

$$\vec{V} = \nabla \phi \tag{1-115}$$

Combining these gives at once

$$\nabla^2 \phi = 0 \tag{1-116}$$

for incompressible, irrotational flows.

We complete our discussion of vector analysis by stating, without proof, two important theorems which will be used in later chapters.

Gauss' Theorem: The volume integral of the divergence of a vector function of position $\vec{V}$ taken over volume τ is equal to the surface integral of $\vec{V}$ taken over the closed surface $\bar{\sigma}$ surrounding the volume τ, or

$$\int_\tau \nabla \cdot \vec{V} \, d\tau = \int_\sigma \vec{V} \cdot d\bar{\sigma} \qquad (1\text{-}117)$$

Stokes' Theorem: The surface integral of the curl of a vector function of position $\vec{V}$ taken over any surface $\bar{\sigma}$ is equal to the line integral of $\vec{V}$ around the border λ of the surface, or

$$\int_\sigma \nabla \times \vec{V} \cdot d\bar{\sigma} = \oint_\lambda \vec{V} \cdot d\lambda \qquad (1\text{-}118)$$

As an exercise, the student should write these theorems in ordinary cartesian form.[12] See Refs. (3) and (4) for additional information on vector analysis.

1-5 Introduction to Complex Variable Theory. In this section a brief introductory treatment of a complex variable theory is presented. Starting with the definition of a complex number, we proceed to a derivation of the Cauchy–Riemann equations and a brief discussion of the Laplacian equation and the solution of this by means of complex variables.

A number of the form

$$z = x + iy \qquad (1\text{-}119)$$

is called a *complex number*. In this, x and y are real numbers and i is the imaginary unit defined by

$$i = \sqrt{-1} \qquad (1\text{-}120)$$

The *real part* of the complex number, z, is x, and the *imaginary part* of the complex number, z, is y. Thus,

$$x = R(z)$$

and

$$y = I(z) \qquad (1\text{-}121)$$

---

[12] Gauss' Theorem (sometimes called the Divergence Theorem) and Stokes' Theorem are extremely important in pure as well as applied mathematics. By giving physical interpretations to $\vec{V}$ (such as fluid velocity or gradient of velocity potential or similar terms) it is possible to derive various fundamental existence, uniqueness, minimum and maximum relations in fluid mechanics and other fields as well. The different forms of the Green's Theorems also follow directly from Gauss' Theorem. Stokes' Theorem may be used to prove, among other things, the necessary and sufficient conditions that a line integral be independent of the path of integration.

Two complex numbers, $a+ib$ and $c+id$ are equal if and only if the real parts are equal and the imaginary parts are equal. That is, if

$$a+ib = c+id \qquad (1\text{-}122)$$

then

$$\left.\begin{aligned} a &= c \\ b &= d \end{aligned}\right\} \qquad (1\text{-}123)$$

Note that a complex number is *zero* only if *both* the real and imaginary parts of the number are zero. Thus, if

$$a+ib = 0 \qquad (1\text{-}124)$$

then

$$\left.\begin{aligned} a &= 0 \\ b &= 0 \end{aligned}\right\} \qquad (1\text{-}125)$$

If, for the complex number

$$a+ib \qquad (1\text{-}126)$$

$b = 0, a \neq 0$, then the number is said to be *pure real*. If, for this number, $a = 0, b \neq 0$, then the number is said to be *pure imaginary*.

The following relations obviously hold for the imaginary unit, i.

$$\left.\begin{aligned} i^2 &= -1 \\ i^3 &= i^2 i = -i \\ i^4 &= i^2 i^2 = 1 \end{aligned}\right\} \qquad (1\text{-}127)$$

and so on.

In adding two complex numbers, we add the real parts and the imaginary parts. Thus,

$$(a+ib)+(c+id) = (a+c)+i(b+d) \qquad (1\text{-}128)$$

Subtraction is defined similarly; thus,

$$(a+ib)-(c+id) = (a-c)+i(b-d) \qquad (1\text{-}129)$$

Multiplication follows the usual rules of algebraic multiplication with the property of the powers of i used as needed. Thus

$$(a+ib)(c+id) = (ac-bd)+i(bc+ad) \qquad (1\text{-}130)$$

To divide one complex number by another we rationalize the denominator. Thus

$$\left.\begin{aligned} \frac{a+ib}{c+id} &= \frac{a+ib}{c+id} \cdot \frac{c-id}{c-id} \\[2mm] &= \left(\frac{ac+bd}{c^2+d^2}\right)+i\left(\frac{bc-ad}{c^2+d^2}\right) \end{aligned}\right\} \qquad (1\text{-}131)$$

If two complex numbers differ in the sign of their imaginary parts, they are said to be *conjugate* to each other. The conjugate of a complex number, such as

$$z = x + iy \tag{1-132}$$

is written as

$$\bar{z} = x - iy \tag{1-133}$$

Note that

$$\left. \begin{aligned} z\bar{z} &= (x+iy)(x-iy) \\ &= x^2 + y^2. \end{aligned} \right\} \tag{1-134}$$

We may represent a complex number in an *Argand plane* (also called the *complex plane* or the *z plane*) by a point P whose abscissa and ordinate are, respectively, the real and imaginary parts of the given complex number, or by the directed line which joins the origin

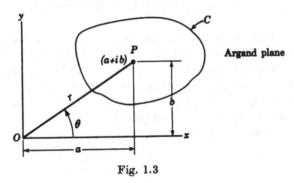

Fig. 1.3

to the point. Thus, in Fig. 1.3, the line OP represents the complex number $a + ib$. Note that the length of OP is given by

$$r = \sqrt{a^2 + b^2} \tag{1-135}$$

and its direction by

$$\theta = \tan^{-1} \frac{b}{a} \tag{1-136}$$

Also, since

$$\left. \begin{aligned} a &= r\cos\theta \\ b &= r\sin\theta \end{aligned} \right\} \tag{1-137}$$

it follows that the complex number $a + ib$ is also given by

$$a + ib = r(\cos\theta + i\sin\theta) = re^{i\theta} \tag{1-138}$$

When given in terms of r and θ, a complex number is said to be given in *polar form*.

The length r is called the *absolute value* or *modulus* of the complex number and the angle θ is called the *amplitude* or *argument* of the complex number.

Given two complex numbers in polar form,

$$z_1 = r_1(\cos\theta_1 + i\sin\theta_1) \tag{1-139}$$

and

$$z_2 = r_2(\cos\theta_2 + i\sin\theta_2) \tag{1-140}$$

then by performing the indicated operations we may show

$$z_1 z_2 = r_1 r_2[\cos(\theta_1+\theta_2)+i\sin(\theta_1+\theta_2)] \tag{1-141}$$

and

$$\frac{z_1}{z_2} = \frac{r_1}{r_2}[\cos(\theta_1-\theta_2)+i\sin(\theta_1-\theta_2)] \tag{1-142}$$

In other words—the product of two complex numbers is a complex number whose modulus is given by the product of the two moduli and whose argument is given by the sum of the two arguments. The quotient of two complex numbers is a complex number whose modulus is given by the quotient of the two moduli and whose argument is given by the difference of the two arguments.

It follows therefore that, given

$$z = r(\cos\theta + i\sin\theta) \tag{1-143}$$

then

$$z^n = r^n(\cos n\theta + i\sin n\theta) \tag{1-144}$$

which, for $r = 1$ is known as *de Moivre's Theorem*.

If z is a complex number given by

$$z = x+iy \tag{1-145}$$

where x and y are real, and if w is a function of z, then w will, in general, be a complex number as well. That is,

$$\left.\begin{aligned} w = f(z) = f(x+iy) \\ = u(x, y)+iv(x, y) \end{aligned}\right\} \tag{1-146}$$

where u and v are real and functions of x and y. As an example, if

$$z = x+iy \tag{1-147}$$

and

$$w = z^2 \tag{1-148}$$

then

$$w = x^2 - y^2 + i2xy$$
$$\left. = u + iv \right\} \tag{1-149}$$

or

$$u = x^2 - y^2$$
$$\left. v = 2xy \right\} \tag{1-150}$$

We now proceed to a definition of differentiation in complex variable theory. That is, we wish to define the derivative of a function w with respect to z. In accordance with our experience with real variables we define this as

$$\frac{dw}{dz} = \lim_{\Delta z \to 0} \frac{f(z + \Delta z) - f(z)}{\Delta z} \tag{1-151}$$

regardless of how $\Delta z \to 0$. But because $z = x + iy$ a fundamental difference from the real variable theory now presents itself. Δz may approach zero in many different ways. For

$$\Delta z = \Delta x + i\Delta y \tag{1-152}$$

and Δz may approach zero, depending upon Δx, Δy, and combinations of these. For example, consider a particular function

$$w = 2x + iy \tag{1-153}$$

in the w plane, Fig. 1.4, and let us determine dw/dz at the point shown if (a) Δz approaches zero along a constant value of y or

$$\Delta z = \Delta x \tag{1-154}$$

(b) Δz approaches zero along a constant value of x, or

$$\Delta z = i\Delta y \tag{1-155}$$

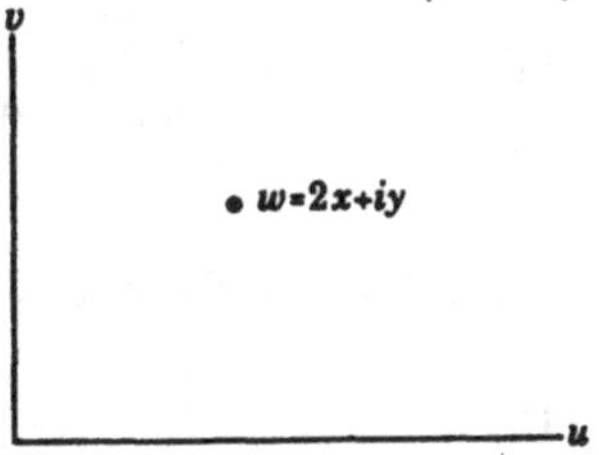

Fig. 1.4

Note that both of the above are possible manners in which Δz may approach zero for the function w at the point shown in the figure. There are an infinite number of other possible routes as well, but the two chosen above bring out the point to be made.

Then

(a)
$$\frac{dw}{dz} = \lim_{\Delta z \to 0} \frac{\Delta w}{\Delta z} = \frac{2\Delta x + i\Delta y}{\Delta x} = 2 \tag{1-156}$$

since $\Delta y = 0$ for constant y.

(b)
$$\frac{dw}{dz} = \lim_{\Delta z \to 0} \frac{\Delta w}{\Delta z} = \frac{2\Delta x + i\Delta y}{i\Delta y} = 1 \tag{1-157}$$

since $\Delta x = 0$ for constant x.

In other words, the derivative of $w = 2x + iy$ with respect to z has different values depending upon how $\Delta z \to 0$. Such a function is called *nonanalytic*.

On the other hand, suppose

$$w = z^2 \tag{1-158}$$

then (neglecting a square of a differential)

$$\left.\begin{aligned}
\Delta w &= (z + \Delta z)^2 - z^2 \\
&= 2z\Delta z \\
&= 2(x + iy)(\Delta x + i\Delta y)
\end{aligned}\right\} \tag{1-159}$$

Now, as before, assume

(a) Δz approaches zero along a constant value of y or

$$\Delta z = \Delta x \tag{1-160}$$

and

(b) Δz approaches zero along a constant value of x or

$$\Delta z = i\Delta y \tag{1-161}$$

Then,

(a)
$$\left.\begin{aligned}
\frac{dw}{dz} &= \lim_{\Delta z \to 0} \frac{\Delta w}{\Delta z} = 2(x + iy)\frac{(\Delta x + i\Delta y)}{\Delta x} \\
&= 2z
\end{aligned}\right\} \tag{1-162}$$

since $\Delta y = 0$ for constant y and

(b)
$$\left.\begin{aligned}
\frac{dw}{dz} &= \lim_{\Delta z \to 0} \frac{\Delta w}{\Delta z} = 2(x + iy)\frac{(\Delta x + i\Delta y)}{i\Delta y} \\
&= 2z
\end{aligned}\right\} \tag{1-163}$$

since $\Delta x = 0$ for constant x.

Note especially that the same value for dw/dz is obtained in both cases, and furthermore that the value of the derivative obtained is the same as obtained in ordinary differentiation of real quantities.

In complex variable theory we will require that our function of w be one whose derivative at any point is a single valued quantity. The function is then an *analytic function* of w, and we extend this requirement to cover the definition of the existence of derivatives of w. That is, we say the derivative dw/dz of a function $w = f(z)$ exists only when

$$\lim_{\Delta z \to 0} \frac{f(z + \Delta z) - f(z)}{\Delta z} \tag{1-164}$$

has the same value for all ways in which Δz can be made to approach zero.

We now derive the conditions on w such that dw/dz exists. These are the Cauchy–Riemann equations, which we state first and then derive.

Theorem: Given
$$w = f(z) = u(x, y) + iv(x, y) \tag{1-165}$$

If dw/dz exists for all values of z in a region R,[13] then everywhere in R

$$\left.\begin{aligned} \frac{\partial u}{\partial x} &= \frac{\partial v}{\partial y} \\[2mm] \frac{\partial u}{\partial y} &= -\frac{\partial v}{\partial x} \end{aligned}\right\} \tag{1-166}$$

and $\partial u/\partial x$, $\partial u/\partial y$, $\partial v/\partial x$, $\partial v/\partial y$ are continuous functions of x and y. The necessity of these conditions may be proven as follows:

$$\left.\begin{aligned} \frac{\Delta w}{\Delta z} &= \frac{\Delta u + i\Delta v}{\Delta z} \\[2mm] &= \frac{u(x+\Delta x, y+\Delta y) + iv(x+\Delta x, y+\Delta y) - u(x, y) - iv(x, y)}{\Delta x + i\Delta y} \end{aligned}\right\} \tag{1-167}$$

First let $\Delta z = \Delta x$, or $\Delta y = 0$, and the above becomes

$$\left.\begin{aligned} \frac{\Delta w}{\Delta z} &= \frac{u(x+\Delta x, y) - u(x, y)}{\Delta x} + \frac{iv(x+\Delta x, y) - iv(x, y)}{\Delta x} \\[4mm] \text{or} \qquad\qquad & \\[2mm] \frac{dw}{dz} &= \frac{\partial u}{\partial x} + i\frac{\partial v}{\partial x} \end{aligned}\right\} \tag{1-168}$$

Now let $\Delta z = i\Delta y$, $\Delta x = 0$, and proceed as above to obtain

$$\left.\begin{aligned} \frac{\Delta w}{\Delta z} &= \frac{u(x, y+\Delta y) - u(x, y)}{i\Delta y} + \frac{v(x, y+\Delta y) - v(x, y)}{\Delta y} \\[2mm] \text{or} \qquad\qquad \frac{dw}{dz} &= -i\frac{\partial u}{\partial y} + \frac{\partial v}{\partial y} \end{aligned}\right\} \tag{1-169}$$

[13] We shall not discuss in any detail what is meant by *region*, and we shall not discuss the related (and from the pure mathematical point of view) extremely important questions concerning *bounded regions*, *simple closed curves*, etc. A more complete discussion of these points will be found in texts on complex variable theory. We shall, in this book, assume the R to be defined with reference to the closed curve C shown in Fig. 1.3 as follows:

1. The region R may be the *unbounded* part of the z plane *exterior* to the closed curve C.
2. The region R may be the *bounded* part of the z plane *interior* to the closed curve C.

By hypothesis these two expressions must be equal. Equating real and imaginary parts, we obtain

(a)
$$\frac{\partial u}{\partial x} = \frac{\partial v}{\partial y}$$

(b)
$$\frac{\partial u}{\partial y} = -\frac{\partial v}{\partial x}$$

$$\left.\right\} \tag{1-170}$$

which are the Cauchy–Riemann equations.

For a proof of the sufficiency condition the reader is referred to Ref. (5).

To illustrate applications of the Cauchy–Riemann equations, let us consider the two functions presented earlier in this section, namely,

(1)
$$w = 2x + iy \tag{1-171}$$

(2)
$$w = z^2 = (x^2 - y^2) + i2xy \tag{1-172}$$

For (1) we have

$$\left.\begin{array}{l} u = 2x \\ v = y \end{array}\right\} \tag{1-173}$$

and

$$\left.\begin{array}{l} \dfrac{\partial u}{\partial x} = 2 \\[2mm] \dfrac{\partial v}{\partial y} = 1 \\[2mm] \dfrac{\partial u}{\partial y} = 0 \\[2mm] \dfrac{\partial v}{\partial x} = 0 \end{array}\right\} \tag{1-174}$$

Therefore,

$$\left.\begin{array}{l} \dfrac{\partial u}{\partial x} \neq \dfrac{\partial u}{\partial y} \\[3mm] 2 \neq 1 \end{array}\right\} \tag{1-175}$$

since

However,

$$\frac{\partial u}{\partial y} = -\frac{\partial v}{\partial x} = 0 \tag{1-176}$$

In other words, the first Cauchy–Riemann equation is not satisfied anywhere and the second is satisfied identically. Since *both* Cauchy–Riemann equations are not satisfied, the function is nonanalytic.

For (2) we have

$$\left.\begin{array}{l} u = x^2 - y^2 \\ v = 2xy \end{array}\right\} \tag{1-177}$$

Therefore,

$$\left.\begin{array}{l} \dfrac{\partial u}{\partial x} = 2x \\[2mm] \dfrac{\partial v}{\partial y} = 2x \\[2mm] \dfrac{\partial u}{\partial y} = -2y \\[2mm] \dfrac{\partial v}{\partial x} = 2y \end{array}\right\} \tag{1-178}$$

and hence the two Cauchy–Riemann equations are identically satisfied everywhere, i.e.,

$$\left.\begin{array}{l} \dfrac{\partial u}{\partial x} = \dfrac{\partial v}{\partial y} \\[3mm] 2x = 2x \end{array}\right\} \tag{1-179}$$

is given by

and

$$\left.\begin{array}{l} \dfrac{\partial u}{\partial y} = -\dfrac{\partial v}{\partial x} \\[3mm] -2y = -2y \end{array}\right\} \tag{1-180}$$

is given by

Thus, the function

$$w = z^2 \tag{1-181}$$

is *analytic* everywhere in the z plane.

We may show now that the real and imaginary parts of any function $w = f(z)$, satisfy the Laplace equation

$$\frac{\partial^2 \phi}{\partial x^2} + \frac{\partial^2 \phi}{\partial y^2} = 0 \tag{1-182}$$

To do this, take $\partial/\partial y$ of Eq. 1-170b and $\partial/\partial x$ of Eq. 1-170a and adding, we obtain

$$\frac{\partial^2 u}{\partial x^2} + \frac{\partial^2 u}{\partial y^2} = 0 \tag{1-183}$$

Similarly, taking $\partial/\partial y$ of 1-170a and $\partial/\partial x$ of 1-170b and subtracting, we have

$$\frac{\partial^2 v}{\partial x^2} + \frac{\partial^2 v}{\partial y^2} = 0 \qquad (1\text{-}184)$$

In other words, the real and imaginary parts of the analytic function $w = f(z)$ are solutions of the Laplace equation.[14] See Refs. (5) and (6) for additional information on complex variable theory.

1-6   Summary. In this chapter the introductory theory of matrix arithmetic (and algebra and calculus), vector analysis, and complex variable theory were presented. The brief treatments given will be sufficient for an understanding of the later sections of the book.

Problems

1. Given

$$A = \begin{pmatrix} 3 & 16 & 9 \\ 2 & 5 & 0 \\ 1 & 7 & 3 \end{pmatrix}$$

$$B = \begin{pmatrix} x^2 & 3zy & y^{-2} \\ xyz & e^x & y^{-3} \\ 0 & 2yx^{-1} & z^3 \end{pmatrix}$$

$$C = (4 \quad 3 \quad 1)$$

$$D = \begin{pmatrix} 2 \\ 9 \\ -6 \end{pmatrix}$$

$$E = \begin{pmatrix} 1 & 0 & 0 \\ 0 & 1 & 0 \\ 0 & 0 & 1 \end{pmatrix}$$

Determine

(a) AB

(b) BA

(c) ABD

(d) CAB

(e) AE

(f) AEB

(g) CE

(h) ED

(i) CED

(j) DC

(k) $\partial B/\partial x$

(l) $\partial B/\partial z$

(m) dB

---

[14] In connection with the Laplace equation, we have assumed the existence and continuity of the second partial derivatives of the functions u and v. It may be shown that an analytic function does in fact possess continuous partial derivatives of all orders in u and v.

2. Show the equations

$$2x + 3y = 16$$
$$12x - 2y = -1$$

in *three* different matrix forms.

3. By interpreting the equality (=) the same as the verb to be (are), show the following statements in matrix form.

> The tall boy and the small girl are tired.
> The fat pig and the brown cat are hungry.
> The tall building and the small house are dirty.
> The fat man and the brown dog are running.

4. Determine A^{-1} using the two methods described in the text if

(a)
$$A = \begin{pmatrix} 6 & -12 \\ -4 & -1 \end{pmatrix}$$

(b)
$$A = \begin{pmatrix} 14 & 3 & -2 \\ 6 & 8 & -1 \\ 0 & 2 & -7 \end{pmatrix}$$

5. Given vectors

$$\bar{U} = i(2x^2 + 3y - z) + j(2 + e^z y) + k(3xzy)$$
$$\bar{V} = i(y \log x) + j(x^2 + 2y^2 z) + k(6x)$$

Determine

(a)	$\bar{U} \cdot \bar{V}$	(e)	$\bar{U}\bar{V}$	(i)	$\nabla \times \bar{U}$
(b)	$\bar{U} \times \bar{V}$	(f)	$\bar{V}\bar{U}$	(j)	$\bar{U} \times \nabla \times \bar{U}$
(c)	$\bar{V} \cdot \bar{V}$	(g)	$\nabla \cdot \bar{V}$	(k)	$\nabla \bar{U} \cdot \bar{V}$
(d)	$\bar{V} \times \bar{U}$	(h)	$\bar{U}\nabla \cdot \bar{V}$	(l)	$\nabla \cdot \nabla(\bar{U} \cdot \bar{V})$

(m) Using the vectors ∇, $\bar{U}$, and $\bar{V}$, show three *vector* operations which are *not* permissible.

6. (a) Prove that the vector $\nabla\phi$, where ϕ is a scalar function of (x, y, z), is normal to curves $\phi = $ constant. Hint: along the curve $\phi = $ constant, $d\phi = 0$.

(b) Hence, if $\bar{R}$ is a unit vector in any direction, show that

$$\bar{R} \cdot \nabla\phi$$

is the *directional* derivative $d\phi/dr$ of ϕ in that direction.

7. Prove that $\bar{A} \cdot \bar{B}$, if $\bar{A}$ and $\bar{B}$ are vectors, is equal to

(a) The product of the magnitude of $\bar{A}$ and the component of $\bar{B}$ along $\bar{A}$.
or
(b) The product of the magnitude of $\bar{B}$ and the component of $\bar{A}$ along $\bar{B}$.
Hint: direct one of the coordinate axes in the direction of the vector.

8. Show that

$$R(z) = \frac{z + \bar{z}}{2}, \quad \text{and} \quad I(z) = \frac{z - \bar{z}}{2i}$$

9. Show that

$$R\left(\frac{z_1}{z_1 + z_2}\right) + R\left(\frac{z_2}{z_1 + z_2}\right) = 1$$

10. Prove

$$|z_1| - |z_2| \leqslant |z_1 + z_2|$$

11. Prove

$$|z_1 + z_2| \leqslant |z_1| + |z_2|$$

12. Prove

$$|\bar{z}| = |z|$$

13. Prove

$$|z_1 z_2| = |z_1| \cdot |z_2|$$

14. Prove

$$\left|\frac{z_1}{z_2}\right| = \frac{|z_1|}{|z_2|} \quad \text{if} \quad z_2 \neq 0$$

15. Show that

$$1 - 2x \cos\theta + x^2 = (1 - xe^{i\theta})(1 - xe^{-i\theta})$$

16. Show that if

$$z = \tanh(u + iv)$$

then

$$x^2 + y^2 = \frac{\cosh 2u - \cos 2v}{\cosh 2u + \cos 2v}$$

17. Prove that the real and imaginary parts of $\cos z$ satisfy the Laplace equation.

18. Is $\overline{e^z}$ equal to $e^{\bar{z}}$?

19. Is $\overline{\cos z}$ equal to $\cos \bar{z}$?

20. (a) Given $w = f(z) = u + iv$. Assume u (or v) is given. How would you determine v (or u)?
    (b) Using the method of (a) obtain v if

$$u = x^3 - 3xy^2 + 3x^2 - 3y^2 + 1$$

21. Show that the real and imaginary parts of the following functions of z satisfy the Cauchy–Riemann equations and also the Laplace equation
    (a) $\tanh z$ (b) ze^z (c) $z^3 + iz + 3$ (d) $\log z$

Chapter 2

TENSORS (OR MATRICES) OF ZERO, FIRST, AND SECOND ORDER

2-1 Introduction. We define a tensor, following Jefferies, Ref. (7), as a quantity having physical significance which satisfies a certain transformation law. There are many possible transformation laws that can be considered. However the transformation law which has the greatest physical significance for our work in applied mechanics is a simple *rotation-of-axes transformation*—i.e., we analyze the behavior of the quantities under study as we rotate the coordinate systems in which they act about the origin of the system.[1] Furthermore, for our purposes and the ordinary engineering usages, it will be sufficient to consider only tensors of order zero, one, and two.[2]

2-2 Zero-order Tensor (Scalar). The tensor of zero order is the *scalar*, that is, a quantity which is independent of orientation of axes. Some examples of scalars are: work, pressure, isotropic modulus of elasticity, and density. Note particularly that we can specify each of these quantities merely by giving a single number denoting its magnitude or value. No reference to a cartesian or other frame is required. In other words, under a rotation of axes, the scalar does not change—

---

[1] Another way to define "tensor" (as we use the expression in this book), which is essentially the equivalent of the above definition, is based on the following argument:

We deal in this book with continuous media, i.e., solids (beams, plates, etc.), fluids, and gases. The extent of a continuous medium, or field, is defined by means of coordinates, such as (x, y, z); and the variation of physical quantities, all of which are tensors, is given in terms of these coordinates. The use of coordinates implies a datum point, or origin, from which the coordinates are measured. The tensor is defined by describing its behaviour or its form or its value or its representation as the coordinate axes are rotated about the origin.

[2] In this text we shall deal exclusively with rectangular Cartesian systems, since these are of primary interest in engineering applications of tensor analysis. With these systems the distinction between so-called covariant and contravariant tensor transformations does not exist and we shall not, therefore, concern ourselves with either the transformations or notations usually associated with these terms. See Ref. (3) for a further discussion of this point.

and this is the transformation law satisfied by the tensor of zero order, the scalar.

2-3   First-order Tensor (Vector). The first-order tensor is the ordinary *vector*, variously represented (see Chapter 1) as

$$\vec{V} = v_x \vec{i} + v_y \vec{j} + v_z \vec{k} \tag{2-1}$$

or (and in general this will be the notation used in this text) either

$$V = (v_x \quad v_y \quad v_z) \tag{2-2}$$

or

$$V = \begin{pmatrix} v_x \\ v_y \\ v_z \end{pmatrix} \tag{2-3}$$

The reason for the notation 2-2 or 2-3 is twofold. First, there is the matter of economy of expression. There is a saving of writing involved in using 2-2 or 2-3 in preference to 2-1. Secondly, the notation of 2-2 or 2-3 is the standard matrix notation for the first-order tensor. Thus, 2-2 is a row matrix (or tensor) and 2-3 is a column matrix (or tensor). From this point on we will call 2-2 and 2-3 vectors, although they are also either matrices or tensors (since we shall invariably be dealing with quantities of physical significance which also satisfy the transformation law for the first-order tensor). In addition, wherever possible, capital letters will be used to represent the complete matrix and small letters with subscripts will be used to represent the elements of the matrix but this will be waived occasionally.

The transformation law satisfied by the vector quantity may be obtained as follows (we do this first for the two-dimensional case—and three-dimensional vector will then be obtained as a generalization).

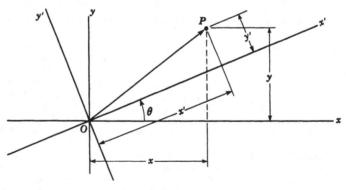

Fig. 2.1

In Fig. 2.1 let OP be a vector, whose coordinates in the O–x–y system are (x, y). Now assume the axes are rotated through an angle θ, about O, to position O–x'–y'. The coordinates of the vector in the O–x'–y' systems are (x', y') and furthermore, as may be verified,

$$\left.\begin{array}{l} x' = x \cos\theta + y \sin\theta \\ y' = -x \sin\theta + y \cos\theta \end{array}\right\} \tag{2-4}$$

or, in our matrix notation

$$\begin{pmatrix} x' \\ y' \end{pmatrix} = \begin{pmatrix} \cos\theta & \sin\theta \\ -\sin\theta & \cos\theta \end{pmatrix} \begin{pmatrix} x \\ y \end{pmatrix} \tag{2-5}$$

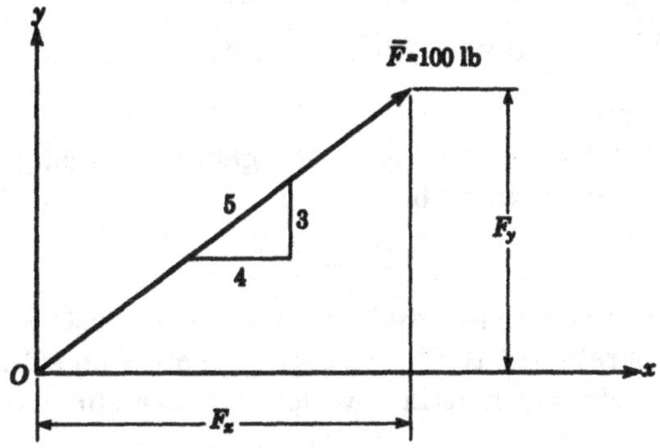

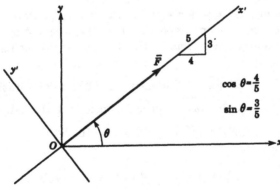

Fig. 2.2

For example, consider a force vector $\bar{F}$, of value 100 lb, directed as shown in Fig. 2.2. It is obvious that for the O–x–y system

$$F = \begin{pmatrix} F_x \\ F_y \end{pmatrix} = \begin{pmatrix} 80 \\ 60 \end{pmatrix} \tag{2-6}$$

Fig. 2.3

Now assume the axis rotated to O–x'–y' as shown in Fig. 2.3. It is equally obvious that

$$F = \begin{pmatrix} F'_x \\ F'_y \end{pmatrix} = \begin{pmatrix} 100 \\ 0 \end{pmatrix} \qquad (2\text{-}7)$$

We verify that this is as it should be by checking the equations

$$\left. \begin{array}{l} F'_x = F_x \cos\theta + F_y \sin\theta \\ F'_y = -F_x \sin\theta + F_y \cos\theta \end{array} \right\} \qquad (2\text{-}8)$$

or

$$\left. \begin{array}{l} F'_x = \tfrac{4}{5}(80) + \tfrac{3}{5}(60) = 64 + 36 = 100 \\ F'_y = -\tfrac{3}{5}(80) + \tfrac{4}{5}(60) = -48 + 48 = 0 \end{array} \right\} \qquad (2\text{-}9)$$

which is as required.

The above relation, Eq. 2-5, is the general transformation law for vectors in two dimensions, that is,

$$V' = RV \qquad (2\text{-}10)$$

in which V' is the vector referred to the O'–x'–y' system of coordinates, V is the vector referred to the O–x–y system of coordinates, and R is the so-called "rotation matrix," which we may show as

$$R = \begin{pmatrix} l_{11} & l_{12} \\ l_{21} & l_{22} \end{pmatrix} \qquad (2\text{-}11)$$

In this representation of R, each element represents the cosine of the angle between the axes corresponding to the subscripts—in which the first subscript is the primed axis and the second is the unprimed, so that, for example,

l_{21} = cosine of the angle between the y' and the x axes   (2-12)

Note that neither x, y nor x', y' in Eq. 2-4 are scalars. That is, the value of each component depends upon the orientation of the axes. But, there is a scalar connected with the vector, this being the squared length. That is,

$$x^2 + y^2 = x'^2 + y'^2 = \text{independent of axial orientation} \qquad (2\text{-}13)$$

For example, for the force vector F considered above in Eq. 2-6

$$F_x^2 + F_y^2 = 80^2 + 60^2 = 10{,}000 \text{ lb} \qquad (2\text{-}14)$$

and, from Eq. 2-7,

$$F_x'^2 + F_y'^2 = 100^2 + 0^2 = 10{,}000 \text{ lb} \qquad (2\text{-}15)$$

which verifies the invariance.

One may show quite easily that the equation

$$V' = RV \qquad (2\text{-}16)$$

is a general one which also holds true in a space of three dimensions. In this, the general rotation matrix element, l_{ij}, is given by

$$l_{ij} = \text{cosine of angle between } i' \text{ and } j \text{ axes.} \qquad (2\text{-}17)$$

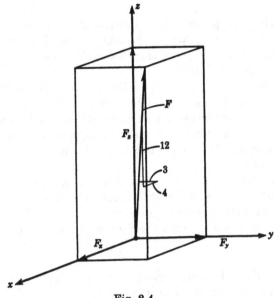

Fig. 2.4

For example, consider the force vector in space, shown in Fig. 2.4. If $F = 260$ lb, then for the direction of F shown in the figure

$$F = \begin{pmatrix} F_x \\ F_y \\ F_z \end{pmatrix} = \begin{pmatrix} 80 \\ 60 \\ 240 \end{pmatrix} \qquad (2\text{-}18)$$

Assume the x' axis coincides with F, and furthermore suppose that the axes are rotated so that x' coincides with x, and y' and z' are perpendicular to it and to each other. Then in the direction cosine table, the elements l_{11}, l_{12}, and l_{13} given in the first row are

$$R = \begin{pmatrix} l_{11} & l_{12} & l_{13} \\ l_{21} & l_{22} & l_{23} \\ l_{31} & l_{32} & l_{33} \end{pmatrix} = \begin{pmatrix} \frac{4}{13} & \frac{3}{13} & \frac{12}{13} \\ - & - & - \\ - & - & - \end{pmatrix} \qquad (2\text{-}19)$$

Hence it follows that F'_x is given by

$$
\left.
\begin{aligned}
F'_x &= l_{11}F_x + l_{12}F_y + l_{13}F_z \\[2mm]
&= \tfrac{4}{13}(80) + \tfrac{3}{13}(60) + \tfrac{12}{13}(240) \\[2mm]
&= 260 \text{ lb}
\end{aligned}
\right\}
\tag{2-20}
$$

as required.

Summarizing the present discussion of zero- and first-order tensors:

(a) A tensor is a quantity having physical significance whose elements satisfy a given transformation law—for our purposes a rotation of axes law.

(b) The tensors of engineering interest may conveniently be represented in matrix form. Hence for our purposes the matrix and the tensor are the same and either name will be used.

(c) The zero-order tensor or scalar is independent of orientation of axes and this is, in fact, the law of transformation for the zero-order tensor.

(d) The first-order tensor is the vector, ordinarily defined as "a quantity having magnitude and direction." We define this quantity more precisely by stating that a first-order tensor is a quantity which satisfies the following transformation law:

$$
V' = RV
$$

where V is the representation of the first-order tensor in the O–x–y–z system, R is the rotation matrix, and V' is the representation of the first-order tensor in the O–x'–y'–z' system.

2-4 The Tensor. The quantity which we shall call a "tensor" is, in reality, a "tensor of the second order." As pointed out previously, scalars and vectors are also part of the tensor family—membership in the family being restricted to physical quantities which satisfy certain transformation laws. All of the quantities of engineering and physical interest are members of this family. We shall in general reserve the name "tensor" or "matrix" for the "tensor of second order" only, and we shall call the first-order tensor a vector and the zero-order tensor a scalar. But the student should understand and appreciate the very intimate connection between all of these fundamental physical quantities.

It will be recalled (see Chapter 1) that in the elementary vector analysis two different products of vectors were defined. One, the scalar product, is given by

$$
A \cdot B
\tag{2-21}
$$

and the second, or the vector product, is given by

$$\bar{A} \times \bar{B} \qquad (2\text{-}22)$$

In Chapter 1 we discussed still another type of product of vectors which is extremely important in engineering and physics, the so-called "dyadic" of the Gibbs vector notation. We repeat, because of its importance at this point, the discussion of certain portions of this subject in Chapter 1. The dyadic is represented in vector notation as

$$\bar{A}\bar{B}$$

and is obtained by carrying through the multiplication in the ordinary algebraic manner for the vectors

$$\left. \begin{array}{l} \bar{A} = a_1 i + a_2 j + a_3 k \\ \bar{B} = b_1 i + b_2 j + b_3 k \end{array} \right\} \qquad (2\text{-}23)$$

so that

$$\begin{aligned} \bar{A}\bar{B} = {} & a_1 b_1 ii + a_1 b_2 ij + a_1 b_3 ik + a_2 b_1 ji + a_2 b_2 jj \\ & + a_2 b_3 jk + a_3 b_1 ki + a_3 b_2 kj + a_3 b_3 kk \end{aligned} \qquad (2\text{-}24)$$

Now, if we assign to i, j, k meaning corresponding to positions in an array (first letter represents row position, second letter represents column position) this can be conveniently represented as

$$\bar{A}\bar{B} = \begin{pmatrix} a_1 b_1 ii & a_1 b_2 ij & a_1 b_3 ik \\ a_2 b_1 ji & a_2 b_2 jj & a_2 b_3 jk \\ a_3 b_1 ki & a_3 b_2 kj & a_3 b_3 kk \end{pmatrix} \qquad (2\text{-}25)$$

or, in line with our desire for simplification of notation, this may also be shown as

$$\bar{A}\bar{B} = \begin{pmatrix} a_1 b_1 & a_1 b_2 & a_1 b_3 \\ a_2 b_1 & a_2 b_2 & a_2 b_3 \\ a_3 b_1 & a_3 b_2 & a_3 b_3 \end{pmatrix} \qquad (2\text{-}26)$$

where the subscripts take over the meaning of the i, j, and k and where now the first subscript corresponds to row position and the second corresponds to column position.

Equation 2-26 is the matrix representation of the dyadic. It is also the matrix representation of the linear vector function[3] and, most important from our point of view, it is the matrix representation of the second-order tensor in the three-dimensional space.

[3] A linear vector function (see Ref. 1) is an operator which produces a vector when it multiplies a vector.

In two dimensions this tensor becomes

$$\tilde{A}B = \begin{pmatrix} a_1b_1 & a_1b_2 \\ a_2b_1 & a_2b_2 \end{pmatrix} \tag{2-27}$$

Thus, for the usual engineering usages, the dyadic, linear vector function, tensor and matrix as we define them herein are identical quantities. We will refer to this quantity as either a matrix or a tensor only. However, in Refs. (1), (8), and (9) will be found discussions dealing with dyadics and linear functions that require only a change in name (and possibly notation) to come within the scope of the present discussion.

The transformation law for the tensor is obtained as follows:
Consider two vectors U and V which are represented by

$$\left. \begin{array}{l} U = \begin{pmatrix} u_x \\ u_y \\ u_z \end{pmatrix} \\[2em] V = \begin{pmatrix} v_x \\ v_y \\ v_z \end{pmatrix} \end{array} \right\} \tag{2-28}$$

Then, in accordance with Eq. 2-26, we define the tensor of the second order as the quantity

$$T = UV^\star \tag{2-29}$$

or, by simply performing the indicated operation,

$$T = \begin{pmatrix} u_x v_x & u_x v_y & u_x v_z \\ u_y v_x & u_y v_y & u_y v_z \\ u_z v_x & u_z v_y & u_z v_z \end{pmatrix} \tag{2-30}$$

This being so, it follows (see Eq. 2-16) that

$$U' = RU \tag{2-31}$$

and (see Eq. 2-16)

$$V' = RV \tag{2-32}$$

so that (see Eq. 1-45)

$$(V')^\star = V^\star R^\star \tag{2-33}$$

and

$$U'V'^\star = RUV^\star R^\star \tag{2-34}$$

or

$$T' = RTR^\star \tag{2-35}$$

and this is the transformation law satisfied by the tensor of the second order.[4] Note that we are, in effect, defining the primed axis tensor as $T' = U'V'^{\star}$, which is consistent with the definition for the unprimed tensor, T.

This transformation law is of paramount importance in engineering applications. It is the essential mathematical requirement behind the well-known Mohr circle construction (which will be described shortly) and shows very clearly that the Mohr circle construction can be applied to any tensor like T, such as the inertia tensor, stress tensor, strain tensor, etc.

It will be instructive to obtain the expanded form of T' in terms of the elements of T, for both the three-dimensional and the two-dimensional forms of T. Thus, if

$$T = \begin{pmatrix} T_{11} & T_{12} & T_{13} \\ T_{21} & T_{22} & T_{23} \\ T_{31} & T_{32} & T_{33} \end{pmatrix} \tag{2-36}$$

then

$$T' = \begin{pmatrix} l_{11} & l_{12} & l_{13} \\ l_{21} & l_{22} & l_{23} \\ l_{31} & l_{32} & l_{33} \end{pmatrix} \begin{pmatrix} T_{11} & T_{12} & T_{13} \\ T_{21} & T_{22} & T_{23} \\ T_{31} & T_{32} & T_{33} \end{pmatrix} \begin{pmatrix} l_{11} & l_{21} & l_{31} \\ l_{12} & l_{22} & l_{32} \\ l_{13} & l_{23} & l_{33} \end{pmatrix} \tag{2-37}$$

where

$$T' = \begin{pmatrix} T'_{11} & T'_{12} & T'_{13} \\ T'_{21} & T'_{22} & T'_{23} \\ T'_{31} & T'_{32} & T'_{33} \end{pmatrix} \tag{2-38}$$

Performing the multiplications indicated, we obtain

$$\begin{aligned}
T'_{11} &= T_{11}l_{11}^2 + T_{12}l_{12}l_{11} + T_{13}l_{13}l_{11} + T_{21}l_{11}l_{12} + T_{22}l_{12}^2 \\
&\quad + T_{23}l_{13}l_{12} + T_{31}l_{11}l_{13} + T_{32}l_{12}l_{13} + T_{33}l_{13}^2 \\[4pt]
T'_{12} &= T_{11}l_{21}l_{11} + T_{12}l_{22}l_{11} + T_{13}l_{23}l_{11} + T_{21}l_{21}l_{12} + T_{22}l_{22}l_{12} \\
&\quad + T_{23}l_{23}l_{12} + T_{31}l_{21}l_{13} + T_{32}l_{22}l_{13} + T_{33}l_{23}l_{13} \\[4pt]
T'_{13} &= T_{11}l_{31}l_{11} + T_{12}l_{32}l_{11} + T_{13}l_{33}l_{11} + T_{21}l_{31}l_{12} + T_{22}l_{32}l_{12} \\
&\quad + T_{23}l_{33}l_{12} + T_{31}l_{31}l_{13} + T_{32}l_{32}l_{13} + T_{33}l_{33}l_{13}
\end{aligned} \tag{2-39}$$

[4] There are various identities, equalities and similar relations which hold for the general tensor of the second order, the unit tensor and the rotation matrix. Some of these are given as problems (Probs. 1–12) at the end of this chapter. The student should refer to these and use them whenever it is convenient or necessary to do so. The following important point must be noted: in order to prove that any quantity is a tensor we must show that it satisfies the relation of Eq. 2-35.

$$T'_{21} = T_{11}l_{11}l_{21} + T_{12}l_{12}l_{21} + T_{13}l_{13}l_{21} + T_{21}l_{11}l_{22} + T_{22}l_{12}l_{22}$$
$$+ T_{23}l_{13}l_{22} + T_{31}l_{11}l_{23} + T_{32}l_{12}l_{23} + T_{33}l_{13}l_{23}$$

$$T'_{22} = T_{11}l_{21}{}^2 + T_{12}l_{22}l_{21} + T_{13}l_{23}l_{21} + T_{21}l_{21}l_{22} + T_{22}l_{22}{}^2$$
$$+ T_{23}l_{23}l_{22} + T_{31}l_{21}l_{23} + T_{32}l_{22}l_{23} + T_{33}l_{23}{}^2$$

$$T'_{23} = T_{11}l_{31}l_{21} + T_{12}l_{32}l_{21} + T_{13}l_{33}l_{21} + T_{21}l_{31}l_{22} + T_{22}l_{31}l_{22}$$
$$+ T_{23}l_{33}l_{22} + T_{31}l_{31}l_{23} + T_{32}l_{32}l_{23} + T_{33}l_{33}l_{23}$$

$$T'_{31} = T_{11}l_{11}l_{31} + T_{12}l_{12}l_{31} + T_{13}l_{13}l_{31} + T_{21}l_{11}l_{32} + T_{22}l_{12}l_{32}$$
$$+ T_{23}l_{13}l_{32} + T_{31}l_{11}l_{33} + T_{32}l_{12}l_{33} + T_{33}l_{13}l_{33}$$

$$T'_{32} = T_{11}l_{21}l_{31} + T_{12}l_{22}l_{31} + T_{13}l_{23}l_{31} + T_{21}l_{21}l_{32} + T_{22}l_{22}l_{32}$$
$$+ T_{23}l_{23}l_{32} + T_{31}l_{21}l_{33} + T_{32}l_{22}l_{33} + T_{33}l_{23}l_{33}$$

$$T'_{33} = T_{11}l_{31}{}^2 + T_{12}l_{32}l_{31} + T_{13}l_{33}l_{31} + T_{21}l_{31}l_{32} + T_{22}l_{32}{}^2$$
$$+ T_{23}l_{33}l_{32} + T_{31}l_{31}l_{33} + T_{32}l_{32}l_{33} + T_{33}l_{33}{}^2$$

$$(2\text{-}39)$$
continued

For the two-dimensional case, this becomes

$$\begin{pmatrix} T'_{11} & T'_{12} \\ T'_{21} & T'_{22} \end{pmatrix} = \begin{pmatrix} \cos\theta & \sin\theta \\ -\sin\theta & \cos\theta \end{pmatrix} \begin{pmatrix} T_{11} & T_{12} \\ T_{21} & T_{22} \end{pmatrix} \begin{pmatrix} \cos\theta & -\sin\theta \\ \sin\theta & \cos\theta \end{pmatrix} \quad (2\text{-}40)$$

which becomes, upon expansion

$$T' = \begin{pmatrix} T_{11}\cos^2\theta + T_{12}\sin\theta\cos\theta & -T_{11}\sin\theta\cos\theta + T_{12}\cos^2\theta \\ \quad + T_{21}\sin\theta\cos\theta + T_{22}\sin^2\theta & \quad - T_{21}\sin^2\theta + T_{22}\sin\theta\cos\theta \\[1em] -T_{11}\sin\theta\cos\theta - T_{12}\sin^2\theta & T_{11}\sin^2\theta - T_{12}\sin\theta\cos\theta \\ \quad + T_{21}\cos^2\theta + T_{22}\sin\theta\cos\theta & \quad - T_{21}\sin\theta\cos\theta + T_{22}\cos^2\theta \end{pmatrix}$$

$$(2\text{-}41)$$

We emphasize that these are perfectly general relations which hold for all tensors of the second order. We shall see later how Eq. 2-41 is the basis of the Mohr circle construction. But before we do this we shall discuss two very important properties of tensors—diagonalization and invariance. These general properties of tensors will be given without proof. The proofs may be found in Refs. (1) and (8).

PROPERTY I: DIAGONALIZATION OF SYMMETRICAL TENSORS.

Given any *symmetrical* tensor, A, whose elements in an $O\text{-}x\text{-}y\text{-}z$ system of coordinates are given by

$$A = \begin{pmatrix} a_{11} & a_{12} & a_{13} \\ a_{21} & a_{22} & a_{23} \\ a_{31} & a_{32} & a_{33} \end{pmatrix} \quad (2\text{-}42)$$

then there is an orthogonal set of axes O–x'–y'–z' (called the *principal axes*) with respect to which the tensor may be given in diagonal form,

$$A' = \begin{pmatrix} a'_{11} & 0 & 0 \\ 0 & a'_{22} & 0 \\ 0 & 0 & a'_{33} \end{pmatrix} \tag{2-43}$$

This property holds also for the 2×2 (that is, the two dimensional) tensor.[5] An example of diagonalization will be presented in the next section in connection with the inertia tensor.

PROPERTY II: INVARIANTS OF THE TENSOR

For any 3×3 tensor A, whose elements in the O–x–y–z system are given by

$$A = \begin{pmatrix} a_{11} & a_{12} & a_{13} \\ a_{21} & a_{22} & a_{23} \\ a_{31} & a_{32} & a_{33} \end{pmatrix} \tag{2-44}$$

there are three "invariants" (or scalars), which are quantities whose values do not change when the coordinate axes are changed to O–x'–y'–z' (a rotation of axes), for which the tensor becomes

$$A' = \begin{pmatrix} a'_{11} & a'_{12} & a'_{13} \\ a'_{21} & a'_{22} & a'_{23} \\ a'_{31} & a'_{32} & a'_{33} \end{pmatrix} \tag{2-45}$$

These invariants are

$$I_1 = \text{the trace} = a_{11} + a_{22} + a_{33} = a'_{11} + a'_{22} + a'_{33} \tag{2-46}$$

---

[5] Indeed, we may "prove" the diagonal property for the 2×2 symmetrical tensor rather simply as follows:

Refer to Eq. 2-40 and 2-41. Assume $T_{11}, T_{12} = T_{21}$, and T_{22} are not zero. Then diagonalization implies that there is an angle θ such that

$$-T_{11} \sin \theta \cos \theta + T_{12}(\cos^2\theta - \sin^2\theta) + T_{22} \sin \theta \cos \theta = 0$$

Dividing through by T_{12} (which is not zero), we obtain

$$\cos^2\theta - \sin^2\theta + K \sin \theta \cos \theta = 0$$

in which $K = (T_{22} - T_{11})/T_{12}$ and we may assume this equation is a continuous one in θ. Now,

for $\theta = 0$ The equation $= +1$
and for $\theta = \pi/2$ The equation $= -1$

Since the equation is continuous and it varies between $+1$ and -1 there is at least one value of θ for which the equation has the value zero. Q.E.D.

The reader should note especially that this simple demonstration is not valid if $T_{12} \neq T_{21}$, i.e., if the tensor is *not* symmetrical.

I_2 = the sum of the principal two-rowed minors,

$$= \begin{vmatrix} a_{22} & a_{23} \\ a_{32} & a_{33} \end{vmatrix} + \begin{vmatrix} a_{11} & a_{13} \\ a_{31} & a_{33} \end{vmatrix} + \begin{vmatrix} a_{11} & a_{12} \\ a_{21} & a_{22} \end{vmatrix}$$

$$= \begin{vmatrix} a'_{22} & a'_{23} \\ a'_{32} & a'_{33} \end{vmatrix} + \begin{vmatrix} a'_{11} & a'_{13} \\ a'_{31} & a'_{33} \end{vmatrix} + \begin{vmatrix} a'_{11} & a'_{12} \\ a'_{21} & a'_{22} \end{vmatrix}$$

(2-47)

I_3 = the determinant of the matrix

$$= \begin{vmatrix} a_{11} & a_{12} & a_{13} \\ a_{21} & a_{22} & a_{23} \\ a_{31} & a_{32} & a_{33} \end{vmatrix} = \begin{vmatrix} a'_{11} & a'_{12} & a'_{13} \\ a'_{21} & a'_{22} & a'_{23} \\ a'_{31} & a'_{32} & a'_{33} \end{vmatrix}$$

(2-48)

For the 2×2 tensor, there are two invariants:[6]

$$I_1 = a_{11} + a_{22} = a'_{11} + a'_{22} \tag{2-49}$$

$$I_2 = \begin{vmatrix} a_{11} & a_{12} \\ a_{21} & a_{22} \end{vmatrix} = \begin{vmatrix} a'_{11} & a'_{12} \\ a'_{21} & a'_{22} \end{vmatrix} \tag{2-50}$$

Examples of invariants will be presented in the next section in connection with the discussion of the inertia tensor.

Another important statement that can be made concerning tensors has to do with the behavior of tensors in equations. The statement can be given in several forms. For our purposes it will be sufficient to state it as:

Any equation expressible in tensor form holds independently of axial orientation and is valid in all systems of coordinates.[7]

For example, if

$$\Phi = K_1\Omega + K_2\Gamma \tag{2-51}$$

is a tensor equation with Φ, Ω, and Γ tensors, then this equation is

---

[6] The two invariance relations for the 2×2 tensor may be verified without difficulty by referring to the general transformed tensor in Eqs. 2-40 and 2-41. It is only necessary to show that for this tensor

$$T_{11} + T_{22} = T'_{11} + T'_{22}$$

and

$$T_{11}T_{22} - T_{12}T_{21} = T'_{11}T'_{22} - T'_{12}T'_{21}$$

[7] Refer to the force vector F of Fig. 2.2 and Eq. 2-6. It is clear that this vector F represents a physical quantity independently of reference to any coordinate system. The *components* of F may differ, depending upon the directions of the x and y axes, but F itself, as a force *vector*, has identity and meaning in the form

$$F = \begin{pmatrix} F_x \\ F_y \end{pmatrix}$$

regardless of which directions we assign to x and y. This is true of tensors generally and because of this we may state, intuitively, that equations given in terms of tensors hold independently of the orientation of the axes. The *elements* of the tensors which are

Footnote continued on page 47.

valid in all coordinate systems. It is only necessary that the appropriate form of the tensor be used for the system in question.

The discussion given above represents sufficient background in the field of matrix-tensor analysis for our present purposes and is, indeed, a sound foundation for additional advanced work in these areas.

In order to illustrate applications of much of the theory presented above we shall now discuss, in some detail, a tensor which is well known to most engineering and science students.

2-5 The Inertia Tensor. Let us consider the following expression, which occurs repeatedly in many problems in dynamics and (as we shall see) also occurs in the fields of engineering elasticity—

$$I = r^2 E - r^{\star} r \tag{2-52}$$

in which[8]

$$r = (x\ y),\ \text{the distance vector}$$
$$r^2 = x^2 + y^2,\ \text{a scalar}$$
$$E = \text{unit matrix or tensor}$$

$$= \begin{pmatrix} 1 & 0 \\ 0 & 1 \end{pmatrix}$$

If we expand the above in the usual way we find

$$I = \begin{pmatrix} yy & -xy \\ -yx & xx \end{pmatrix} \tag{2-53}$$

and (see Fig. 2.5), if we multiply the right-hand side of Eq. 2-53 by dA and integrate over the entire area, then each element represents a moment of inertia (the off-diagonal elements being the "products of inertia") and we have the very important second-order tensor, the *inertia tensor*, given by

$$I = \begin{pmatrix} I_{xx} & -I_{xy} \\ -I_{yx} & I_{yy} \end{pmatrix} \tag{2-54}$$

Footnote continued from page 46.

contained in the equations will alter or change form depending upon the orientation of axes or depending upon the coordinate system being used. But the equations themselves, as statements of tensor identities, are unchanging. See also the discussion of this point in Ref. (3).

Another way of saying this:

The equations of mathematical physics are given, quantitatively, in the form of partial differential equations. These equations which express laws or relations of nature must, in their fundamental forms, be given in an expression which is independent of the coordinate system used. If this were not so then different investigators, using different permissible coordinate systems, would arrive at different solutions to the (essentially) same problem. Such a situation goes against reason and cannot be admitted in our scientific philosophy.

[8] Note that I is a tensor since it is made up of the difference two terms which are tensors, i.e., satisfy the relation $T' = RTR^*$. See Prob. 12 at the end of this chapter.

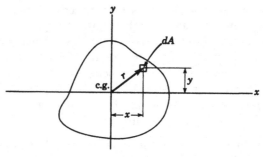

Fig. 2.5

Note that this is a *symmetric* tensor, since obviously $I_{xy} = I_{yx}$.
In this expression,

$$I_{xx} = \text{area moment of inertia with respect to } x\text{-}x \text{ axes.}$$
$$I_{yy} = \text{area moment of inertia with respect to } y\text{-}y \text{ axes.}$$
$$I_{xy} = I_{yx} = \text{area product of inertia with respect to } x\text{-}y \text{ axes.}$$

To illustrate a simple moment of inertia calculation, let us determine I_{xx} and I_{yy} for the rectangle about its centroidal axes, see Fig. 2.6.

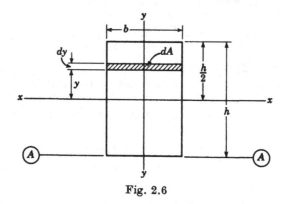

Fig. 2.6

Consider I_{xx} first. We have

$$\left. \begin{aligned}
I_{xx} &= \int_{-(h/2)}^{+(h/2)} y^2 \, dA \\
&= \int_{-(h/2)}^{+(h/2)} y^2 b \, dy \\
I_{xx} &= \frac{bh^3}{12}
\end{aligned} \right\} \tag{2-55}$$

In the same way, we would find

$$
\left.\begin{aligned}
I_{yy} &= \int_{-(b/2)}^{+(b/2)} x^2 \, dA \\
&= \int_{-(b/2)}^{+(b/2)} x^2 h \, dx \\
I_{yy} &= \frac{hb^3}{12}
\end{aligned}\right\} \tag{2-56}
$$

Also it is obvious that $I_{xy} = 0$, since for every positive contribution to this term, there is an equal negative one.

If we wish, we can determine I_{AA} where A–A is an axis through the base. In this case we have

$$
\left.\begin{aligned}
I_{AA} &= \int_0^h y^2 \, dA \\
&= \int_0^h y^2 b \, dy \\
I_{AA} &= \frac{bh^3}{3}
\end{aligned}\right\} \tag{2-57}
$$

There are other moments of inertia besides *area moments of inertia*. For example, in certain fields the concept of *line moment of inertia* is important. This is essentially a one-dimensional counterpart of the area moment of inertia. We may speak also of *mass moments of inertia* which is a three-dimensional form of the inertia tensor. In fact, one may generalize this concept to an n-dimensional space.

In the example cited above we have I_{xx} and I_{yy}, the moments of inertia with respect to the y and x axes respectively. It is also possible to define a moment of inertia, I_r, the so-called *polar moment* of inertia which is the moment of inertia with respect to an axis through the origin and normal to the area—that is (see Fig. 2.5)

$$
\left.\begin{aligned}
I_r &= \int_A r^2 \, dA \\
&= \int_A (x^2 + y^2) \, dA \\
&= \int_A x^2 \, dA + \int_A y^2 \, dA
\end{aligned}\right\} \tag{2-58}
$$

or, in general,

$$
I_r = I_{xx} + I_{yy} \tag{2-59}
$$

Also, we may show that if x and y are the centroidal axes, then given

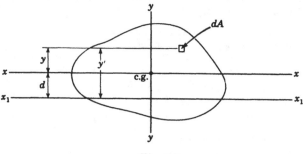

Fig. 2.7

an axis x_1 parallel to x and at, say, a distance d from it, see Fig. 2.7,

$$
\left.
\begin{aligned}
I_{x_1x_1} &= \int_A y_1^2 \, dA \\[2mm]
&= \int_A (y+d)^2 \, dA \\[2mm]
&= \int_A y^2 \, dA + 2d \int_A y \, dA + d^2 \int_A dA
\end{aligned}
\right\}
\tag{2-60}
$$

and since the x axis is a centroidal axis, the middle integral equals zero, or

$$
I_{x_1x_1} = I_{xx} + Ad^2
\tag{2-61}
$$

This is the very important Parallel Axis Theorem, and it is stated in words as follows:

The moment of inertia of an area with respect to any axis exceeds its moment of inertia with respect to a parallel axis drawn through its centroid by the product of the area by the square of the distance between the parallel axes.

The student should verify the theorem for the rectangle example given earlier in the section.

A similar parallel axis theorem may be given for lines and for volumes.

Also, a parallel axis theorem for products of inertia may be obtained in a manner similar to that used in Eq. (2.60). The student should do this as an exercise.

Going back to the tensor representation of the inertia it will be instructive to determine I', the above tensor in an O–x'–y' system that makes an angle θ with the O–x–y system. This is given at once (see Eq. 2-41) by

$$
I' = \begin{pmatrix} I_{x'x'} & -I_{x'y'} \\ -I_{y'x'} & I_{y'y'} \end{pmatrix}
$$

$$= \begin{pmatrix} I_{xx} \cos^2\theta + I_{yy} \sin^2\theta & (I_{yy} - I_{xx}) \sin\theta \cos\theta \\ -2I_{xy} \sin\theta \cos\theta & -(\cos^2\theta - \sin^2\theta)I_{xy} \\ (I_{yy} - I_{xx}) \sin\theta \cos\theta & I_{xx} \sin^2\theta + I_{yy} \cos^2\theta \\ -(\cos^2\theta - \sin^2\theta)I_{xy} & +2I_{xy} \sin\theta \cos\theta \end{pmatrix} \qquad (2\text{-}62)$$

The above represents the rotation of axes expressions for the area moment of inertia. For example, for the rectangle of Fig. 2.6 we had, about the centroidal x and y axes,

$$\left. \begin{aligned} I_{xx} &= \frac{bh^3}{12} \\ I_{yy} &= \frac{hb^3}{12} \\ I_{xy} &= I_{yx} = 0 \end{aligned} \right\} \qquad (2\text{-}63)$$

Suppose we have axes x' and y' making $45°$ with the x and y axes respectively. What will $I_{x'x'}$, $I_{y'y'}$ and $I_{x'y'}$ be given by?

We have, since $\cos^2\theta = \sin^2\theta = \sin\theta \cos\theta = \frac{1}{2}$, from Eq. 2-62, the following:

$$\left. \begin{aligned} I_{x'x'} &= \frac{I_{xx}}{2} + \frac{I_{yy}}{2} - I_{xy} = \frac{bh^3}{24} + \frac{hb^3}{24} \\ I_{y'y'} &= \frac{I_{xx}}{2} + \frac{I_{yy}}{2} + I_{xy} = \frac{bh^3}{24} + \frac{hb^3}{24} \\ I_{x'y'} &= \frac{I_{yy} - I_{xx}}{2} = \frac{hb^3}{24} - \frac{bh^3}{24} \end{aligned} \right\} \qquad (2\text{-}64)$$

Let us now determine the invariants of the area moment of inertia tensor. These are given at once (see Eqs. 2-49 and 2-50) by

$$\mathscr{I}_1 = I_{xx} + I_{yy} = I_{x'x'} + I_{y'y'} \qquad (2\text{-}65)$$

and

$$\mathscr{I}_2 = I_{xx}I_{yy} - I_{xy}I_{yx} = I_{x'x'}I_{y'y'} - I_{x'y'}I_{y'x'} \qquad (2\text{-}66)$$

We use the symbol $\mathscr{I}$ for invariant instead of I to distinguish the invariant from the inertia element.

The invariant $\mathscr{I}_1$ is usually discussed in elementary courses of statics. The invariant $\mathscr{I}_2$ is rarely mentioned; it is, however, an invariant just as $\mathscr{I}_1$ is, and in some ways it is of more interest, particularly since it is an invariant which contains all elements of the inertia tensor.

We may verify the invariance relations for the rectangle considered above as follows:

$$I_{xx} + I_{yy} = \frac{bh^3}{12} + \frac{hb^3}{12} \tag{2-67}$$

and (from Eq. 2-64)

$$I_{x'x'} + I_{y'y'} = \frac{bh^3}{12} + \frac{hb^3}{12} \tag{2-68}$$

which verifies the relation for $\mathscr{I}_1$

For $\mathscr{I}_2$ we have

$$I_{xx}I_{yy} - I_{xy}{}^2 = \frac{b^4 h^4}{144} \tag{2-69}$$

and

$$\left.\begin{array}{l} I_{x'x'}I_{y'y'} - I_{x'y'}{}^2 = \left(\dfrac{bh^3}{24} + \dfrac{hb^3}{24}\right)^2 - \left(\dfrac{bh^3}{24} - \dfrac{hb^3}{24}\right)^2 \\[2mm] \qquad = \dfrac{b^4 h^4}{144} \end{array}\right\} \tag{2-70}$$

which verifies the relation for $\mathscr{I}_2$.

We may note also that since the tensor Eq. 2-54 is a symmetric one, then by virtue of the theorem stated in Art. 2-4, it follows that this tensor may be put in diagonal form. That is, there is a set of axes, $O\text{-}x_p\text{-}y_p$, the *principal axes*, such that the tensor becomes

$$I_p = \begin{pmatrix} I_{x_p x_p} & 0 \\ 0 & I_{y_p y_p} \end{pmatrix} \tag{2-71}$$

Thus, the *products of inertia are zero about the principal axes.*

It is clear therefore, that for the rectangle of Fig. 2.6, the x and y axes are the principal axes. In fact, it must follow that axes of symmetry are *always* principal axes, since for an axis of symmetry I_{xy} will equal zero.

Also, we may show that $I_{x_p x_p}$ and $I_{y_p y_p}$ are either the maximum or minimum values of the moment of inertia about any set of orthogonal axes through O. This is done as follows: Assume $I_{xx} > I_{yy}$. This is no real restriction, since we can always label the axes so that this is so. We have, by virtue of the invariance relations,

$$\left.\begin{array}{l} I_{x_p x_p} + I_{y_p y_p} = I_{xx} + I_{yy} \\ I_{x_p x_p} I_{y_p y_p} = I_{xx} I_{yy} - I_{xy}{}^2 \end{array}\right\} \tag{2-72}$$

where the left-hand sides are the principal axes values and the right-hand sides are the values for any other set of orthogonal axes through the origin. Then

$$I_{y_p y_p} = I_{xx} + I_{yy} - I_{x_p x_p} \tag{2-73}$$

and, substituting this in the second of the above, we have

$$I_{x_p x_p}(I_{xx} + I_{yy} - I_{x_p x_p}) = I_{xx} I_{yy} - I_{xy}{}^2 \tag{2-74}$$

Solving for $I_{x_p x_p}$, we have

$$\left. \begin{aligned} I_{x_p x_p} &= \frac{(I_{xx} + I_{yy}) \pm \sqrt{(I_{xx} + I_{yy})^2 - 4(I_{xx} I_{yy} - I_{xy}{}^2)}}{2} \\[2mm] &= \frac{(I_{xx} + I_{yy}) \pm \sqrt{(I_{xx} - I_{yy})^2 + 4 I_{xy}{}^2}}{2} \\[2mm] &= \frac{(I_{xx} + I_{yy}) \pm [(I_{xx} - I_{yy}) + \epsilon]}{2} \end{aligned} \right\} \tag{2-75}$$

in which ϵ is some positive quantity. Hence

$$I_{x_p x_p} = \begin{cases} I_{xx} + \dfrac{\epsilon}{2} \\[2mm] \text{or} \\[2mm] I_{yy} - \dfrac{\epsilon}{2} \end{cases} \tag{2-76}$$

In other words, $I_{x_p x_p}$ is either greater or less than the moment of inertia about any other set of. axes. This may also be demonstrated for $I_{y_p y_p}$, in the same way.

Hence, in order for the equality of the first invariance to be valid, it follows that $I_{x_p x_p}$ and $I_{y_p y_p}$ are the $\begin{Bmatrix} \text{maximum} \\ \text{minimum} \end{Bmatrix}$ or $\begin{Bmatrix} \text{minimum} \\ \text{maximum} \end{Bmatrix}$ value of the moment of inertia about any set of orthogonal axes through the origin.

Equation 2-62 has a most suggestive form. Whenever equations are given in terms of the circular functions (sine, cosine, etc.), one is led to consider the possibility of constructing a graphical solution of the equations using circular arcs. Such a construction does, in fact, exist for these equations—and therefore for all other tensors of the second order, in two dimensions. This construction is the Mohr circle, which we describe next.

We consider Eq. 2-54 and assume that for a given set of rectangular axes the moments of inertia I_{xx}, I_{yy}, and I_{xy} are known. Then, for any other rectangular set of axes making an angle θ with the initial set, we have the primed quantities as given by Eq. 2-62. From this equation we obtain

$$I_{x'x'} = \frac{I_{xx}+I_{yy}}{2} + \frac{I_{xx}-I_{yy}}{2}\cos 2\theta - I_{xy}\sin 2\theta$$

$$I_{x'y'} = \frac{I_{xx}-I_{yy}}{2}\sin 2\theta + I_{xy}\cos 2\theta \qquad (2\text{-}77)$$

or

$$\left(I_{x'x'} - \frac{I_{x'x'}+I_{y'y'}}{2}\right)^2 + I_{x'y'}{}^2 = \left(\frac{I_{xx}-I_{yy}}{2}\right)^2 + I_{xy}{}^2$$

$$= \text{constant for any } \theta$$

But Eq. 2-77 is the equation of a circle in an I plane, center at $[(I_{xx}+I_{yy})/2, O]$ of radius equal to $\sqrt{[(I_{xx}-I_{yy})/2]^2 + I_{xy}{}^2}$. This circle may most easily be constructed as follows (see Fig. 2.8):

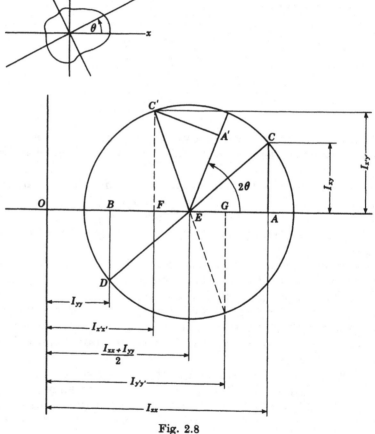

Fig. 2.8

We choose the x direction such that $I_{xx} > I_{yy}$. Lay off abscissae equal to I_{xx} ($= OA$) and I_{yy} ($= OB$). Lay off ordinates (assumed positive) I_{xy} ($= AC, BD$) as shown. From center E draw a circle through C and D. This is the Mohr circle.

Now consider $\triangle ECA$ as a movable indicator. To find $I_{x'x'}$, $I_{y'y'}$, $I_{x'y'}$ at the angle θ, rotate ECA through 2θ in the same direction as θ to position $EA'C'$. Then $OF = I_{x'x'}$, $OG = I_{y'y'}$ and $C'F = I_{x'y'}$.

In this way, a complete graphical solution of the inertia (and other tensors) may be obtained. For example, the student should verify the following relations, by referring to the Mohr circle:

(a) Principal moments of inertia always occur on planes perpendicular to each other.

(b) $(I_p)_{\max}$ = the maximum principal I

$$= \frac{I_{xx}+I_{yy}}{2} + (I_{xy})_{\max}$$

and

$$(I_p)_{\min} = \frac{I_{xx}+I_{yy}}{2} - (I_{xy})_{\max}$$

(c) The plane of $(I_{xy})_{\max}$ is at $45°$ to the principal planes.

In a similar way, other general relations concerning the elements of the second-order tensor can be obtained from the Mohr circle.

The three-dimensional form of the inertia tensor may be obtained at once from the general equation, Eq. 2-52, in its three-dimensional form. This leads to

$$I = \begin{pmatrix} I_{xx} & -I_{xy} & -I_{xz} \\ -I_{yx} & I_{yy} & -I_{yz} \\ -I_{zx} & -I_{zy} & I_{zz} \end{pmatrix} \tag{2-78}$$

This tensor also is symmetric, hence has principal axes, also it has the three invariants which are common to all tensors of this order and, of course, satisfies the standard transformation of axes relations.

Also, it may be noted, a Mohr circle construction has been developed for the three-dimensional case, see Ref. (11).

Summarizing the above discussion of tensors of the second order:

(a) If

$$U = \begin{pmatrix} u_x \\ u_y \\ u_z \end{pmatrix}$$

$$V = \begin{pmatrix} v_x \\ v_y \\ v_z \end{pmatrix}$$

are tensors of the first order, then

$$T = UV^\star$$

is a tensor of the second order.

(b) The transformation law satisfied by the second-order tensor is

$$T' = RTR^\star$$

where R is the direction cosine tensor defined by Eq. 2-17, and $T' = U'V'^\star$.

(c) As an example of the second-order tensor, we considered a tensor defined by

$$I = r^2 E - r^\star r$$

and showed that physically this corresponded to the moment of inertia tensor, and also that this was a symmetric tensor.

(d) Applying the transformation law of step (b) above to this tensor, we obtained the well-known rotation of axes relations for moments of inertia and, in addition, the two invariants of the two-dimensional form were obtained.

(e) The Mohr circle construction for two-dimensional tensors of the second order was described in detail.

(f) The three-dimensional form of the inertia was briefly discussed.

2-6 The Indicial or Subscript Tensor Notation. In the previous articles we have discussed the topic which is called "matrix-tensor analysis" in this book. The hyphenated title is used to emphasize that (1) the subject deals with tensors, i.e., the family name of physical quantities which behave in certain fixed ways when the coordinate axes are rotated about the origin; (2) these tensors may be represented by matrices.

Now, there is another very common representation of tensors—the *indicial* or *subscript* notation. Although we shall use the matrix notation throughout this book, it may be of interest to discuss briefly the indicial notation. The basis of this notation is the summation convention which we discuss first.

The expression

$$\sum_{i=1,2,3} a_{ii} \tag{2-79}$$

represents

$$a_{11} + a_{22} + a_{33} \qquad (2\text{-}80)$$

We can define a_{ii} (without the summation sign or index numbers) as the equivalent of the above, if we introduce a *summation convention*, namely,

A recurring suffix letter indicates the sum must be formed of all terms obtainable by assigning to the suffix the values of the indices. (The indices in Eq. 2-79 are $1, 2, 3$.) Thus the expression[9]

$$a_{ij}b_{ij}, \qquad i = 1, 2, 3 \qquad j = 1, 2, 3 \qquad (2\text{-}81)$$

means that the summation must be carried out for all values, $1, 2, 3$ of i and j or[9]

$$a_{ij}b_{ij} = a_{11}b_{11} + a_{22}b_{22} + a_{33}b_{33} + 2(a_{23}b_{23} + a_{31}b_{31} + a_{12}b_{12}) \quad (2\text{-}82)$$

Note, in the above expression it was assumed that

$$\left. \begin{array}{l} a_{ij} = a_{ji} \\ b_{ij} = b_{ji} \end{array} \right\} \qquad (2\text{-}83)$$

If this is not true then the expression must be modified accordingly.

The repeated letter indexes are called *dummy* indexes. The non-repeated indexes are called *free* indexes. Thus

$$a_{ij}a_{jk} = a_{i1}a_{1k} + a_{i2}a_{2k} + a_{i3}a_{3k} \qquad (2\text{-}84)$$

and j is the dummy index whereas i and k are the free indexes. Thus, a dummy index can always be replaced by any other letter which does not occur in the same term.

The use of brackets is important in this notation. Thus, for $i = 1, 2, 3$

$$(a_{ii})^2 = (a_{11} + a_{22} + a_{33})^2 \qquad (2\text{-}85)$$

whereas

$$a_{ii}^2 = a_{11}^2 + a_{22}^2 + a_{33}^2 \qquad (2\text{-}86)$$

In such a term[10] as

$$a_{ij}b_{jk} = c_{ik} \qquad (2\text{-}87)$$

the free indexes (i and k) must be the same on both sides of the equation and the equation holds for all values of the free suffixes. Thus, Eq. 2-87 stands for *nine* equations, of which two typical ones are

$$\left. \begin{array}{ll} a_{11}b_{11} + a_{12}b_{21} + a_{13}b_{31} = c_{11} & (i = 1, k = 1) \\ a_{21}b_{13} + a_{22}b_{23} + a_{23}b_{33} = c_{23} & (i = 2, k = 3) \end{array} \right\} \qquad (2\text{-}88)$$

[9] The reader should note that this is also equal to $b_{ij}a_{ij}$—in other words, the order of terms in the indicial notation multiplication may be interchanged.
[10] Note: this is just the expression $AB = C$ of the matrix notation.

The summation convention also holds for derivative and differential expressions. Thus, for $i = 1, 2, 3$

$$\frac{\partial u_i}{\partial x_i} = \frac{\partial u_1}{\partial x_1} + \frac{\partial u_2}{\partial x_2} + \frac{\partial u_3}{\partial x_3} \tag{2-89}$$

and the expression

$$\frac{\partial \sigma_{ij}}{\partial x_j} = 0 \tag{2-90}$$

is *three* equations

$$\left. \begin{array}{l} \dfrac{\partial \sigma_{11}}{\partial x_1} + \dfrac{\partial \sigma_{12}}{\partial x_2} + \dfrac{\partial \sigma_{13}}{\partial x_3} = 0 \\[2mm] \dfrac{\partial \sigma_{21}}{\partial x_1} + \dfrac{\partial \sigma_{22}}{\partial x_2} + \dfrac{\partial \sigma_{23}}{\partial x_3} = 0 \\[2mm] \dfrac{\partial \sigma_{31}}{\partial x_1} + \dfrac{\partial \sigma_{32}}{\partial x_2} + \dfrac{\partial \sigma_{33}}{\partial x_3} = 0 \end{array} \right\} \tag{2-91}$$

and in Chapter 4 it will be shown that these are the "equilibrium equations" of elasticity—body forces neglected.

An important symbol in the indicial notation is the δ symbol, which is the equivalent of E, the unit matrix. δ is defined as

$$\left. \begin{array}{l} \delta_{ij} = 1 \quad \text{if} \quad i = j \\ \delta_{ij} = 0 \quad \text{if} \quad i \neq j \end{array} \right\} \tag{2-92}$$

As an example of the use of this, the reader should verify that

$$a_{ij} - \tfrac{1}{3}\delta_{ij}a_{kk} \tag{2-93}$$

is equivalent to the following matrix, which appears in plasticity theory, Chapter 9.

$$\begin{pmatrix} \dfrac{2a_{11} - a_{22} - a_{33}}{3} & a_{12} & a_{13} \\[3mm] a_{21} & \dfrac{2a_{22} - a_{33} - a_{11}}{3} & a_{23} \\[3mm] a_{31} & a_{32} & \dfrac{2a_{33} - a_{11} - a_{22}}{3} \end{pmatrix} \tag{2-94}$$

We shall now discuss the connection between *tensors* and the indicial notation.

In Eq. 2-10 we noted that the transformation law for the first-order tensor (i.e., vector) is

$$V' = RV \tag{2-95}$$

In the indicial notation this is just given by

$$v'_i = a_{ij}v_j, \quad i = 1, 2, 3 \tag{2-96}$$

in which

v'_i is the vector in the primed system,

v_j is the vector in the unprimed system, and

a_{ij} is the rotation matrix.

Thus, the x component (i.e., 1 component) of v'_i is given by

$$v'_1 = a_{11}v_1 + a_{12}v_2 + a_{13}v_3 \tag{2-97}$$

and so on for the others.

In Eq. 2-35 we pointed out that a tensor of the second order satisfies the following transformation law

$$P' = RPR^\star \tag{2-98}$$

This is identical, in the indicial notation,[11] to

$$P'_{ij} = a_{ik}a_{jl}P_{kl} \tag{2-99}$$

Thus—for $i = 1, j = 1$, the reader should verify that both expressions give the following:

$$\begin{aligned}
P'_{11} = {} & a_{11}a_{11}P_{11} + a_{11}a_{12}P_{12} + a_{11}a_{13}P_{13} \\
& + a_{12}a_{11}P_{21} + a_{12}a_{12}P_{22} + a_{12}a_{13}P_{23} \\
& + a_{13}a_{11}P_{31} + a_{13}a_{12}P_{32} + a_{13}a_{13}P_{33}
\end{aligned} \tag{2-100}$$

In the above expression, obviously a_{ik} and a_{jl} represented the direction cosine matrices R and $R^\star$. Hence, see Probs. 1 and 2 at the end of this chapter,

$$a_{ik}a_{jk} = \delta_{ij} \tag{2-101}$$

and

$$a_{ki}a_{kj} = \delta_{ij} \tag{2-102}$$

A general property of tensors that is of interest and that follows quite simply from the indicial notation is the following:

If $f(P_{ij})$ is a function of a tensor, P_{ij} and if $q_{ij} = \partial f/\partial P_{ij}$, then we can show that q_{ij} is also a tensor (in which the transformed or primed axes form for q is given by $q'_{ij} = \partial f/\partial P'_{ij}$).

We prove this property as follows:

It may be shown (the reader should verify this by expanding the expressions and noting the application of the chain-rule in partial

[11] Note that in this expression also the order of terms on the right-hand side may be interchanged without affecting the result. See footnote, p. 57. Also we use the letter P to represent the indicial notation tensor, whereas T is used for the matrix-tensor.

differentiation), that

$$\frac{\partial f(P_{ij})}{\partial P'_{ij}} = \frac{\partial f(P_{ij})}{\partial P_{kl}} \frac{\partial P_{kl}}{\partial P'_{ij}} \tag{2-103}$$

which is also given by

$$\frac{\partial f(P_{ij})}{\partial P'_{ij}} = \frac{\partial f(P_{ij})}{\partial P_{kl}} a_{ik} a_{jl} \tag{2-104}$$

since (see Eq. 2-99)

$$P_{kl} = a_{ik} a_{jl} P'_{ij} \tag{2-105}$$

in which the role of the primed and unprimed has been interchanged.[12]
Therefore,

$$\left. \begin{aligned} q'_{ij} &= a_{ik} a_{jl} \frac{\partial f(P_{ij})}{\partial P_{kl}} \\ &= a_{ik} a_{jl} q_{kl} \end{aligned} \right\} \tag{2-106}$$

and this is just the required tensor transformation.

We may also obtain the invariants of tensors rather directly utilizing the indicial notation. Thus, as was stated in Art. 2.4, there are three invariants for the second-order tensor in the three-dimensional space. This may be verified as follows:

We shall prove that

and

$$\left. \begin{aligned} &P_{ii} \\ &P_{ij} P_{ji} \\ \\ &P_{ij} P_{jk} P_{ki} \end{aligned} \right\} \tag{2-107}$$

are invariants. That is, we shall prove that

$$P_{ii} = P'_{ii} \tag{2-108}$$

$$P_{ij} P_{ji} = P'_{ij} P'_{ji} \tag{2-109}$$

$$P_{ij} P_{jk} P_{ki} = P'_{ij} P'_{jk} P'_{ki} \tag{2-110}$$

To prove Eq. 2-108, we have (see Eq. 2-99)

$$P'_{ii} = a_{ik} a_{il} P_{kl} \tag{2-111}$$

and from Eq. 2-102, this becomes

$$P'_{ii} = \delta_{kl} P_{kl} \tag{2-112}$$

[12] This result follows at once by multiplying both sides of Equation 2-99 by $a_{ik} a_{jl}$ and noting that

$$a_{ik} a_{ik} = 1$$
$$a_{jl} a_{jl} = 1$$

or (see Eq. 2-92)

$$P'_{ii} = P_{kk} = P_{ii} \qquad \text{Q.E.D.} \qquad (2\text{-}113)$$

To prove Eq. 2-109, we have (using once again Eqs. 2-99, 2-102, and 2-92),[13]

$$
\left.
\begin{aligned}
P'_{ij}P'_{ji} &= (a_{ik}a_{jl}P_{kl})(a_{jm}a_{in}P_{mn}) \\
&= a_{ik}a_{in}P_{kl}a_{jl}a_{jm}P_{mn} \\
&= \delta_{kn}P_{kl}\delta_{lm}P_{mn} \\
&= P_{nl}P_{ln} = P_{ij}P_{ji} \quad \text{Q.E.D.}
\end{aligned}
\right\} \qquad (2\text{-}114)
$$

Finally, we prove Eq. 2-110, as follows

$$
\left.
\begin{aligned}
P'_{ij}P'_{jk}P'_{ki} &= (a_{il}a_{jm}P_{lm})(a_{jn}a_{kr}P_{nr})(a_{ks}a_{it}P_{st}) \\
&= (a_{il}a_{it}P_{lm})(a_{jm}a_{jn}P_{nr})(a_{kr}a_{ks}P_{st}) \\
&= \delta_{lt}P_{lm}\delta_{mn}P_{nr}\delta_{rs}P_{st} \\
&= P_{tm}P_{mr}P_{rt} = P_{ij}P_{jk}P_{ki} \quad \text{Q.E.D.}
\end{aligned}
\right\} \qquad (2\text{-}115)
$$

The reader should compare the invariants given in Eqs. 2-108, 2-109, and 2-110 with those given previously for the matrix tensor and should satisfy himself that the three sets are *equivalent*, although the second and third invariants are not identically the same in both forms.

2-7 Summary. The general definition was given of a tensor as a quantity having physical significance and satisfying a certain transformation law. This led at once to a discussion of the scalar, vector, and tensor, all of which are part of the family of "tensors."

Some general properties of tensors as well as the transformation laws were next discussed. Following this, much of the preceding was illustrated by referring to a specific tensor—the inertia tensor. This in turn led to a discussion of the Mohr circle.

Finally, the indicial-subscript tensor notation was briefly described.

Problems

1. Show that
   $R^{\star}R = E$. Hint: use for V the unit vectors and then use the scalar and vector product properties of normal vectors.
2. Show that

$$RR^{\star} = E$$

[13] In this and the following proof we rearrange several terms in the expressions. This can be done without changing the value of the expressions, in the indicial notation. See previous footnote on p. 57.

3. Show that E is a second-order tensor, with

$$E = E' = \begin{pmatrix} 1 & 0 & 0 \\ 0 & 1 & 0 \\ 0 & 0 & 1 \end{pmatrix}$$

4. Prove that
$$R = (R^\star)^{-1}$$

5. Prove that
$$R^\star = R^{-1}$$

6. Prove that
$$\det R = |R| = 1.$$

7. Prove that
$$A^{-1}A = E.$$

8. Show that
$$A^{\star\prime} = A^{\prime\star}.$$

9. If A is a second-order tensor and B is a second-order tensor, show that AB is also a second-order tensor. Hint: it is necessary to demonstrate that $A'B' = RABR^\star$.

10. If A is a second-order tensor, show that A^{-1} is also a second-order tensor.

11. If A is a second-order tensor, show that $A^\star$ is also a second-order tensor.

12. If A is a second-order tensor and B is a second-order tensor, show that $A+B$ is also a second-order tensor. Show the same for $A-B$.

13. If, for a given area and set of axes

$$I_{xx} = 1000$$

$$I_{yy} = 500$$

$$I_{xy} = 250$$

draw the Mohr circle and determine

$(I_{xx})_{\max}$

$(I_{yy})_{\max}$

$(I_{xy})_{\max}$

and determine the angles between the principal axes and the x–y axes.

14. (a) Determine I_{xx}, I_{yy} and I_{xy} for the figure shown.
  (b) Draw the Mohr circle, determine the maximum and minimum moment of inertia, the maximum product of inertia, and the angle between the principal axes and the x and y axes.

15. Verify by direct substitution of the

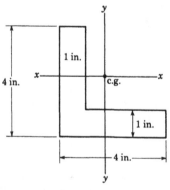

Prob. Fig. 2.14

transformed plane terms the two invariants of the two-dimensional inertia tensor.

16. (a) List the three invariants of the stress tensor;

$$T = \begin{pmatrix} \sigma_x & \tau_{xy} & \tau_{xz} \\ \tau_{yx} & \sigma_y & \tau_{yz} \\ \tau_{zx} & \tau_{zy} & \sigma_z \end{pmatrix}$$

(b) Explain by means of a sketch on a differential cube element the physical significance of the two invariants of the two-dimensional form of this tensor.

17. The following equations are given in indicial tensor form. In all cases assume a three-dimensional space. Write out the cartesian forms of the equations.

(a)
$$\frac{\partial u_i}{\partial x_i}$$

(b)
$$\frac{\partial u_i}{\partial t} + u_k \frac{\partial u_i}{\partial x_k} = g_i - \frac{1}{\rho}\frac{\partial p}{\partial x_i} + \frac{\nu \partial^2 u_i}{\partial x_k \partial x_k}$$

(c)
$$\sigma_{ij} = K\left(\frac{\partial w}{\partial v_{ij}} + \frac{\partial w}{\partial v_{ji}}\right)$$

(d)
$$e_{rs} = \frac{\partial \theta_i}{\partial x_r}\frac{\partial \theta_j}{\partial x_s}v_{ij}$$

18. Write each of the following using the summation convention:

(a)
$$d\phi = \frac{\partial \phi}{\partial x}dx + \frac{\partial \phi}{\partial y}dy + \frac{\partial \phi}{\partial z}dz$$

(b)
$$\frac{dA}{dt} = \frac{\partial A}{\partial x}\frac{dx}{dt} + \frac{\partial A}{\partial y}\frac{dy}{dt} + \frac{\partial A}{\partial z}\frac{dz}{dt}$$

(c)
$$ds^2 = g_{11}(dx_1)^2 + g_{22}(dx_2)^2 + g_{33}(dx_3)^2$$

(d)
$$\sum_{p=1}^{3}\sum_{q=1}^{3} g_{pq}\,dx_p dx_q$$

19. Solve Probs. 1 through 12 of this chapter using the indicial summation tensor forms.

Chapter 3

CURVILINEAR COORDINATES

3-1 Introduction. The fundamental equations of mathematical physics (and therefore of engineering) are, of necessity, given in a tensor form, i.e., in a form independent of coordinate systems. For example, a basic term which appears in many equations is the Laplacian of a scalar function ϕ, which in tensor (i.e., invariant) form is given by

$$\nabla^2 \phi \tag{3-1}$$

and which, in the (x, y) system of coordinates is given by

$$\frac{\partial^2 \phi}{\partial x^2} + \frac{\partial^2 \phi}{\partial y^2} \tag{3-2}$$

In the differential equations that this term occurs in, not only must the equation be solved but it is also necessary that the boundary conditions be satisfied. It is, in fact, this requirement which makes it possible to obtain unique solutions to given problems. If the boundary is a straight wall or a right-angle corner it will generally be most convenient to use rectangular (x, y) coordinates. However if the boundary is curved—say, circular, or elliptical, or some other shape—then the solution can generally be best obtained in circular or elliptical or other *curvilinear* coordinates, since the boundary condition will invariably be most simply stated in this coordinate system.

For example, the two-dimensional circular coordinates are the *polar coordinates* (r, θ). That is, lines (or curves) of r = constant, θ = constant, take the place of the rectangular coordinates, x = constant, y = constant, as shown in Fig. 3.1.

An essential and extremely important similarity in both systems of coordinates shown in Fig. 3.1 is that they are both "*orthogonal*" coordinate systems, i.e., in both sets of coordinates, the curves of constant coordinates are normal to each other at all points of intersection. This will be true of all coordinate systems we consider, although it is not required that this be so, in general. We call such coordinates *orthogonal curvilinear coordinates*, and we shall study in this chapter the methods

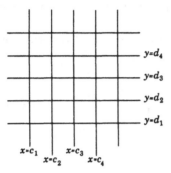
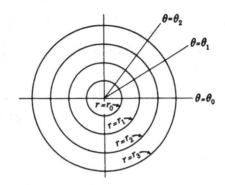

(a) Rectangular coordinates   (b) Polar (circular) coordinates

Fig. 3.1

by which equations may be transformed into the infinite number of possible curvilinear coordinate systems.[1]

The following discussion of curvilinear coordinates, which is based almost entirely on matrix-tensor arguments, follows the treatment of this topic given by Murnaghan, Ref. (1).

3-2   The Rotation Matrix and the Magnification Factors. We review some preliminary ideas first. If

$$V = \begin{pmatrix} v_x \\ v_y \\ v_z \end{pmatrix} \tag{3-3}$$

is a first-order tensor in the O-x-y-z system (see Art. 2-3, and note that the following expression is obtained by simply multiplying both sides of Eq. 2.16 by $R^\star$ and recalling that $R^\star R = E$), then

$$V = R^\star V' \tag{3-4}$$

where

$$V' = \begin{pmatrix} v'_x \\ v'_y \\ v'_z \end{pmatrix} \tag{3-5}$$

[1]The equations are given initially in terms of x, y, z, (and not the more general rotated x', y', z') coordinates. We wish to give these equations in terms of α, β, γ orthogonal coordinates (and not in terms of more general rotated α', β', γ' coordinates). This is accomplished by assuming that at all points the directions of the α, β, γ curves coincide with the directions of the x, y, z curves and this is the basic assumption in the technique which will be described here. See also the footnote on p. 69–70.

is this vector in the O–x'–y'–z' system and R is the rotation (direction cosine) matrix.

If we introduce coordinates (α, β, γ) related to (x, y, z) through the equations[2]

$$
\left.
\begin{aligned}
x &= x\,(\alpha, \beta, \gamma) \\
y &= y\,(\alpha, \beta, \gamma) \\
z &= z\,(\alpha, \beta, \gamma)
\end{aligned}
\right\}
\tag{3-6}
$$

and

$$
\left.
\begin{aligned}
\alpha &= \alpha\,(x, y, z) \\
\beta &= \beta\,(x, y, z) \\
\gamma &= \gamma\,(x, y, z)
\end{aligned}
\right\}
\tag{3-7}
$$

then the curves

$$
\left.
\begin{aligned}
x &= \text{constant} \\
y &= \text{constant} \\
z &= \text{constant}
\end{aligned}
\right\}
\tag{3-8}
$$

form a system of orthogonal curvilinear coordinates (actually rectangular coordinates for this special case), and the curves

$$
\left.
\begin{aligned}
\alpha &= \text{constant} \\
\beta &= \text{constant} \\
\gamma &= \text{constant}
\end{aligned}
\right\}
\tag{3-9}
$$

will also form a system of curvilinear coordinates. In particular, if the curves of Eq. 3-9 are mutually perpendicular at every point, then these are also an orthogonal curvilinear set.

Because of Eq. 3-6 we have, from the chain rule of differentiation,

$$
dx = \frac{\partial x}{\partial \alpha}d\alpha + \frac{\partial x}{\partial \beta}d\beta + \frac{\partial x}{\partial \gamma}d\gamma
\tag{3-10}
$$

[2] In Eqs. 3-6 and 3-7 x, y, z are mutually independent and likewise α, β, γ are mutually independent. Therefore neither of the following Jacobian determinants can vanish identically, i.e.,

$$
J\left|\frac{x,\,y,\,z}{\alpha,\,\beta,\,\gamma}\right| =
\begin{vmatrix}
\dfrac{\partial x}{\partial \alpha} & \dfrac{\partial x}{\partial \beta} & \dfrac{\partial x}{\partial \gamma} \\[2ex]
\dfrac{\partial y}{\partial \alpha} & \dfrac{\partial y}{\partial \beta} & \dfrac{\partial y}{\partial \gamma} \\[2ex]
\dfrac{\partial z}{\partial \alpha} & \dfrac{\partial z}{\partial \beta} & \dfrac{\partial z}{\partial \gamma}
\end{vmatrix} \neq 0
\qquad
J\left|\frac{\alpha,\,\beta,\,\gamma}{x,\,y,\,z}\right| =
\begin{vmatrix}
\dfrac{\partial \alpha}{\partial x} & \dfrac{\partial \alpha}{\partial y} & \dfrac{\partial \alpha}{\partial z} \\[2ex]
\dfrac{\partial \beta}{\partial x} & \dfrac{\partial \beta}{\partial y} & \dfrac{\partial \beta}{\partial z} \\[2ex]
\dfrac{\partial \gamma}{\partial x} & \dfrac{\partial \gamma}{\partial y} & \dfrac{\partial \gamma}{\partial z}
\end{vmatrix} \neq 0
$$

or, for the three components dx, dy, and dz,

$$
\begin{pmatrix} dx \\ dy \\ dz \end{pmatrix} = \begin{pmatrix} \dfrac{\partial x}{\partial \alpha} & \dfrac{\partial x}{\partial \beta} & \dfrac{\partial x}{\partial \gamma} \\[6pt] \dfrac{\partial y}{\partial \alpha} & \dfrac{\partial y}{\partial \beta} & \dfrac{\partial y}{\partial \gamma} \\[6pt] \dfrac{\partial z}{\partial \alpha} & \dfrac{\partial z}{\partial \beta} & \dfrac{\partial z}{\partial \gamma} \end{pmatrix} \begin{pmatrix} d\alpha \\ d\beta \\ d\gamma \end{pmatrix}
\tag{3-11}
$$

or

$$
dx = A_1 \, d\alpha
$$

where

$$
A_1 = \begin{pmatrix} \dfrac{\partial x}{\partial \alpha} & \dfrac{\partial x}{\partial \beta} & \dfrac{\partial x}{\partial \gamma} \\[6pt] \dfrac{\partial y}{\partial \alpha} & \dfrac{\partial y}{\partial \beta} & \dfrac{\partial y}{\partial \gamma} \\[6pt] \dfrac{\partial z}{\partial \alpha} & \dfrac{\partial z}{\partial \beta} & \dfrac{\partial z}{\partial \gamma} \end{pmatrix}
\tag{3-12}
$$

and similarly, from Eq. 3-7,[3]

$$
\begin{pmatrix} d\alpha \\ d\beta \\ d\gamma \end{pmatrix} = \begin{pmatrix} \dfrac{\partial \alpha}{\partial x} & \dfrac{\partial \alpha}{\partial y} & \dfrac{\partial \alpha}{\partial z} \\[6pt] \dfrac{\partial \beta}{\partial x} & \dfrac{\partial \beta}{\partial y} & \dfrac{\partial \beta}{\partial z} \\[6pt] \dfrac{\partial \gamma}{\partial x} & \dfrac{\partial \gamma}{\partial y} & \dfrac{\partial \gamma}{\partial z} \end{pmatrix} \begin{pmatrix} dx \\ dy \\ dz \end{pmatrix}
\tag{3-13}
$$

or

$$
d\alpha = A_2 \, dx
\tag{3-14}
$$

[3] Note that the first matrices on the right-hand side of both Eqs. 3-11 and 3-13 are the Jacobian *matrices*, defined by

$$
J\!\left(\frac{x,\, y,\, z}{\alpha,\, \beta,\, \gamma}\right) = \begin{pmatrix} \dfrac{\partial x}{\partial \alpha} & \dfrac{\partial x}{\partial \beta} & \dfrac{\partial x}{\partial \gamma} \\[6pt] \dfrac{\partial y}{\partial \alpha} & \dfrac{\partial y}{\partial \beta} & \dfrac{\partial y}{\partial \gamma} \\[6pt] \dfrac{\partial z}{\partial \alpha} & \dfrac{\partial z}{\partial \beta} & \dfrac{\partial z}{\partial \gamma} \end{pmatrix} \qquad J\!\left(\frac{\alpha,\, \beta,\, \gamma}{x,\, y,\, z}\right) = \begin{pmatrix} \dfrac{\partial \alpha}{\partial x} & \dfrac{\partial \alpha}{\partial y} & \dfrac{\partial \alpha}{\partial z} \\[6pt] \dfrac{\partial \beta}{\partial x} & \dfrac{\partial \beta}{\partial y} & \dfrac{\partial \beta}{\partial z} \\[6pt] \dfrac{\partial \gamma}{\partial x} & \dfrac{\partial \gamma}{\partial y} & \dfrac{\partial \gamma}{\partial z} \end{pmatrix}
$$

Compare with footnote on p. 66 in which Jacobian *determinants* are discussed.

in which

$$A_2 = \begin{pmatrix} \dfrac{\partial \alpha}{\partial x} & \dfrac{\partial \alpha}{\partial y} & \dfrac{\partial \alpha}{\partial z} \\[2mm] \dfrac{\partial \beta}{\partial x} & \dfrac{\partial \beta}{\partial y} & \dfrac{\partial \beta}{\partial z} \\[2mm] \dfrac{\partial \gamma}{\partial x} & \dfrac{\partial \gamma}{\partial y} & \dfrac{\partial \gamma}{\partial z} \end{pmatrix}$$

(3-15)

Now

$$A_1 A_2 = \begin{pmatrix} 1 & 0 & 0 \\ 0 & 1 & 0 \\ 0 & 0 & 1 \end{pmatrix}$$

(3-16)

since, for example, considering the term of the first row, second column, of $A_1 A_2$ we have, again utilizing the chain rule of differentiation,

$$\frac{\partial x}{\partial \alpha}\frac{\partial \alpha}{\partial y} + \frac{\partial x}{\partial \beta}\frac{\partial \beta}{\partial y} + \frac{\partial x}{\partial \gamma}\frac{\partial \gamma}{\partial y} = \frac{\partial x}{\partial y} = 0$$

(3-17)

(since x and y are independent of each other) and for the first row, first column, term we have

$$\frac{\partial x}{\partial \alpha}\frac{\partial \alpha}{\partial x} + \frac{\partial x}{\partial \beta}\frac{\partial \beta}{\partial x} + \frac{\partial x}{\partial \gamma}\frac{\partial \gamma}{\partial x} = \frac{\partial x}{\partial x} = 1$$

(3-18)

The other seven terms of $A_1 A_2$ are obtained in a similar manner.

Therefore A_1 and A_2 are inverse to each other.

In discussing curvilinear coordinates and the transformations related thereto, it is necessary that we obtain a measure of the relative lengths of the curvilinear coordinates in comparison to the lengths of the rectangular coordinates. That is, if (α, β, γ) are the curvilinear coordinates, then it is desirable that we have a *measure* of these in terms in (x, y, z). Furthermore, since we are primarily concerned with differential equations given originally in terms of rectangular coordinates (x, y, z) and the transformation of these in terms of curvilinear coordinates (α, β, γ) we shall require a comparison of the lengths of $(d\alpha, d\beta, d\gamma)$ in terms of (dx, dy, dz). This comparison is most directly effected if at each point in the field we assume the tangents to the curves of constant (α, β, γ) of the orthogonal curvilinear coordinate system coincide with the curves of constant (x, y, z) of the orthogonal rectangular system; this will, in fact, be the basis of our discussion of transformation theory. In other words, we shall essentially assume that at each point in the field the (α,β,γ) axes are rotated so that they coincide with the (x,y,z) axes and the rotation matrix will be evaluated accordingly.

Note that this is equivalent to the statement that at all points we shall assume the matrix components $\begin{pmatrix} d\alpha \\ d\beta \\ d\gamma \end{pmatrix}$ have the same directions as the corresponding components $\begin{pmatrix} dx \\ dy \\ dz \end{pmatrix}$. The reader should note, however, that *dimensionally* $d\alpha$, $d\beta$, and $d\gamma$ may not be the same as dx, dy, or dz. This dimensional homogeneity is introduced by means of the magnification factors which we discuss next.

If at a given point, $\begin{pmatrix} \alpha \\ \beta \\ \gamma \end{pmatrix}$ has the direction $\begin{pmatrix} x \\ y \\ z \end{pmatrix}$, then Eq. 3-11 takes the form

$$\begin{pmatrix} dx \\ dy \\ dz \end{pmatrix} = \begin{pmatrix} \dfrac{1}{h_1} & 0 & 0 \\ 0 & \dfrac{1}{h_2} & 0 \\ 0 & 0 & \dfrac{1}{h_3} \end{pmatrix} \begin{pmatrix} d\alpha \\ d\beta \\ d\gamma \end{pmatrix} \tag{3-19}$$

so that (h_1, h_2, h_3) are *magnification factors* for the curvilinear matrix components $(d\alpha, d\beta, d\gamma)$. We see from Eq. 3-19, for example, that

$$ds^2 = dx^2 + dy^2 + dz^2 = \frac{d\alpha^2}{h_1{}^2} + \frac{d\beta^2}{h_2{}^2} + \frac{d\gamma^2}{h_3{}^2} \tag{3-20}$$

and this is a useful form for determining (h_1, h_2, h_3) in any given case.[4]

---

[4] In the most general case (see Ref. 46),

$$(ds)^2 = g_{ij}\, dx_i\, dx_j \qquad i = 1, 2, 3 \qquad j = 1, 2, 3$$

in which (x_i, x_j) are the curvilinear coordinates and the g_{ij} are called metric coefficients which may be represented in matrix form as

$$\begin{pmatrix} g_{11} & g_{12} & g_{13} \\ g_{21} & g_{22} & g_{23} \\ g_{31} & g_{32} & g_{33} \end{pmatrix}$$

For the special case of *orthogonal* coordinate systems—and *these are the only systems that we shall consider herein*—it may be shown that the matrix takes the form

$$\begin{pmatrix} g_{11} & 0 & 0 \\ 0 & g_{22} & 0 \\ 0 & 0 & g_{33} \end{pmatrix}$$

Footnote continued on page 70.

Also, in view of Eq. 3-16 and because of Eq. 3-19, it follows that the matrix A_2 of Eqs. 3-14 and 3.15 is given by

$$\begin{pmatrix} h_1 & 0 & 0 \\ 0 & h_2 & 0 \\ 0 & 0 & h_3 \end{pmatrix} \tag{3-21}$$

when the directions of $\begin{pmatrix} d\alpha \\ d\beta \\ d\gamma \end{pmatrix}$ coincide with the directions of $\begin{pmatrix} dx \\ dy \\ dz \end{pmatrix}$

EXAMPLE OF POLAR COORDINATES

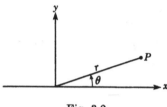

Fig. 3.2

A simple application of the above results will help to fix ideas. Consider the plane polar coordinates (r, θ). These correspond to (α, β) of the above notation. Also (see Fig. 3.2), for any point P we have

$$\left. \begin{array}{l} x = r \cos \theta \\ y = r \sin \theta \end{array} \right\} \tag{3-22}$$

so that from Eqs. 3-10 and 3-11

$$\begin{pmatrix} dx \\ dy \end{pmatrix} = \begin{pmatrix} \cos \theta & -r \sin \theta \\ \sin \theta & r \cos \theta \end{pmatrix} \begin{pmatrix} dr \\ d\theta \end{pmatrix} \tag{3-23}$$

or

$$dx = A_1 \, dr \tag{3-24}$$

Footnote continued from page 69.

For our purposes, it is deemed best to arrive at this form as a consequence of the requirement that, at a given point, the components $(d\alpha \ d\beta \ d\gamma)$ have the same direction as $(dx \ dy \ dz)$—which is, in fact, the physical interpretation which one may give to the above diagonal form of the matrix.

Furthermore we see that $1/h_1$, $1/h_2$, and $1/h_3$ are quantities which *transform* $d\alpha$, $d\beta$, and $d\gamma$ into the same dimensional form as dx, dy, and dz. It follows, therefore, that,

$$\begin{pmatrix} \dfrac{1}{h_1} d\alpha \\ \dfrac{1}{h_2} d\beta \\ \dfrac{1}{h_3} d\gamma \end{pmatrix}$$ have the same dimensions as $\begin{pmatrix} dx \\ dy \\ dz \end{pmatrix}$ and insofar as $\begin{pmatrix} dx \\ dy \\ dz \end{pmatrix}$ may be thought of as

components of a vector, then also $\begin{pmatrix} \dfrac{1}{h_1} d\alpha \\ \dfrac{1}{h_2} d\beta \\ \dfrac{1}{h_3} d\gamma \end{pmatrix}$ may be considered as the components of the

same vector in the α, β, γ system of curvilinear coordinates.

Also, since

$$r = (x^2 + y^2)^{1/2}$$
$$\theta = \tan^{-1}\frac{y}{x}$$

$$(3\text{-}25)$$

we have

$$\frac{\partial r}{\partial x} = \frac{x}{(x^2+y^2)^{1/2}} = \cos\theta$$

$$\frac{\partial r}{\partial y} = \frac{y}{(x^2+y^2)^{1/2}} = \sin\theta$$

$$\frac{\partial \theta}{\partial x} = -\frac{y}{(x^2+y^2)} = -\frac{\sin\theta}{r}$$

$$\frac{\partial \theta}{\partial y} = \frac{x}{(x^2+y^2)} = \frac{\cos\theta}{r}$$

$$(3\text{-}26)$$

so that $A_1 A_2$ which is given by

$$\begin{pmatrix} \dfrac{\partial x}{\partial r} & \dfrac{\partial x}{\partial \theta} \\[2mm] \dfrac{\partial y}{\partial r} & \dfrac{\partial y}{\partial \theta} \end{pmatrix} \begin{pmatrix} \dfrac{\partial r}{\partial x} & \dfrac{\partial r}{\partial y} \\[2mm] \dfrac{\partial \theta}{\partial x} & \dfrac{\partial \theta}{\partial y} \end{pmatrix} \qquad (3\text{-}27)$$

is equal to

$$E_2 = \begin{pmatrix} 1 & 0 \\ 0 & 1 \end{pmatrix} \qquad (3\text{-}28)$$

as the reader may verify.

When $\begin{pmatrix} dr \\ d\theta \end{pmatrix}$ have the directions $\begin{pmatrix} dx \\ dy \end{pmatrix}$, then obviously $\theta = 0$, and Eq. 3-23 becomes

$$\begin{pmatrix} dx \\ dy \end{pmatrix} = \begin{pmatrix} 1 & 0 \\ 0 & r \end{pmatrix} \begin{pmatrix} dr \\ d\theta \end{pmatrix} \qquad (3\text{-}29)$$

so that (see Eq. 3-19)

$$\frac{1}{h_1} = 1$$
$$\frac{1}{h_2} = r$$

$$(3\text{-}30)$$

which may have been determined independently from Eq. 3-20, since

$$ds^2 = dx^2 + dy^2 = dr^2 + r^2\,d\theta^2 \qquad (3\text{-}31)$$

Because of Eq. 3-23, we see that, in the general case,

$$\begin{pmatrix} dx \\ dy \end{pmatrix} = \begin{pmatrix} \cos\theta & -\sin\theta \\ \sin\theta & \cos\theta \end{pmatrix} \begin{pmatrix} dr \\ r\,d\theta \end{pmatrix} \qquad (3\text{-}32)$$

We see therefore that the *vector* $\begin{pmatrix} dr \\ r\,d\theta \end{pmatrix}$ has the same dimensions as $\begin{pmatrix} dx \\ dy \end{pmatrix}$, and the equation may be interpreted as the transformation equation for a two-dimensional vector (see Eq. 2-11), with

$$R^{\star} = \begin{pmatrix} \cos\theta & -\sin\theta \\ \sin\theta & \cos\theta \end{pmatrix} \tag{3-33}$$

as required.

For the case where $\theta = 0$

$$R^{\star} = \begin{pmatrix} 1 & 0 \\ 0 & 1 \end{pmatrix} \tag{3-34}$$

That is, $R^{\star}$ becomes the unit matrix.

Returning now to the general three-dimensional case, we note that since

$$\frac{\partial}{\partial x} = \frac{\partial}{\partial\alpha}\frac{\partial\alpha}{\partial x} + \frac{\partial}{\partial\beta}\frac{\partial\beta}{\partial x} + \frac{\partial}{\partial\gamma}\frac{\partial\gamma}{\partial x} \tag{3-35}$$

then for the case where the directions of the vectors coincide, we have, from Eqs. 3-13, 3-14, and 3-21,

$$\frac{\partial}{\partial x} = h_1\frac{\partial}{\partial\alpha} \tag{3-36}$$

and similarly,

$$\frac{\partial}{\partial y} = h_2\frac{\partial}{\partial\beta} \tag{3-37}$$

and

$$\frac{\partial}{\partial z} = h_3\frac{\partial}{\partial\gamma} \tag{3-38}$$

These relations enable us to obtain the generalized form of the rotation matrix, $R^{\star}$, in curvilinear coordinates, in the following way.

Note that $\begin{pmatrix} dx \\ dy \\ dz \end{pmatrix}$ is a vector, corresponding to V of Eq. 3-3; $\begin{pmatrix} d\alpha \\ d\beta \\ d\gamma \end{pmatrix}$, however, are differential coordinates (not necessarily vector components). Hence A_1 of Eq. 3-11 may not, directly, be taken as the rotation matrix. However, the factors $1/h_1$, $1/h_2$, $1/h_3$ are just the terms which transform $\begin{pmatrix} d\alpha \\ d\beta \\ d\gamma \end{pmatrix}$ into vector components

$$\begin{pmatrix} \dfrac{1}{h_1}d\alpha \\[2ex] \dfrac{1}{h_2}d\beta \\[2ex] \dfrac{1}{h_3}d\gamma \end{pmatrix}$$

in accordance with the transformation Eqs. 3-4 and 3-5. It is clear, therefore, that Eq. 3-4 will be the required vector transformation equation if $R^\star$ is defined, in a generalized form, as

$$R^\star = \begin{pmatrix} h_1\dfrac{\partial x}{\partial \alpha} & h_2\dfrac{\partial x}{\partial \beta} & h_3\dfrac{\partial x}{\partial \gamma} \\[2ex] h_1\dfrac{\partial y}{\partial \alpha} & h_2\dfrac{\partial y}{\partial \beta} & h_3\dfrac{\partial y}{\partial \gamma} \\[2ex] h_1\dfrac{\partial z}{\partial \alpha} & h_2\dfrac{\partial z}{\partial \beta} & h_3\dfrac{\partial z}{\partial \gamma} \end{pmatrix} \qquad (3\text{-}39)$$

so that

$$\begin{pmatrix} dx \\ dy \\ dz \end{pmatrix} = \begin{pmatrix} h_1\dfrac{\partial x}{\partial \alpha} & h_2\dfrac{\partial x}{\partial \beta} & h_3\dfrac{\partial x}{\partial \gamma} \\[2ex] h_1\dfrac{\partial y}{\partial \alpha} & h_2\dfrac{\partial y}{\partial \beta} & h_3\dfrac{\partial y}{\partial \gamma} \\[2ex] h_1\dfrac{\partial z}{\partial \alpha} & h_2\dfrac{\partial z}{\partial \beta} & h_3\dfrac{\partial z}{\partial \gamma} \end{pmatrix} \begin{pmatrix} \dfrac{1}{h_1}d\alpha \\[2ex] \dfrac{1}{h_2}d\beta \\[2ex] \dfrac{1}{h_3}d\gamma \end{pmatrix} \qquad (3\text{-}40)$$

$R^\star$ in this form is consistent with all the previous expressions and is just the rotation matrix of tensor theory in orthogonal curvilinear coordinates.

Furthermore, at any point when the directions of constant (α, β, γ) coincide with those of constant (x, y, z), then $R^\star$ becomes the unit matrix (see Eq. 3-19)

$$R^\star = \begin{pmatrix} 1 & 0 & 0 \\ 0 & 1 & 0 \\ 0 & 0 & 1 \end{pmatrix}$$

However, it is important to note that $R^\star$ *varies* from point to point. It is only for the special condition, at any point, that (α, β, γ) is assumed to be in the directions (x, y, z) that $R^\star$ becomes the unit matrix.

We may now determine the general curvilinear form of one of the expressions which appears in the equations of mathematical physics and engineering—the gradient of a scalar field.

3-3 The Gradient of a Scalar Field in Curvilinear Coordinates.

By *definition*, the gradient of a scalar field in curvilinear coordinates is the value of the gradient when the curvilinear axes are coincident with the cartesian (x, y, z) axes. Then

$$
\begin{aligned}
\operatorname{grad} f &= \begin{pmatrix} \dfrac{\partial f}{\partial x} \\[2ex] \dfrac{\partial f}{\partial y} \\[2ex] \dfrac{\partial f}{\partial z} \end{pmatrix} \\[4ex]
&= \begin{pmatrix} \dfrac{\partial \alpha}{\partial x} & \dfrac{\partial \beta}{\partial x} & \dfrac{\partial \gamma}{\partial x} \\[2ex] \dfrac{\partial \alpha}{\partial y} & \dfrac{\partial \beta}{\partial y} & \dfrac{\partial \gamma}{\partial y} \\[2ex] \dfrac{\partial \alpha}{\partial z} & \dfrac{\partial \beta}{\partial z} & \dfrac{\partial \gamma}{\partial z} \end{pmatrix} \begin{pmatrix} \dfrac{\partial f}{\partial \alpha} \\[2ex] \dfrac{\partial f}{\partial \beta} \\[2ex] \dfrac{\partial f}{\partial \gamma} \end{pmatrix}
\end{aligned} \tag{3-41}
$$

or, from Eqs. 3-13 and 3-21,

$$
\begin{pmatrix} \dfrac{\partial f}{\partial x} \\[2ex] \dfrac{\partial f}{\partial y} \\[2ex] \dfrac{\partial f}{\partial z} \end{pmatrix} = \begin{pmatrix} h_1 & 0 & 0 \\ 0 & h_2 & 0 \\ 0 & 0 & h_3 \end{pmatrix} \begin{pmatrix} \dfrac{\partial f}{\partial \alpha} \\[2ex] \dfrac{\partial f}{\partial \beta} \\[2ex] \dfrac{\partial f}{\partial \gamma} \end{pmatrix} \tag{3-42}
$$

so that, in curvilinear coordinates,

$$
\operatorname{grad} f = \begin{pmatrix} h_1 \dfrac{\partial f}{\partial \alpha} \\[3ex] h_2 \dfrac{\partial f}{\partial \beta} \\[3ex] h_3 \dfrac{\partial f}{\partial \gamma} \end{pmatrix} \tag{3-43}
$$

This result could have been written down at once utilizing Eqs. 3-36, 3-37, and 3-38. In particular, for plane polar coordinates,

$$
\operatorname{grad} f = \begin{pmatrix} \dfrac{\partial f}{\partial r} \\[3ex] \dfrac{1}{r} \dfrac{\partial f}{\partial \theta} \end{pmatrix} \tag{3-44}
$$

In order to obtain the more complicated expressions in curvilinear coordinates, it is necessary that we obtain the curvilinear forms for the derivatives of $R^\star$, and this is done next.

3-4 The Derivatives of R in Curvilinear Coordinates.

Now consider the case

$$V = R^\star A \tag{3-45}$$

where V is a vector in the $(O\text{–}x\text{–}y\text{–}z)$ system and A is the same vector in curvilinear coordinates (α, β, γ), $R^\star$ being the rotation matrix. In general, the components of A are rotated from those of V.

Although at any point we shall assume that the directions of constant (α, β, γ) are along constant (x, y, z) so that $R^\star$ becomes the unit matrix, if we differentiate $R^\star$ at any point, the result is not equal to zero, since $R^\star$ varies from point to point in the curvilinear coordinate field. Thus, if we take derivatives with respect to x of the above equation, we have

$$\frac{\partial V}{\partial x} = \frac{\partial R^\star}{\partial x}A + R^\star \frac{\partial A}{\partial x} \tag{3-46}$$

or, because of Eq. 3-36,

$$\frac{\partial V}{\partial x} = \frac{\partial R^\star}{\partial x}A + R^\star h_1 \frac{\partial A}{\partial \alpha} \tag{3-47}$$

where, in the general case (see Eq. 3-40),

$$R^\star = \begin{pmatrix} h_1\dfrac{\partial x}{\partial \alpha} & h_2\dfrac{\partial x}{\partial \beta} & h_3\dfrac{\partial x}{\partial \gamma} \\[2ex] h_1\dfrac{\partial y}{\partial \alpha} & h_2\dfrac{\partial y}{\partial \beta} & h_3\dfrac{\partial y}{\partial \gamma} \\[2ex] h_1\dfrac{\partial z}{\partial \alpha} & h_2\dfrac{\partial z}{\partial \beta} & h_3\dfrac{\partial z}{\partial \gamma} \end{pmatrix} \tag{3-48}$$

Let us represent V and A as

$$V = \begin{pmatrix} v_x \\ v_y \\ v_z \end{pmatrix} \tag{3-49}$$

and

$$A = \begin{pmatrix} a_\alpha \\ a_\beta \\ a_\gamma \end{pmatrix} \tag{3-50}$$

Now $\partial R^\star/\partial x$ is an antisymmetric tensor. This follows from the fact that everywhere

$$RR^\star = E \tag{3-51}$$

so that

$$\frac{\partial R}{\partial x}R^\star + R\frac{\partial R^\star}{\partial x} = 0 \qquad (3\text{-}52)$$

everywhere, and in particular, when the axes coincide $R = R^\star =$ unit tensor or

$$\frac{\partial R}{\partial x} + \frac{\partial R^\star}{\partial x} = 0 \qquad (3\text{-}53)$$

which proves the antisymmetry property. Therefore, the diagonal elements of $\partial R^\star/\partial x$ are all zero. In addition, it may be shown that if r_{jk} is the element of the jth row and k^{th} column of $R^\star$, then

$$\frac{\partial r_{jk}}{\partial x_l} = 0 \quad \text{if} \quad l \neq j \neq k \qquad (3\text{-}54)$$

and, using the fact that the squared sum of the columns and rows of $R^\star$ are equal to unity, we obtain, finally, the following very important relations (the reader is referred to Ref. (1) for details of the development):

$$\frac{\partial R^\star}{\partial x} = h_1 \begin{pmatrix} 0 & h_2\dfrac{\partial(1/h_1)}{\partial\beta} & h_3\dfrac{\partial(1/h_1)}{\partial\gamma} \\[2ex] -h_2\dfrac{\partial(1/h_1)}{\partial\beta} & 0 & 0 \\[2ex] -h_3\dfrac{\partial(1/h_1)}{\partial\gamma} & 0 & 0 \end{pmatrix}$$

$$\frac{\partial R^\star}{\partial y} = h_2 \begin{pmatrix} 0 & -h_1\dfrac{\partial(1/h_2)}{\partial\alpha} & 0 \\[2ex] h_1\dfrac{\partial(1/h_2)}{\partial\alpha} & 0 & h_3\dfrac{\partial(1/h_2)}{\partial\gamma} \\[2ex] 0 & -h_3\dfrac{\partial(1/h_2)}{\partial\gamma} & 0 \end{pmatrix}$$

$$\frac{\partial R^\star}{\partial z} = h_3 \begin{pmatrix} 0 & 0 & -h_1\dfrac{\partial(1/h_3)}{\partial\alpha} \\[2ex] 0 & 0 & -h_2\dfrac{\partial(1/h_3)}{\partial\beta} \\[2ex] h_1\dfrac{\partial(1/h_3)}{\partial\alpha} & h_2\dfrac{\partial(1/h_3)}{\partial\beta} & 0 \end{pmatrix} \qquad (3\text{-}55)$$

Equation 3-55 and those which precede it enable us to determine in any curvilinear system the various scalar, vector, and second-order tensor relations which occur in the fields of engineering and physics. Several examples will suffice to bring out the usefulness of these relations. As a preliminary step we consider two special types of space curvilinear coordinates which are frequently encountered in practical problems—cylindrical and spherical coordinates. We shall then obtain the various illustrative expressions in general terms and also in terms of these special coordinate systems.

3-5 Cylindrical Coordinates. Cylindrical coordinates (ρ, ϕ, z)

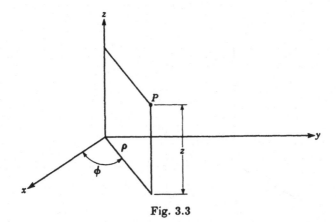

Fig. 3.3

are related to rectangular coordinates (x, y, z) by means of the relations (see Fig. 3.3)

$$\left.\begin{array}{l} x = \rho \cos \phi \\ y = \rho \sin \phi \\ z = z \end{array}\right\} \tag{3-56}$$

so that Eq. 3-11 becomes

$$\begin{pmatrix} dx \\ dy \\ dz \end{pmatrix} = \begin{pmatrix} \cos \phi & -\rho \sin \phi & 0 \\ \sin \phi & \rho \cos \phi & 0 \\ 0 & 0 & 1 \end{pmatrix} \begin{pmatrix} d\rho \\ d\phi \\ dz \end{pmatrix} \tag{3-57}$$

and, using Eq. 3-20,

$$ds^2 = dx^2 + dy^2 + dz^2 = d\rho^2 + \rho^2 d\phi^2 + dz^2 \tag{3-58}$$

from which

$$\left.\begin{array}{l} h_1 = 1 \\[2mm] h_2 = \dfrac{1}{\rho} \\[2mm] h_3 = 1 \end{array}\right\} \qquad (3\text{-}59)$$

3-6 Spherical Coordinates. For spherical coordinates (r, ϕ, θ) we have (see Fig. 3.4)

$$\left.\begin{array}{l} x = r \sin\theta \cos\phi \\ y = r \sin\theta \sin\phi \\ z = r \cos\theta \end{array}\right\} \qquad (3\text{-}60)$$

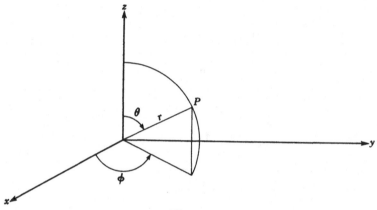

Fig. 3.4

so that

$$\begin{pmatrix} dx \\ dy \\ dz \end{pmatrix} = \begin{pmatrix} \sin\theta\cos\phi & -r\sin\theta\sin\phi & r\cos\theta\cos\phi \\ \sin\theta\sin\phi & r\sin\theta\cos\phi & r\cos\theta\sin\phi \\ \cos\theta & 0 & -r\sin\theta \end{pmatrix} \begin{pmatrix} dr \\ d\phi \\ d\theta \end{pmatrix} \qquad (3\text{-}61)$$

and

$$ds^2 = dx^2 + dy^2 + dz^2 = dr^2 + (r\sin\theta)^2\,d\phi^2 + r^2\,d\theta^2 \qquad (3\text{-}62)$$

or

$$\left.\begin{array}{l} h_1 = 1 \\[2mm] h_2 = \dfrac{1}{r\sin\theta} \\[2mm] h_3 = \dfrac{1}{r} \end{array}\right\} \qquad (3\text{-}63)$$

3-7 Div $\vec{V}$ in Curvilinear Coordinates. Going back to Eq. 3-55 and the following equations, let us determine

$$\text{div } V = \frac{\partial v_x}{\partial x} + \frac{\partial v_y}{\partial y} + \frac{\partial v_z}{\partial z}$$

first in general curvilinear coordinates, and then in cylindrical and spherical coordinates. We have

$$V = R^\star A \tag{3-64}$$

or

$$\begin{pmatrix} v_x \\ v_y \\ v_z \end{pmatrix} = R^\star \begin{pmatrix} a_\alpha \\ a_\beta \\ a_\gamma \end{pmatrix} \tag{3-65}$$

so that

$$\frac{\partial}{\partial x}\begin{pmatrix} v_x \\ v_y \\ v_z \end{pmatrix} = \frac{\partial}{\partial x}\left[R^\star \begin{pmatrix} a_\alpha \\ a_\beta \\ a_\gamma \end{pmatrix} \right] = \frac{\partial R^\star}{\partial x}A + R^\star \frac{\partial A}{\partial x} \tag{3-66}$$

or, $\partial v_x/\partial x$ corresponds to the *first row* of the matrix

$$\frac{\partial R^\star}{\partial x}A + R^\star \frac{\partial A}{\partial x}$$

$\partial v_y/\partial y$ and $\partial v_z/\partial z$ are obtained in a similar manner.

Hence, from Eq. 3-36, 3-37, 3-38 and the three parts of Eq. 3-55, and because $R^\star$ is the unit matrix, we have

$$\frac{\partial v_x}{\partial x} + \frac{\partial v_y}{\partial y} + \frac{\partial v_z}{\partial z} = \overbrace{\frac{\partial R^\star}{\partial x}A + R^\star \frac{\partial A}{\partial x}}^{\text{1st row of}} + \overbrace{\frac{\partial R^\star}{\partial y}A + R^\star \frac{\partial A}{\partial y}}^{\text{2nd row of}} + \overbrace{\frac{\partial R^\star}{\partial z}A + R^\star \frac{\partial A}{\partial z}}^{\text{3rd row of}}$$

$$= \left[h_1\frac{\partial a_\alpha}{\partial \alpha} + h_1 h_2\frac{\partial(1/h_1)}{\partial \beta}a_\beta \atop h_1 h_3\frac{\partial(1/h_1)}{\partial \gamma}a_\gamma \right] + \left[h_2\frac{\partial a_\beta}{\partial \beta} + h_1 h_2\frac{\partial(1/h_2)}{\partial \alpha}a_\alpha \atop + h_2 h_3\frac{\partial(1/h_2)}{\partial \gamma}a_\gamma \right]$$

$$+ \left[h_3\frac{\partial a_\gamma}{\partial \gamma} + h_1 h_3\frac{\partial(1/h_3)}{\partial \alpha}a_\alpha \atop + h_2 h_3\frac{\partial(1/h_3)}{\partial \beta}a_\beta \right]$$

$$= h_1 h_2 h_3\left[\frac{\partial}{\partial \alpha}\left(\frac{a_\alpha}{h_2 h_3}\right) + \frac{\partial}{\partial \beta}\left(\frac{a_\beta}{h_3 h_1}\right) + \frac{\partial}{\partial \gamma}\left(\frac{a_\gamma}{h_1 h_2}\right) \right] \tag{3-67}$$

Now, using Eqs. 3-59 and 3-63, we find that in cylindrical coordinates

$$\text{div } \vec{V} = \frac{\partial v_x}{\partial x} + \frac{\partial v_y}{\partial y} + \frac{\partial v_z}{\partial z} = \frac{\partial(a_\rho)}{\partial \rho} + \frac{1}{\rho}a_\rho + \frac{1}{\rho}\frac{\partial(a_\phi)}{\partial \phi} + \frac{\partial(a_z)}{\partial z} \tag{3-68}$$

and in spherical coordinates

$$\text{div } \vec{V} = \frac{\partial v_x}{\partial x} + \frac{\partial v_y}{\partial y} + \frac{\partial v_z}{\partial z} = \frac{\partial(a_r)}{\partial r} + \frac{2}{r}a_r + \frac{1}{r}\frac{\partial(a_\theta)}{\partial \theta} + \frac{\cot\theta}{r}a_\theta + \frac{1}{r\sin\theta}\frac{\partial(a_\phi)}{\partial \phi}$$

$$\tag{3-69}$$

3-8   The Laplacian f in Curvilinear Coordinates. If we want the Laplacian of a scalar function in curvilinear coordinates, we use Eq. 3-43, noting that

$$\left.\begin{aligned}\nabla^2 f &= \text{div grad } f \\ &= \nabla \cdot (\nabla f)\end{aligned}\right\} \tag{3-70}$$

and (see Eq. 3-43)

$$\text{grad } f = \nabla f = \begin{pmatrix} h_1\dfrac{\partial f}{\partial \alpha} \\[2ex] h_2\dfrac{\partial f}{\partial \beta} \\[2ex] h_3\dfrac{\partial f}{\partial \gamma} \end{pmatrix} \tag{3-71}$$

from which, with

$$\begin{pmatrix} a_\alpha \\ a_\beta \\ a_\gamma \end{pmatrix} = \begin{pmatrix} h_1\dfrac{\partial f}{\partial \alpha} \\[2ex] h_2\dfrac{\partial f}{\partial \beta} \\[2ex] h_3\dfrac{\partial f}{\partial \gamma} \end{pmatrix} \tag{3-72}$$

we have at once from Eq. 3-67

$$\nabla^2 f = h_1 h_2 h_3 \left\{ \frac{\partial}{\partial \alpha}\left[\frac{h_1(\partial f/\partial \alpha)}{h_2 h_3}\right] + \frac{\partial}{\partial \beta}\left[\frac{h_2(\partial f/\partial \beta)}{h_1 h_3}\right] + \frac{\partial}{\partial \gamma}\left[\frac{h_3(\partial f/\partial \gamma)}{h_1 h_2}\right] \right\} \tag{3-73}$$

and in cylindrical coordinates

$$\nabla^2 f = \frac{\partial^2 f}{\partial \rho^2} + \frac{1}{\rho}\frac{\partial f}{\partial \rho} + \frac{1}{\rho^2}\frac{\partial^2 f}{\partial \phi^2} + \frac{\partial^2 f}{\partial z^2} \tag{3-74}$$

while in spherical coordinates

$$\nabla^2 f = \frac{\partial^2 f}{\partial r^2} + \frac{2}{r}\frac{\partial f}{\partial r} + \frac{1}{r^2}\frac{\partial^2 f}{\partial \theta^2} + \frac{\cot\theta}{r^2}\frac{\partial f}{\partial \theta} + \frac{1}{r^2\sin^2\theta}\frac{\partial^2 f}{\partial \phi^2} \tag{3-75}$$

3-9 The Second-order Tensor in Curvilinear Coordinates.

As another example we obtain the general curvilinear form of a second-order tensor which occurs repeatedly in problems of elasticity, fluids, structures, plates and shells, and other fields of applied physics and engineering. That is, we wish to determine the general curvilinear form of the tensor,

$$\Lambda = \begin{pmatrix} \dfrac{\partial v_x}{\partial x} & \dfrac{\partial v_x}{\partial y} & \dfrac{\partial v_x}{\partial z} \\[2ex] \dfrac{\partial v_y}{\partial x} & \dfrac{\partial v_y}{\partial y} & \dfrac{\partial v_y}{\partial z} \\[2ex] \dfrac{\partial v_z}{\partial x} & \dfrac{\partial v_z}{\partial y} & \dfrac{\partial v_z}{\partial z} \end{pmatrix} \tag{3-76}$$

Knowing this tensor in general curvilinear form we can then obtain, in any coordinate system, the following tensors which we shall discuss in later sections of the text:

1. The deformation tensor used in elasticity theory. This is obtained by simply substituting the deformations (u, v, w) for the general components (v_x, v_y, v_z) (see footnote, p. 91).
2. The *time rate* of deformation tensor used in viscous flow and plasticity theory. Substitute the *velocity* (u, v, w) for (v_x, v_y, v_z) (see Eq. 8-22).
3. The *strain* tensor, η, used in elasticity, plates, etc., since (see Eq. 4-18)

$$\eta = \frac{\Lambda + \Lambda^\star}{2}$$

in which the deformations (u, v, w) are substituted for (v_x, v_y, v_z).
4. The *time rate of strain* tensor, Φ, of viscous flow theory, see Eq. 8-24
5. The rotation or curl tensor, Ω, see Eq. 8-25, which is obtained from

$$\Omega = \frac{\Lambda - \Lambda^\star}{2}$$

6. The curvature tensor of plate and shell theory, Eq. 7-6. Simply substitute $[(\partial w/\partial x), (\partial w/\partial y)]$ for (v_x, v_y).

Other tensors of these types, appearing in these and other fields are obtained similarly.

We obtain the general curvilinear form of Eq. 3-76 by using the results previously obtained. See Eqs. 3-45 through 3-55. For (with R

equal to the unit matrix), typical terms are given by the following:

$$
\begin{aligned}
\frac{\partial v_x}{\partial x} &= \overbrace{\frac{\partial R^\star}{\partial x}A + R^\star\frac{\partial A}{\partial x}}^{\text{1st row of}} \\
&= \left[h_1 h_2 \frac{\partial(1/h_1)}{\partial\beta}a_\beta + h_1 h_3 \frac{\partial(1/h_1)}{\partial\gamma}a_{\gamma.} \right] + h_1 \frac{\partial a_\alpha}{\partial\alpha}
\end{aligned}
\left.\rule{0pt}{5.5em}\right\}
\tag{3-77}
$$

and

$$
\begin{aligned}
\frac{\partial v_z}{\partial y} &= \overbrace{\frac{\partial R^\star}{\partial y}A + R^\star\frac{\partial A}{\partial y}}^{\text{3rd row of}} \\
&= -h_2 h_3 \frac{\partial(1/h_2)}{\partial\gamma}a_\beta + h_2 \frac{\partial a_\gamma}{\partial\beta}
\end{aligned}
\left.\rule{0pt}{5.5em}\right\}
\tag{3-78}
$$

The other seven terms are obtained in the same way, giving the final result

$$
\left(
\begin{aligned}
&h_1\frac{\partial(a_\alpha)}{\partial\alpha} + h_1 h_2 \frac{\partial(1/h_1)}{\partial\beta}a_\beta + h_1 h_3 \frac{\partial(1/h_1)}{\partial\gamma}a_\gamma \\[1em]
&h_1\frac{\partial(a_\beta)}{\partial\alpha} - h_1 h_2 \frac{\partial(1/h_1)}{\partial\beta}a_\alpha \\[1em]
&h_1\frac{\partial(a_\gamma)}{\partial\alpha} - h_1 h_3 \frac{\partial(1/h_1)}{\partial\gamma}a_\alpha \\[1em]
&\qquad h_2\frac{\partial(a_\alpha)}{\partial\beta} - h_2 h_1 \frac{\partial(1/h_2)}{\partial\alpha}a_\beta \\[1em]
&\qquad h_2\frac{\partial(a_\beta)}{\partial\beta} + h_2 h_3 \frac{\partial(1/h_2)}{\partial\gamma}a_\gamma + h_2 h_1 \frac{\partial(1/h_2)}{\partial\alpha}a_\alpha \\[1em]
&\qquad h_2\frac{\partial(a_\gamma)}{\partial\beta} - h_2 h_3 \frac{\partial(1/h_2)}{\partial\gamma}a_\beta \\[1em]
&\qquad\qquad h_3\frac{\partial(a_\alpha)}{\partial\gamma} - h_3 h_1 \frac{\partial(1/h_3)}{\partial\alpha}a_\gamma \\[1em]
&\qquad\qquad h_3\frac{\partial(a_\beta)}{\partial\gamma} - h_3 h_2 \frac{\partial(1/h_3)}{\partial\beta}a_\gamma \\[1em]
&\qquad\qquad h_3\frac{\partial(a_\gamma)}{\partial\gamma} + h_3 h_1 \frac{\partial(1/h_3)}{\partial\alpha}a_\alpha + h_3 h_2 \frac{\partial(1/h_3)}{\partial\beta}a_\beta
\end{aligned}
\right)
\tag{3-79}
$$

3-10   The Divergence of a Tensor, Div T. The second-order tensor transformation is given (see Eq. 2-35) by

$$T' = RTR^\star \tag{3-80}$$

in which the unprimed refers to the x, y, and z axes and the primed to the rotated axes. Since $RR^\star = R^\star R = E_3$, the above expression is equivalent to

$$'T = R^\star T'R \tag{3-81}$$

and this form is more convenient for our purposes. Thus, if

$$T = \begin{pmatrix} \widehat{xx} & \widehat{xy} & \widehat{xz} \\ \widehat{yx} & \widehat{yy} & \widehat{yz} \\ \widehat{zx} & \widehat{zy} & \widehat{zz} \end{pmatrix} \tag{3-82}$$

where $\widehat{xy}$ is a typical element of the tensor in the x–y–z system, and if

$$T' = \begin{pmatrix} \widehat{\alpha\alpha} & \widehat{\alpha\beta} & \widehat{\alpha\gamma} \\ \widehat{\beta\alpha} & \widehat{\beta\beta} & \widehat{\beta\gamma} \\ \widehat{\gamma\alpha} & \widehat{\gamma\beta} & \widehat{\gamma\gamma} \end{pmatrix} \tag{3-83}$$

represents this tensor in the α–β–γ system with $\widehat{\alpha\beta}$ the corresponding element in this system, then the divergence of the tensor (which occurs in many applications) is defined by

$$\operatorname{div} T = \left(\frac{\partial}{\partial x} \frac{\partial}{\partial y} \frac{\partial}{\partial z} \right) \begin{pmatrix} \widehat{xx} & \widehat{xy} & \widehat{xz} \\ \widehat{yx} & \widehat{yy} & \widehat{yz} \\ \widehat{zx} & \widehat{zy} & \widehat{zz} \end{pmatrix} \tag{3-84}$$

and, for example, using Eq. 3-55, noting that $R = R^\star = E$

$$
\left.
\begin{aligned}
&\overset{\text{1st row—1st column element of}}{\frac{\partial}{\partial x}\widehat{xx} = \overbrace{\frac{\partial}{\partial x}(R^\star T'R)}} \\[2mm]
&\qquad = \overbrace{\frac{\partial R^\star}{\partial x}T'R + R^\star \frac{\partial T'}{\partial x}R + R^\star T'\frac{\partial R}{\partial x}}^{\text{1st row—1st column element of}} \\[2mm]
&\qquad = \overbrace{\frac{\partial R^\star}{\partial x}T' + h_1\frac{\partial T'}{\partial \alpha} + T'\frac{\partial R}{\partial x}}^{\text{1st row—1st column element of}} \\[2mm]
&\qquad = \left[h_1 h_2 \frac{\partial(1/h_1)}{\partial \beta}\widehat{\beta\alpha} + h_1 h_3 \frac{\partial(1/h_1)}{\partial \gamma}\widehat{\gamma\alpha} \right] + \left[h_1\frac{\partial\widehat{\alpha\alpha}}{\partial\alpha} \right] \\[2mm]
&\qquad\qquad + \left[\widehat{\alpha\beta}h_1 h_2 \frac{\partial(1/h_1)}{\partial\beta} + \widehat{\alpha\gamma}h_1 h_3 \frac{\partial(1/h_1)}{\partial\gamma} \right]
\end{aligned}
\right\} \tag{3-85}
$$

The other terms are obtained similarly, and we obtain finally for

$$\frac{\partial \widehat{xx}}{\partial x} + \frac{\partial (\widehat{yx})}{\partial y} + \frac{\partial (\widehat{zx})}{\partial z} \tag{3-86}$$

in the curvilinear coordinates, the expression

$$h_1 h_2 h_3 \left[\frac{\partial}{\partial \alpha} \left(\frac{\widehat{\alpha\alpha}}{h_2 h_3} \right) + \frac{\partial}{\partial \beta} \left(\frac{\widehat{\beta\alpha}}{h_3 h_1} \right) + \frac{\partial}{\partial \gamma} \left(\frac{\widehat{\gamma\alpha}}{h_1 h_2} \right) \right] + h_1 h_2 \frac{\partial}{\partial \beta} \left(\frac{1}{h_1} \right) \widehat{\alpha\beta}$$
$$+ h_1 h_3 \frac{\partial}{\partial \gamma} \left(\frac{1}{h_1} \right) \widehat{\alpha\gamma} - h_1 h_2 \frac{\partial}{\partial \alpha} \left(\frac{1}{h_2} \right) \widehat{\beta\beta} - h_1 h_3 \frac{\partial}{\partial \alpha} \left(\frac{1}{h_3} \right) \widehat{\gamma\gamma} \tag{3-87}$$

The second and third equations are obtained from this by cyclical interchange.

The reader should note that the three expressions referred to here are just the equilibrium equations of elasticity with, for example,

$$\sigma_x = \widehat{xx} \tag{3-88}$$

See Eq. 4-85.

3-11 Summary.

1. We defined the general curvilinear coordinate system and the more special orthogonal curvilinear coordinate system.

2. By defining our various quantities for the special case in which the curvilinear system is coincident with the rectangular cartesian system at each point, we obtained simple matrix expressions for the various curvilinear coordinate relations. We saw that the rotation matrix R played a fundamental role in the theory of curvilinear coordinates.

3. The magnification factors h_1, h_2, and h_3, of the curvilinear coordinates were obtained, the significance of these was discussed, and a simple method for determining these was described, based upon the invariance of the squared differential length.

4. The curvilinear form of the gradient of a scalar was obtained.

5. The important quantities $\partial R^\star / \partial x$, $\partial R^\star / \partial y$, and $\partial R^\star / \partial z$ were obtained, and the usefulness of these in certain vector and tensor curvilinear forms was described.

6. Expressions for gradient and Laplacian were obtained for the special cylindrical and spherical curvilinear coordinates.

7. The second-order tensor curvilinear coordinate transformations were obtained.

Problems

1. The equations of equilibrium in elasticity are (body forces neglected)

$$\operatorname{div} T = 0$$

where T is the stress tensor, which in cartesian (x, y, z) form is given by

$$\begin{pmatrix} \sigma_x & \tau_{xy} & \tau_{xz} \\ \tau_{yx} & \sigma_y & \tau_{yz} \\ \tau_{zx} & \tau_{zy} & \sigma_z \end{pmatrix}$$

(a) Write the expanded form of these equations in the (x, y, z) system.
(b) Write the expanded form of these equations in general (α, β, γ) curvilinear coordinates.
(c) Write the expanded form of these in the space polar coordinate system, in terms of the corresponding stresses, of which σ_ρ and $\tau_{\rho\phi}$ are two typical stresses.
(d) Do the same for the spherical system in terms of σ_r, $\tau_{\theta\phi}$ etc.

2. Write the expression for

$$\nabla f = \operatorname{grad} f$$

in general (α, β, γ) form and then in the cylindrical, and spherical, coordinate systems.

3. The tensor

$$A = \begin{pmatrix} \dfrac{\partial}{\partial x} \\[2ex] \dfrac{\partial}{\partial y} \\[2ex] \dfrac{\partial}{\partial z} \end{pmatrix} (u \quad v \quad w)$$

may be called the *deformation* tensor. The symmetrical part of this is the *strain* tensor (which will be discussed in Chapter 4) and the antisymmetrical part is the rotation tensor (which has components $\nabla \times \vec{V}$, the curl of vector analysis).

(a) Show the tensor A in its expanded cartesian form.
(b) Show the symmetrical and antisymmetrical parts of A in their expanded cartesian forms.
(c) Show the two tensors of (b) in general (α, β, γ) curvilinear form.
(d) Show the three tensors of (a) and (b) in cylindrical (ρ, ϕ, z) form.
(e) Show the three tensors of (a) and (b) in spherical (r, ϕ, θ) form.

4. The tensor

$$\begin{pmatrix} \dfrac{\partial}{\partial x} \\[2ex] \dfrac{\partial}{\partial y} \\[2ex] \dfrac{\partial}{\partial z} \end{pmatrix} \begin{pmatrix} \dfrac{\partial f}{\partial x} & \dfrac{\partial f}{\partial y} & \dfrac{\partial f}{\partial z} \end{pmatrix}$$

occurs in many physical problems.

(a) Show the expanded cartesian form of this tensor.
(b) Write the general (α, β, γ) curvilinear form of this tensor.
(c) Show the cylindrical (ρ, ϕ, z) form of this tensor.
(d) Show the spherical (r, ϕ, θ) form of this tensor.

5. Prove that in any two-dimensional curvilinear coordinate system (α, β) where

$$\alpha + i\beta = f(z)$$

$$z = x + iy$$

$$h_1 = h_2 = \sqrt{\left(\frac{\partial \alpha}{\partial x}\right)^2 + \left(\frac{\partial \alpha}{\partial y}\right)^2} = \sqrt{\left(\frac{\partial \beta}{\partial x}\right)^2 + \left(\frac{\partial \beta}{\partial y}\right)^2}$$

6. (a) Show that

$$\alpha = \tfrac{1}{2}(x^2 - y^2)$$

$$\beta = xy$$

is a system described in Prob. 5.

(b) Show that curves $\begin{cases} x = \text{constant} \\ y = \text{constant} \end{cases}$ plot as confocal parabolas (with a common axis) in the α–β plane. These are *parabolic coordinates*.

7. (a) Show that

$$\alpha = a \cosh x \cos y$$

$$\beta = a \sinh x \sin y$$

is a system described in Prob. 5.

(b) Show that the curves $\begin{cases} x = \text{constant} \\ y = \text{constant} \end{cases}$ plot as confocal $\begin{cases} \text{ellipses} \\ \text{hyperbolas} \end{cases}$ in the α–β plane. These are *elliptic coordinates*.

8. Prolate spheroidal coordinates (ξ, η, ϕ) are defined by

$$x = a \sinh \xi \sin \eta \cos \phi$$

$$y = a \sinh \xi \sin \eta \sin \phi$$

$$z = a \cosh \xi \cos \eta$$

(a) Obtain h_1, h_2 and h_3 for these coordinates.

9. Oblate spheroidal coordinates (ξ, η, ϕ) are defined by

$$x = a \cosh \xi \cos \eta \cos \phi$$

$$y = a \cosh \xi \cos \eta \sin \phi$$

$$z = a \sinh \xi \sin \eta$$

(a) Obtain h_1, h_2, and h_3 for these coordinates.

10. (a) Show that

$$\alpha = \frac{a \sinh y}{\cosh y - \cos x}$$

$$\beta = \frac{a \sin x}{\cosh y - \cos x}$$

is a system described in Prob. 5.

(b) Show that curves $\begin{cases} x = \text{constant} \\ y = \text{constant} \end{cases}$ plot as bipolar circles in the α–β plane. These are *bipolar coordinates*.

11. For the special coordinate systems of Probs. 6 to 10, obtain the curvilinear coordinate forms for

   (a) div $\vec{V} = \nabla \cdot \vec{V}$,
   (b) grad $f = \nabla f$,
   (c) Laplacian $f = \nabla^2 f$,
   (d) The tensors of Probs. 3 and 4.

12. The Laplacian of a vector, $\nabla^2 \vec{V}$, is given by

$$\text{div } T$$

where $\vec{V} = (u\ v\ w)$ and $T = \begin{pmatrix} \dfrac{\partial}{\partial x} \\[2mm] \dfrac{\partial}{\partial y} \\[2mm] \dfrac{\partial}{\partial z} \end{pmatrix} (u\quad v\quad w)$

Using the results given in Arts. 3-9 and 3-10, obtain the cylindrical and spherical forms for $\nabla^2 \vec{V}$.

13. (a) Show that the differential volume element in an orthogonal coordinate system is given by

$$dV = \frac{1}{h_1 h_2 h_3} \, d\alpha \, d\beta \, d\gamma$$

   (b) Determine the expressions for dV in cylindrical coordinates, in spherical coordinates, and for the coordinates of Probs. 8 and 9.

14. (a) Determine the elements of the differential area vector in orthogonal curvilinear coordinates.

   (b) Determine the forms of this vector for the coordinate systems of Prob. 13b.

15. If the vector V is given by $(2z\ -x\ 2y)$ determine this vector, A, in cylindrical coordinates.

Chapter 4

INTRODUCTION TO THEORY OF ELASTICITY

4-1 Introduction. In this chapter a relatively brief treatment is given of the mathematical theory of elasticity. The more comprehensive treatment is beyond the scope of this book and may be found in Refs. 12 through 15. The present treatment differs from that of the references in several important respects. First, in accordance with the stated purpose of the text, wherever possible we base derivations on matrix arguments. Secondly, we derive a relation which holds for large strains. The fundamental derivations are based upon the work of F. D. Murnaghan (Ref. 16).

We first derive the expression for the strain matrix. Following this the equations of equilibrium are obtained. To do this the stress tensor is introduced. Hooke's Law is then derived and also the compatibility conditions. From this point on the treatment follows along classical lines, and in the next chapter we apply the results obtained herein to the solution of the bending problem and the torsion problem.

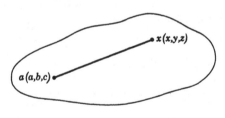

Fig. 4.1

4-2 The Strain Matrix (Tensor). Consider a body, Fig. 4.1, initially unstressed. At any point a, having coordinates (a, b, c), a differential element has a squared length given by

$$\left. \begin{aligned} dS_0{}^2 &= da^2 + db^2 + dc^2 \\ &= da^\star da \end{aligned} \right\} \quad (4\text{-}1)$$

where

$$da = \begin{pmatrix} da \\ db \\ dc \end{pmatrix} \quad (4\text{-}2)$$

88

Now assume that the body is strained due to the application of forces or for some other reason The point a moves to the point x, having coordinates (x, y, z), where

$$\left. \begin{aligned} x &= a+u \\ y &= b+v \\ z &= c+w \end{aligned} \right\} \qquad (4\text{-}3)$$

and the differential element da is now a differential element dx (u, v and w are the *deformations* in the x,y,z directions, respectively). We characterize the "state of strain" in the body at a point by noting that the effect of the strain was to change the differential length of the fiber originally at a, so that

$$\text{state of strain} = dx^{\star}dx - da^{\star}da \qquad (4\text{-}4)$$

Now,

$$x = x(a) \qquad (4\text{-}5)$$

(Note that this statement is an abbreviation for the following *three* statements—we use this notation consistently in this and later chapters

$$\left. \begin{aligned} x &= x(a,b,c) \\ y &= y(a,b,c) \\ z &= z(a,b,c) \end{aligned} \right\} \qquad (4\text{-}6)$$

these being the usual statements of functional dependence.)

Thus

$$dx = J\,da \qquad (4\text{-}7)$$

where

$$J = \frac{\partial(x, y, z)}{\partial(a, b, c)} = \begin{pmatrix} \dfrac{\partial x}{\partial a} & \dfrac{\partial x}{\partial b} & \dfrac{\partial x}{\partial c} \\[2mm] \dfrac{\partial y}{\partial a} & \dfrac{\partial y}{\partial b} & \dfrac{\partial y}{\partial c} \\[2mm] \dfrac{\partial z}{\partial a} & \dfrac{\partial z}{\partial b} & \dfrac{\partial z}{\partial c} \end{pmatrix} \qquad (4\text{-}8)$$

is the Jacobian matrix of (x, a)

Now (see Eq. 1-45)

$$dx^{\star} = da^{\star}J^{\star} \qquad (4\text{-}9)$$

and therefore Eq. 4-4 becomes

$$\left. \begin{aligned} \text{state of strain} &= da^{\star}J^{\star}J\,da - da^{\star}E_3\,da \\ &= da^{\star}(J^{\star}J - E_3)\,da \end{aligned} \right\} \qquad (4\text{-}10)$$

where E_3 is the 3×3 unit matrix.

J and $J^\star$ are certainly tensors, since (for example)

$$J^\star = \begin{pmatrix} \dfrac{\partial}{\partial a} \\[2mm] \dfrac{\partial}{\partial b} \\[2mm] \dfrac{\partial}{\partial c} \end{pmatrix} (x \quad y \quad z) \tag{4-11}$$

which is just the defining expression for the tensor (see Art. 2-4).

Therefore (see Prob. 11 at end of Chapter 2), J is also a tensor. Hence (see Eq. 2-35)

$$\left. \begin{aligned} J' &= RJR^\star \\ J^{\star\prime} &= RJ^\star R^\star \end{aligned} \right\} \tag{4-12}$$

and

$$\left. \begin{aligned} J^{\star\prime}J' &= RJ^\star R^\star RJR^\star \\ &= RJ^\star JR^\star \end{aligned} \right\} \tag{4-13}$$

since

$$R^\star R = E_3 \tag{4-14}$$

Therefore $J^\star J$ is also a tensor. And since E_3 is a tensor it follows that

$$J^\star J - E_3 \tag{4-15}$$

is also a tensor.

The quantity $J^\star J - E_3$ is one of the more important ones in the theory of elasticity. A large part of the subject may be built upon this term as a foundation. We call

$$J^\star J - E_3 = 2\eta \tag{4-16}$$

where η is the strain matrix. It is important to note that at no point was any restriction made as to "smallness" of strains. In other words, the above deformation matrix holds for all values of the deformation. Also, as noted above, η is a tensor.

It is left as an exercise to the reader to expand Eq. 4-16 and verify that

$$\begin{pmatrix} \eta_{aa} & \eta_{ab} & \eta_{ac} \\ \eta_{ba} & \eta_{bb} & \eta_{bc} \\ \eta_{ca} & \eta_{cb} & \eta_{cc} \end{pmatrix} = \begin{pmatrix} \dfrac{\partial u}{\partial a} & \dfrac{1}{2}\left(\dfrac{\partial u}{\partial b}+\dfrac{\partial v}{\partial a}\right) & \dfrac{1}{2}\left(\dfrac{\partial u}{\partial c}+\dfrac{\partial w}{\partial a}\right) \\[3mm] \dfrac{1}{2}\left(\dfrac{\partial u}{\partial b}+\dfrac{\partial v}{\partial a}\right) & \dfrac{\partial v}{\partial b} & \dfrac{1}{2}\left(\dfrac{\partial v}{\partial c}+\dfrac{\partial w}{\partial b}\right) \\[3mm] \dfrac{1}{2}\left(\dfrac{\partial u}{\partial c}+\dfrac{\partial w}{\partial a}\right) & \dfrac{1}{2}\left(\dfrac{\partial v}{\partial c}+\dfrac{\partial w}{\partial b}\right) & \dfrac{\partial w}{\partial c} \end{pmatrix}$$

(*This term continued on next page*).

$$+\frac{1}{2}\begin{pmatrix} \left(\frac{\partial u}{\partial a}\right)^2 + \left(\frac{\partial v}{\partial a}\right)^2 + \left(\frac{\partial w}{\partial a}\right)^2 & \frac{\partial u}{\partial a}\frac{\partial u}{\partial b} + \frac{\partial v}{\partial a}\frac{\partial v}{\partial b} + \frac{\partial w}{\partial a}\frac{\partial w}{\partial b} \\ \frac{\partial u}{\partial b}\frac{\partial u}{\partial a} + \frac{\partial v}{\partial b}\frac{\partial v}{\partial a} + \frac{\partial w}{\partial b}\frac{\partial w}{\partial a} & \left(\frac{\partial u}{\partial b}\right)^2 + \left(\frac{\partial v}{\partial b}\right)^2 + \left(\frac{\partial w}{\partial b}\right)^2 \\ \frac{\partial u}{\partial c}\frac{\partial u}{\partial a} + \frac{\partial v}{\partial c}\frac{\partial v}{\partial a} + \frac{\partial w}{\partial c}\frac{\partial w}{\partial a} & \frac{\partial u}{\partial c}\frac{\partial u}{\partial b} + \frac{\partial v}{\partial c}\frac{\partial v}{\partial b} + \frac{\partial w}{\partial c}\frac{\partial w}{\partial b} \end{pmatrix}$$

$$\left.\begin{matrix} \frac{\partial u}{\partial a}\frac{\partial u}{\partial c} + \frac{\partial v}{\partial a}\frac{\partial v}{\partial c} + \frac{\partial w}{\partial a}\frac{\partial w}{\partial c} \\ \frac{\partial u}{\partial b}\frac{\partial u}{\partial c} + \frac{\partial v}{\partial b}\frac{\partial v}{\partial c} + \frac{\partial w}{\partial b}\frac{\partial w}{\partial c} \\ \left(\frac{\partial u}{\partial c}\right)^2 + \left(\frac{\partial v}{\partial c}\right)^2 + \left(\frac{\partial w}{\partial c}\right)^2 \end{matrix}\right) \tag{4-17}$$

The matrix η given by Eq. 4-17 is the general strain tensor of elasticity including higher-order terms and not restricted to small deformations.

In the usual approximate theory of elasticity, the above expressions are approximated in two very important ways.

1. It is assumed that a is essentially the same as x. Therefore derivatives with respect to a are replaced by derivatives with respect to x. That is, we change from Lagrangian coordinates to Eulerian coordinates.[1]

2. Because of assumed small deformations, the higher-order terms are neglected and therefore the second matrix on the right-hand side of Eq. 4-17 is neglected.

With these two assumptions, the strain tensor takes the form,[2]

[1] Physically, the difference between the Lagrangian and Eulerian coordinate approaches is the following:

1. In the Lagrange coordinate system we are essentially attempting to follow the movement of each particle in the body. In other words, if the initial coordinates of a particle were (a, b, c) then the coordinates of *this same particle*, which in its deformed position is at (x, y, z), are given by $x(a, b, c)$, $y(a, b, c)$ and $z(a, b, c)$. That is, the final positions of all particles are functions of the initial positions of these particles.

2. The Euler approach, on the other hand, specifies conditions at each point in the *deformed body*. In other words, attention is focused on what has happened at a particular point rather than what has happened to a particular particle. Thus, all quantities considered in this formulation are functions of the points being considered, i.e., of x, y, and z.

[2] The expression

$$\begin{pmatrix} \frac{\partial}{\partial x} \\ \frac{\partial}{\partial y} \\ \frac{\partial}{\partial z} \end{pmatrix} (u \quad v \quad w) = \begin{pmatrix} \frac{\partial u}{\partial x} & \frac{\partial v}{\partial x} & \frac{\partial w}{\partial x} \\ \frac{\partial u}{\partial y} & \frac{\partial v}{\partial y} & \frac{\partial w}{\partial y} \\ \frac{\partial u}{\partial z} & \frac{\partial v}{\partial z} & \frac{\partial w}{\partial z} \end{pmatrix}$$

Footnote continued on page 92.

$$\eta = \begin{pmatrix} \dfrac{\partial u}{\partial x} & \dfrac{1}{2}\left(\dfrac{\partial u}{\partial y}+\dfrac{\partial v}{\partial x}\right) & \dfrac{1}{2}\left(\dfrac{\partial u}{\partial z}+\dfrac{\partial w}{\partial x}\right) \\[2ex] \dfrac{1}{2}\left(\dfrac{\partial v}{\partial x}+\dfrac{\partial u}{\partial y}\right) & \dfrac{\partial v}{\partial y} & \dfrac{1}{2}\left(\dfrac{\partial v}{\partial z}+\dfrac{\partial w}{\partial y}\right) \\[2ex] \dfrac{1}{2}\left(\dfrac{\partial w}{\partial x}+\dfrac{\partial u}{\partial z}\right) & \dfrac{1}{2}\left(\dfrac{\partial w}{\partial y}+\dfrac{\partial v}{\partial z}\right) & \dfrac{\partial w}{\partial z} \end{pmatrix}$$

$$= \begin{pmatrix} \eta xx & \eta xy & \eta xz \\ \eta yx & \eta yy & \eta yz \\ \eta zx & \eta zy & \eta zz \end{pmatrix} \tag{4-18}$$

Note that this tensor is a *symmetric* tensor.

The physical interpretation of the elements of this matrix and their connection with the deformation of an elastic body will now be obtained.

We point out first that we are concerned, in elasticity, with deformable bodies, i.e., with bodies which

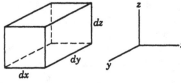

Fig. 4.2

elongate or shorten or otherwise deform under the application of forces, loads or other effects. If we consider the elemental cube shown in Fig. 4.2, we see that there are basically, two types of deformation which this body can be subject to in that any deformation, no matter how complicated, can be given as a combination of these. These are (1) pure elongation (see Fig. 4.3a) or contraction, and (2) pure sliding or shearing action (see Fig. 4.3b).

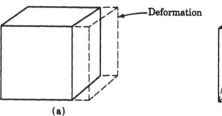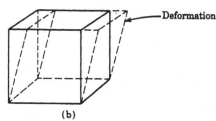

(a)   (b)

Fig. 4.3

Footnote continued from page 91.

which is obviously a tensor, may be called the "deformation tensor." The transpose of the above is certainly also a tensor, and it too may be called the "deformation tensor." The reader should verify that the symmetrical part of this tensor is just the strain tensor η of Eq. 4–18. See also the Appendix discussion following Chap. 10.

As a physical example of the first type of deformation consider a bar subjected to a simple tension load, Fig. 4.4. Consider conditions

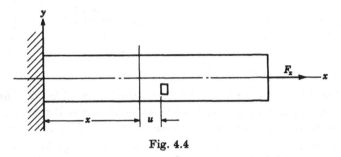

Fig. 4.4

at a plane originally at a distance x from the left end, which is fixed in a wall. At this point the deformation is u. Now let us consider a differential element of the body originally at this point, x.

We obtain a measure of the first type of deformation mentioned above as follows: We consider the projection of the element on an x–y plane and consider first only deformations in the x direction. The deformations in the y and z directions then follow directly from this (see Fig. 4.5).

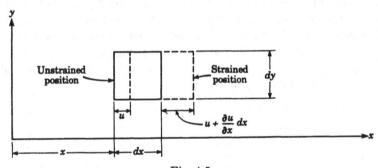

Fig. 4.5

The deformation in the x direction of the face closest to the origin is u. Then, since the deformation changes over the length, dx, of the element, we have for the far face, a movement given by $u + (du/\partial x)\, dx$.

This means,

$$\text{net total elongation of element} = \left(u + \frac{\partial u}{\partial x} dx\right) - u \left. \vphantom{\frac{\partial u}{\partial x}} \right\} $$
$$= \frac{\partial u}{\partial x} dx \qquad (4\text{-}19)$$

$$\left.\begin{aligned}
\text{and, unit elongation} &= \frac{\text{total elongation}}{\text{original length}} \\[2mm]
&= \frac{(\partial u/\partial x)\,dx}{dx} \\[2mm]
&= \frac{\partial u}{\partial x}
\end{aligned}\right\} \quad (4\text{-}20)$$

In exactly the same way, we can find the unit elongations (or contractions) in the y and z directions. Thus, if v is the movement in the y direction, and w is the movement in the z direction, then

$$\left.\begin{aligned}
\text{unit elongation in the } y \text{ direction} &= \frac{\partial v}{\partial y} \\[2mm]
\text{unit elongation in the } z \text{ direction} &= \frac{\partial w}{\partial z}
\end{aligned}\right\} \quad (4\text{-}21)$$

Note: although the word "elongation" is used in the above expressions, they also hold for "contractions," in which case they are simply negative quantities.

Note also that the quantities $\partial u/\partial x$, $\partial v/\partial y$, and $\partial w/\partial z$ are just the main diagonal elements of the strain tensor. In other words, the main diagonal elements of the strain tensor are the unit elongations (or contractions) corresponding to small normal deformations.

Now let us consider the sliding or shearing deformations, and let us obtain a measure of this quantity. Once again we consider first a projection on the x–y plane of the $dx\,dy\,dz$ body, undergoing a sliding or shearing deformation, see Fig. 4.6.

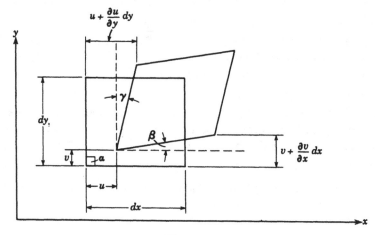

Fig. 4.6

We take as a measure of the shearing or sliding deformation the change in the original right angle, α, of the element. Thus,

$$\left. \begin{array}{l} \text{shearing unit deformation} \\ \qquad \text{in the } x\text{–}y \text{ plane} \end{array} \right\} \begin{array}{l} = \beta + \gamma \\[2mm] = \dfrac{\partial v}{\partial x} + \dfrac{\partial u}{\partial y} \end{array} \right\} \qquad (4\text{-}22)$$

In the same way, we have for the y–z and z–y planes,

$$\left. \begin{array}{ll} \begin{array}{l} \text{shearing unit deformation} \\ \qquad \text{in the } y\text{–}z \text{ plane} \cdot \end{array} & = \dfrac{\partial w}{\partial y} + \dfrac{\partial v}{\partial z} \\[4mm] \begin{array}{l} \text{and shearing unit deformation} \\ \qquad \text{in the } z\text{–}x \text{ plane} \end{array} & = \dfrac{\partial u}{\partial z} + \dfrac{\partial w}{\partial x} \end{array} \right\} \qquad (4\text{-}23)$$

Note that these measures of shearing deformations,

$$\frac{\partial v}{\partial x} + \frac{\partial u}{\partial y}$$

$$\frac{\partial w}{\partial y} + \frac{\partial v}{\partial z}$$

$$\frac{\partial u}{\partial z} + \frac{\partial w}{\partial x}$$

are equal to *twice* the off-diagonal elements of the strain tensor. The one-half values in the tensor are essential—without them, the quantity of Eq. 4-18 will *not* be a tensor and hence will not behave in accordance with the known properties of tensors.

The two-dimensional form of the strain tensor, for the x–y system becomes (also shown is the American notation—see the table on page 96—which we will use in this text)

$$\begin{pmatrix} e_x & \tfrac{1}{2}\gamma_{xy} \\ \tfrac{1}{2}\gamma_{yx} & e_y \end{pmatrix} = \begin{pmatrix} \dfrac{\partial u}{\partial x} & \dfrac{1}{2}\left(\dfrac{\partial u}{\partial y} + \dfrac{\partial v}{\partial x}\right) \\[4mm] \dfrac{1}{2}\left(\dfrac{\partial v}{\partial x} + \dfrac{\partial u}{\partial y}\right) & \dfrac{\partial v}{\partial y} \end{pmatrix} \qquad (4\text{-}24)$$

Because this is a tensor of the second order, it satisfies the transformation laws for this tensor (Eq. 2-40) and furthermore the Mohr circle construction of Fig. 2.8 applies to the elements of this tensor.

We repeat, for convenience, the six fundamental unit deformations of the theory of elasticity, and we add, furthermore, some of the common European and American notations for these:

TABLE 4.1

u = deformation in x direction
v = deformation in y direction
w = deformation in z direction

Unit elongation or contraction in x direction	$\dfrac{\partial u}{\partial x}$	e_x	e_{xx}
Unit elongation or contraction in y direction	$\dfrac{\partial v}{\partial y}$	e_y	e_{yy}
Unit elongation or contraction in z direction	$\dfrac{\partial w}{\partial z}$	e_z	e_{zz}
Unit shearing strain in x–y plane	$\dfrac{\partial u}{\partial y}+\dfrac{\partial v}{\partial x}$	γ_{xy}	e_{xy}
Unit shearing strain in y–z plane	$\dfrac{\partial v}{\partial z}+\dfrac{\partial w}{\partial y}$	γ_{yz}	e_{yz}
Unit shearing strain in z–x plane	$\dfrac{\partial w}{\partial x}+\dfrac{\partial u}{\partial z}$	γ_{zx}	e_{zx}
		American Practice (See Ref. 13)	British and Russian Practice (See Refs. 12 and 18)

Summary up to this point:

1. The state of strain for an elastic body was defined.

2. Utilizing (1), the large deformation strain tensor, including higher order terms, was derived.

3. The two usual approximations of the small deformation theory were then introduced, and these led to

4. The small deformation strain tensor, η.

5. The physical connection of the elements of η with the strains of the deformed body were derived.

6. It was emphasized that the off-diagonal elements of the strain tensor do not represent shear strains directly, but are related to the shear strains as follows:

$$\text{shear strain} = 2 \ (\text{tensor element})$$

4-3 The Stress Tensor. The last of the fundamental tensors of elasticity which we shall consider in this chapter is the stress tensor. We define a stress as the force per unit area, and on the basis of the pure elongation (or shortening) strains and sliding (shearing) strains, we look for related forces and hence stresses. Thus, a type of force which will cause a pure elongation is the one shown in Fig. 4.4, a small portion of which is shown in Fig. 4.7. This is a force normal to the

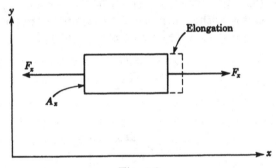

Fig. 4.7

area, A_x, hence it is a normal force, and this leads to a normal stress, σ_x, defined by

$$\sigma_x = \frac{\partial F_x}{\partial A_x} \tag{4-25}$$

in which A_x is the area (normal to x axis) on which the force F_x acts. In a similar manner we have

and

$$\left. \begin{aligned} \sigma_y &= \frac{\partial F_y}{\partial A_y} \\ \sigma_z &= \frac{\partial F_z}{\partial A_z} \end{aligned} \right\} \tag{4-26}$$

The stresses σ_x, σ_y, and σ_z given above are the only normal stresses

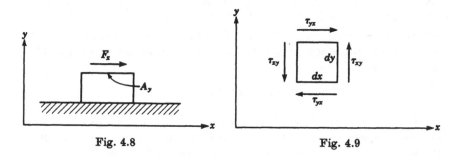

Fig. 4.8   Fig. 4.9

which can act on the faces of an elemental volume $dx\,dy\,dz$ having faces normal to the orthogonal axes x, y, and z.

The shearing or sliding effect is obviously caused by a force acting as shown in Fig. 4.8. Defining the shearing stress τ_{yx} (note the subscripts—a stress in the x direction acting on a plane perpendicular to the y axis) as a force divided by an area, we have

$$\tau_{yx} = \frac{\partial F_x}{\partial A_y} \tag{4-27}$$

and, similarly,

$$\left.\begin{aligned}\tau_{zy} &= \frac{\partial F_y}{\partial A_z} \\[2mm] \tau_{xz} &= \frac{\partial F_z}{\partial A_x}\end{aligned}\right\} \tag{4-28}$$

Corresponding to the stress τ_{yx} we have a companion stress τ_{xy} (see Fig. 4.9).

It is clear that in order for the element shown to be in static equilibrium[3]

$$\tau_{xy} = \tau_{yx} \tag{4-29}$$

since moment balance requires

$$(\tau_{xy}\,dy)\,dx = (\tau_{yx}\,dx)\,dy \tag{4-30}$$

---

[3] Although we have not discussed static equilibrium up to this point, it is desirable to prove the equality of shear stresses at this time, in the simplest possible manner, namely, by using the equilibrium relation. See Art. 4-4 for a more complete statement of the equilibrium relations and also for a different proof of the equality of shear stresses.

In the same way it may be shown that there are stresses τ_{yz} and τ_{zx}, which are given by the equalities

$$\left.\begin{array}{c} \tau_{zy} = \tau_{yz} \\ \tau_{xz} = \tau_{zx} \end{array}\right\} \tag{4-30a}$$

This equality of shear stresses on mutually perpendicular planes is a general property which is always true. We shall utilize this property of shear stresses in later portions of the book.

The nine stresses,

$$\begin{array}{cc} \sigma_x & \tau_{xy} = \tau_{yx} \\ \sigma_y & \tau_{yz} = \tau_{zy} \\ \sigma_z & \tau_{zx} = \tau_{xz} \end{array}$$

are the only possible independent stresses which can act on the faces of the body $dx\,dy\,dz$. This is so because a force in any direction on any of the faces can always be resolved into components corresponding to the normal and shear stresses.

These stresses, in matrix form are shown as

$$T = \begin{pmatrix} \sigma_x & \tau_{xy} & \tau_{xz} \\ \tau_{yx} & \sigma_y & \tau_{yz} \\ \tau_{zx} & \tau_{zy} & \sigma_z \end{pmatrix} \tag{4-31}$$

and we may prove that this is a tensor of the second order, as follows: Obviously

$$T = \begin{pmatrix} \dfrac{\partial}{\partial A_x} \\[2ex] \dfrac{\partial}{\partial A_y} \\[2ex] \dfrac{\partial}{\partial A_z} \end{pmatrix} \begin{pmatrix} F_x & F_y & F_z \end{pmatrix} \tag{4-32}$$

Also, $(F_x\ F_y\ F_z)$ is a vector.
Hence, if

$$\begin{pmatrix} \dfrac{\partial}{\partial A_x} \\[2ex] \dfrac{\partial}{\partial A_y} \\[2ex] \dfrac{\partial}{\partial A_z} \end{pmatrix} \tag{4-33}$$

is a vector (i.e., satisfies the transformation $V' = RV$, see Eq. 2-16, or $v'_i = a_{ij}v_j$, see Eq. 2-96) then T is certainly a tensor of the second order, since it is then obtained as a dyadic product of two vectors. We show that this is actually so, as follows:

We know that the area is a vector. Let us denote this area, in the indicial notation, by

$$A_i = \begin{pmatrix} A_x \\ A_y \\ A_z \end{pmatrix} \tag{4-34}$$

Then (see Eq. 2-96),

$$A'_i = a_{ij}A_j \tag{4-35}$$

and

$$\frac{\partial}{\partial A'_i} = \frac{\partial}{\partial A_j}\frac{\partial A_j}{\partial A'_i} \tag{4-36}$$

or

$$\frac{\partial}{\partial A'_i} = a_{ij}\frac{\partial}{\partial A_j} \quad \text{Q.E.D.} \tag{4-37}$$

in which the identity $a_{ij}a_{ij} = 1$ was used. Thus the stress matrix is a tensor.

In the standard American notation this tensor is shown in matrix form as follows,

$$T = \begin{pmatrix} \sigma_x & \tau_{xy} & \tau_{xz} \\ \tau_{yx} & \sigma_y & \tau_{yz} \\ \tau_{zx} & \tau_{zy} & \sigma_z \end{pmatrix} \tag{4-38}$$

In a typical Russian or British notation this tensor is shown as

$$\begin{pmatrix} x_x & x_y & x_z \\ y_x & y_y & y_z \\ z_x & z_y & z_z \end{pmatrix}$$

In Fig. 4.10 we show these stresses acting on a typical element.

In the two-dimensional x–y space, the stress tensor becomes

$$T = \begin{pmatrix} \sigma_x & \tau_{xy} \\ \tau_{yx} & \sigma_y \end{pmatrix} \tag{4-39}$$

and because this is a symmetrical tensor it may be put in diagonal form, i.e., has principal stresses. Also, because it is a tensor of the second order, it may be shown graphically by means of the Mohr circle.

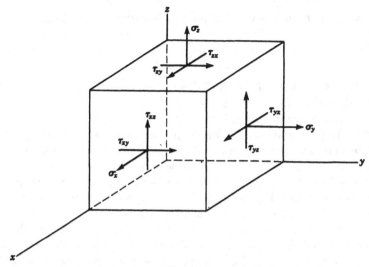

Fig. 4.10

Regarding the signs of the stresses: these are governed, in the mathematical theory, by the directions of the axes. However, in applied (structural) analysis many conventions are possible and are used. For example, we may define the sign of the stress with reference to clockwise or counter-clockwise action, or we may define it with reference to the type of deformation (elongation or contraction) it causes. See Chap. 6 and Ref. 10.

Summarizing the discussion of the stress tensor:

(a) based upon the known strain tensor elements, corresponding stresses were assumed, these being normal and shear stresses.

(b) It was shown that these stresses (nine in all, three normal and six shear) are the elements of a *stress tensor*.

(c) The main diagonal elements of this tensor are the normal (tension or compression) stresses, and the off-diagonal elements are the shear stresses.

(d) It was shown that the tensor is symmetric, that is, the corresponding off-diagonal shear stresses are equal to each other.

4-4 The Equations of Static Equilibrium. The static equilibrium equations of elasticity can be obtained in various ways. Basically, all of them start with the Newton Laws,

$$\left. \begin{aligned} \Sigma \bar{F} &= m\bar{a} \\ \Sigma \bar{M} &= I\bar{\omega} \end{aligned} \right\} \qquad (4\text{-}40)$$

from which it follows that if a body is in *static* equilibrium,

$$\begin{aligned} \Sigma \bar{F} &= 0 \\ \Sigma \bar{M} &= 0 \end{aligned} \Bigg\} \qquad (4\text{-}41)$$

In Eq. 4.40, $\bar{F}$ and $\bar{M}$ are the force and moment vectors, m and I are the mass and mass moment-of-inertia; $\bar{a}$ and $\bar{\omega}$ are the linear and angular acceleration vectors.

Hence the *net* sum of the forces in the three directions, x, y, and z, must be zero and the *net* moments of these forces about the x, y, and z axes must be zero, if a body is to be in static equilibrium.

We shall use this technique to determine the boundary conditions on the stresses (see Eq. 4-73) and, in fact, we already used Eq. 4-41 to prove the symmetry of the stress tensor (see Eq. 4-30).

However, in view of the emphasis on matrix-tensor methods in this text, it will be instructive to derive the static equilibrium equations (and also the proof of the stress tensor symmetry) in another manner. We shall utilize energy methods, particularly the law of conservation of energy in a form especially suitable for our purposes. To do this it will be necessary that we speak very briefly about a variational notation.

We introduce the notion of "variation of ( )" which is represented symbolically by

$$\delta(\quad)$$

The term in the bracket may be any function of (x, y, z) and hence also of (a, b, c).

We introduce this notation at this point since it is customary to define as the *virtual displacement* or *virtual deformation*, the matrix

$$\delta x = \begin{pmatrix} \delta x \\ \delta y \\ \delta z \end{pmatrix} \qquad (4\text{-}42)$$

in which each element of the matrix is given as a function of x, y and z and, in general, δx is very small. Hence δx may be assumed as a differential quantity. Physical interpretations of the quantities will be given later in this section. Furthermore, if we assume continuous second derivatives and because the variables are independent it follows that (typically),

$$\delta \left(\frac{\partial f}{\partial x} \right) = \frac{\partial}{\partial x} (\delta f) \qquad (4\text{-}43)$$

We assume furthermore (in analogy to differentiation) that

$$\delta(A + B) = \delta A + \delta B \qquad (4\text{-}44\text{a})$$

and

$$\delta(AB) = (\delta A)B + A(\delta B) \qquad (4\text{-}44\text{b})$$

in which A and B may be matrices and the order of operation in Eq. 4-44b must be maintained.

The above identities are all that are required for our later developments.

In the following discussion we revert to the Lagrangian system of coordinates. We had (see Eq. 4-16)

$$\eta = \tfrac{1}{2}(J^\star J - E) \qquad (4\text{-}45)$$

In connection with our derivation of the equilibrium and related equations it will be necessary that we obtain $\delta\eta$. This is given by

$$\delta\eta = \tfrac{1}{2}[(\delta J^\star)J + J^\star(\delta J)] \qquad (4\text{-}46)$$

since $\delta E = 0$

Now[4]

$$\left.\begin{aligned}
\delta J &= \delta\left(\frac{\partial x}{\partial a}\right) \\[2mm]
&= \frac{\partial(\delta x)}{\partial a}
\end{aligned}\right\} \qquad (4\text{-}47)$$

[4] Since

$$J = \begin{pmatrix} \dfrac{\partial x}{\partial a} & \dfrac{\partial x}{\partial b} & \dfrac{\partial x}{\partial c} \\[2mm] \dfrac{\partial y}{\partial a} & \dfrac{\partial y}{\partial b} & \dfrac{\partial y}{\partial c} \\[2mm] \dfrac{\partial z}{\partial a} & \dfrac{\partial z}{\partial b} & \dfrac{\partial z}{\partial c} \end{pmatrix}$$

and from the definition $\delta(\text{---})$, it follows that

$$\delta J = \begin{pmatrix} \delta\left(\dfrac{\partial x}{\partial a}\right) & \delta\left(\dfrac{\partial x}{\partial b}\right) & \delta\left(\dfrac{\partial x}{\partial c}\right) \\[2mm] \delta\left(\dfrac{\partial y}{\partial a}\right) & \delta\left(\dfrac{\partial y}{\partial b}\right) & \delta\left(\dfrac{\partial y}{\partial c}\right) \\[2mm] \delta\left(\dfrac{\partial z}{\partial a}\right) & \delta\left(\dfrac{\partial z}{\partial b}\right) & \delta\left(\dfrac{\partial z}{\partial c}\right) \end{pmatrix}$$

and because of Eq. 4-43, this is also given by

$$\delta J = \begin{pmatrix} \dfrac{\partial}{\partial a}(\delta x) & \dfrac{\partial}{\partial b}(\delta x) & \dfrac{\partial}{\partial c}(\delta x) \\[2mm] \dfrac{\partial}{\partial a}(\delta y) & \dfrac{\partial}{\partial b}(\delta y) & \dfrac{\partial}{\partial c}(\delta y) \\[2mm] \dfrac{\partial}{\partial a}(\delta z) & \dfrac{\partial}{\partial b}(\delta z) & \dfrac{\partial}{\partial c}(\delta z) \end{pmatrix}$$

all of which is shown in the shorthand notation of Eq. 4-47.

and

$$\frac{\partial(\delta x)}{\partial a} = \frac{\partial(\delta x)}{\partial x}\frac{\partial x}{\partial a} + \frac{\partial(\delta x)}{\partial y}\frac{\partial y}{\partial a} + \frac{\partial(\delta x)}{\partial z}\frac{\partial z}{\partial a} \qquad (4\text{-}48)$$

plus eight more similar equations; or (the student should verify this), from Eq. 4-48,

$$\delta J = \frac{\partial(\delta x)}{\partial x}J \qquad (4\text{-}49)$$

From the properties of the transpose,

$$\delta J^\star = (\delta J)^\star = J^\star \left[\frac{\partial(\delta x)}{\partial x}\right]^\star \qquad (4\text{-}50)$$

so that

$$\delta\eta = J^\star D J \qquad (4\text{-}51)$$

where D = *variational deformation matrix.*

$$= \frac{1}{2}\left\{\left[\frac{\partial(\delta x)}{\partial x}\right]^\star + \left[\frac{\partial(\delta x)}{\partial x}\right]\right\} \qquad (4\text{-}52)$$

Now

$$J = E_3 + \frac{\partial u}{\partial a} \qquad (4\text{-}53)$$

and, for the small deformation theory, with

$$1 \gg \frac{\partial u}{\partial a} \qquad (4\text{-}54)$$

we have approximately

$$J = E_3 \qquad (4\text{-}55)$$

so that

$$\delta\eta \cong D \qquad (4\text{-}56)$$

However, for more exact work, the form of Eq. 4-51 must be used.

We next introduce the concept of *virtual displacement* and state the Principle of Virtual Displacements. This is the form of the Law of Conservation of Energy that we shall use.

By a virtual displacement is meant any arbitrary (usually very small) displacement that is compatible with the geometrical constraints (such as supports) of a structure. As noted earlier, we designate a virtual displacement as δx, that is, the variation or differential change of the position coordinate x.

The principle of virtual displacements is a fundamental law of mechanics. It may be stated in the following form.

If the virtual work done by all forces acting on a structure during a rigid virtual deformation is zero, then these forces are in equilibrium.

To illustrate this law in a most elementary fashion consider the simple beam shown in Fig. 4.11.

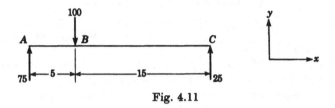

Fig. 4.11

Suppose the beam is moved vertically upward as a rigid body an amount δy (this is the virtual deformation), then the *net* work done by the forces in this virtual deformation is

$$\delta w = -100(\delta y) + 75(\delta y) + 25(\delta y)$$
$$= 0 \tag{4-57}$$

as required for equilibrium. The negative sign follows, since the deflection in this case is *opposite* the direction of the force.

Suppose further that the beam is given a virtual rigid body rotation about point B, with point A moving *up* an amount δy. Then, from the geometry, point C must move *down* an amount $3\delta y$, and we have, for the net work done by the external forces,

$$\delta w = 75(\delta y) - 25(3\delta y)$$
$$= 0 \tag{4-58}$$

again as required for equilibrium.[5]

We had (Eqs. 4-10 and 4-16)

$$da^\star(2\eta)\,da = dS^2 - dS_0^2 \tag{4-59}$$

so that[6]

$$\delta(dS^2 - dS_0^2) = \delta(dS^2)$$
$$= da^\star(2\delta\eta)\,da \tag{4-60}$$

---

[5]. It should be noted that *both* rigid virtual motions (linear and angular) must be considered in specifying equilibrium since, for example, equilibrium under the vertical upward linear rigid body motion could have been indicated by forces of 50 at each of A and C. However these would not have satisfied the angular rigid body virtual rotation requirement.

[6] Note: in these equations, dS_0 and da are the fixed initial lengths squared and length. Hence the variation of these quantities is zero. Variation effects are included in the final or deformed values.

or, using Eq. 4-51,

$$\delta \text{ (state of strain)} = 2da^\star (J^\star DJ)\, da \qquad (4\text{-}61)$$

If we assume the approximations corresponding to assumed small displacements, this becomes

$$\delta \text{ (state of strain)} = 2dx^\star D\, dx \qquad (4\text{-}62)$$

It is obvious in both 4–61 and 4–62 that the variation of the "state of strain" is zero if $D = 0$.

If the variation of the *state of strain* is zero, we are considering, in effect, a condition corresponding to a rigid body deformation, since a rigid body deformation, physically, is just one which does not change the state of strain of a body. To verify this, let us check Eq. 4-52 for the following two cases (the results immediately following will also be used in later developments in this chapter):

1. Assume the virtual displacement is a rigid linear one, so that every particle of the body has the same virtual displacement, say

$$\left.\begin{array}{l} \delta x = 6 \\ \delta y = 2 \\ \delta z = 4.2 \end{array}\right\} \qquad (4\text{-}63)$$

2. Assume the virtual displacement is a rigid body rotation $\delta\theta$ about the z axis, so that (see Fig. 4.12)

$$\left.\begin{array}{l} \delta x = -y\delta\theta \\ \delta y = x\delta\theta \end{array}\right\} \qquad (4\text{-}64)$$

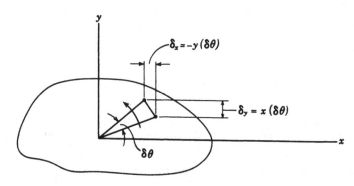

Fig. 4.12

Let us determine D for these two cases. For case (1), with

$$D = \begin{pmatrix} \dfrac{\partial(\delta x)}{\partial x} & \dfrac{1}{2}\left[\dfrac{\partial(\delta x)}{\partial y}+\dfrac{\partial(\delta y)}{\partial x}\right] & \dfrac{1}{2}\left[\dfrac{\partial(\delta x)}{\partial z}+\dfrac{\partial(\delta z)}{\partial x}\right] \\[3mm] \dfrac{1}{2}\left[\dfrac{\partial(\delta y)}{\partial x}+\dfrac{\partial(\delta x)}{\partial y}\right] & \dfrac{\partial(\delta y)}{\partial y} & \dfrac{1}{2}\left[\dfrac{\partial(\delta y)}{\partial z}+\dfrac{\partial(\delta z)}{\partial y}\right] \\[3mm] \dfrac{1}{2}\left[\dfrac{\partial(\delta z)}{\partial x}+\dfrac{\partial(\delta x)}{\partial z}\right] & \dfrac{1}{2}\left[\dfrac{\partial(\delta z)}{\partial y}+\dfrac{\partial(\delta y)}{\partial z}\right] & \dfrac{\partial(\delta z)}{\partial z} \end{pmatrix}$$

$$\text{(4-65)}$$

we have (obviously)

$$D_1 = 0 \qquad\qquad\qquad (4\text{-}66)$$

For case (2), Eq. 4-65 becomes

$$D_2 = \begin{pmatrix} 0 & \frac{1}{2}(\delta\theta - \delta\theta) & 0 \\ \frac{1}{2}(-\delta\theta + \delta\theta) & 0 & 0 \\ 0 & 0 & 0 \end{pmatrix} \qquad (4\text{-}67)$$

Therefore the variation of the state of strain is zero for (1) a rigid virtual translation of the entire body and (2) a rigid virtual rotation of the body.

We may draw the following conclusions from the above:

A rigid linear motion of the entire body and a rigid rotation of the entire body correspond to no deformation of the body. In effect, a suitable transformation of axes would bring the body to the original equilibrium position. Furthermore, the variation of the state of strain is zero, (i.e., $D = 0$) for both the linear rigid body virtual motion and for the rigid body virtual rotation. If for these cases the body is in equilibrium, then the virtual work done by the forces acting on the body is zero.

We are now ready to proceed to a consideration of the forces acting on the body.

There are in general, two types of forces which may act on a deformable body: (a) surface forces, that is forces that are proportional to areas, and (b) body forces that are proportional to the mass.

Thus, if F is a body force per unit mass, ρ being the mass per unit volume, and if

$$F = \begin{pmatrix} F_x \\ F_y \\ F_z \end{pmatrix} \qquad\qquad (4\text{-}68)$$

then the virtual work done by these forces during the virtual displacement δ given by

$$\delta = \begin{pmatrix} \delta x \\ \delta y \\ \delta z \end{pmatrix} \qquad (4\text{-}69)$$

is given by

$$\delta w_B = \int_{\text{E.V.}} \rho F^\star \delta \, dV = \int_{\text{E.V.}} \rho(\delta x F_x + \delta y F_y + \delta z F_z) \, dV \quad (4\text{-}70)$$

(E.V. = entire volume.)

To obtain the virtual work done by the surface forces, we must consider the stress tensor (see Art. 4-3).

It was shown in Art. 4-3 that there are acting on differential elements at all points in a stressed body nine stresses, or forces per unit area,

$$\left. \begin{array}{l} \sigma_x \\ \sigma_y \\ \sigma_z \end{array} \right\} \text{the normal stresses} \qquad \left. \begin{array}{l} \tau_{xy} = \tau_{yx} \\ \tau_{yz} = \tau_{zy} \\ \tau_{zx} = \tau_{xz} \end{array} \right\} \text{the shear stresses}$$

The stresses, when multiplied by the area on which they act, result in the *surface* forces mentioned above.

In order to account for the effect of these in the present analysis, we consider an element of the stressed body at the surface or boundary of the body. This is shown most generally in Fig. 4.13, where BCD represents the surface, inclined as shown to the axes x, y, and z, the origin O being inside the body. The tetrahedron $OBCD$ represents a point (differential volume) in the stressed body.

Let $\bar{x}$, $\bar{y}$, and $\bar{z}$ be the stress components acting on the boundary BCD in the directions x, y, and z, and let the stresses within the body be represented by the stress symbols previously described. Conditions are then as shown in Fig. 4.13, where the stress components are assumed as acting at the centroids of the differential areas shown.

Now, if A = area of BCD, and if $N(l, m, n)$ is the normal to this plane, then

$$\left. \begin{array}{l} \cos(N, x) = l \\ \cos(N, y) = m \\ \cos(N, z) = n \end{array} \right\} \qquad (4\text{-}71)$$

and the areas of the three faces of the tetrahedron (perpendicular to the x, y, and z axes respectively) are Al, Am, and An.

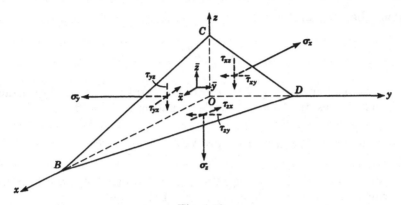

Fig. 4.13

Consider Newton's Law in the form given in Eq. 4-41. Then for the tetrahedron to be in equilibrium under the forces acting on it (summing forces in the x, y, and z directions),

$$\left.\begin{array}{l} A\bar{x} - Al\sigma_x - Am\tau_{yx} - An\tau_{zx} = 0 \\ A\bar{y} - Al\tau_{xy} - Am\sigma_y - An\tau_{zy} = 0 \\ A\bar{z} - Al\tau_{xz} - Am\tau_{yz} - An\sigma_z = 0 \end{array}\right\} \qquad (4\text{-}72)$$

These are the *boundary conditions satisfied by the stresses*. This relation is equivalent to

$$A(\bar{x}\ \bar{y}\ \bar{z}) = A(l\ m\ n)\begin{pmatrix} \sigma_x & \tau_{xy} & \tau_{xz} \\ \tau_{yx} & \sigma_y & \tau_{yz} \\ \tau_{zx} & \tau_{zy} & \sigma_z \end{pmatrix} \qquad (4\text{-}73)$$

or, in matrix notation,

$$A\bar{x} = ANT$$

or

$$\bar{x} = NT \qquad (4\text{-}74)$$

where T is the stress tensor defined in Eq. 4-31.
    Then if

$$\begin{array}{l} dS = \text{area vector} \\ \quad = A(l\ m\ n) \end{array} \qquad (4\text{-}75)$$

the surface force may be represented by

$$A(\bar{x}\ \bar{y}\ \bar{z}) = dST \qquad (4\text{-}76)$$

Having the expression for the surface force in Eq. 4-76, we can say that the virtual work done by the surface forces in going through a virtual displacement δ, Eq. 4-69, is given by

$$\int_{\text{E.S.}} dST\delta$$

(E.S. = entire surface) and therefore the total work done by surface and body forces due to a virtual deformation is given by

$$w_{\text{virtual}} = \int_{\text{E.V.}} \rho F^\star \delta \, dV + \int_{\text{E.S.}} dST\delta \qquad (4\text{-}77)$$

and for a body which is in equilibrium and subject to a rigid virtual deformation this will be equal to zero. If this is so, then

$$\int_{\text{E.V.}} \rho F^\star \delta \, dV + \int_{\text{E.S.}} dST\delta = 0 \qquad (4\text{-}78)$$

We now apply Gauss' Theorem (see Eq. 1-117) for transforming a surface integral into a volume integral. That is, we use the identity[7]

$$\int_{\text{E.S.}} dS(T\delta) = \int_{\text{E.V.}} \text{div}(T\delta) \, dV \qquad (4\text{-}79)$$

in which div represents "divergence," and in this equation the term $\text{div}(T\delta)$ is a scalar and in expanded form is shown as

$$\left(\frac{\partial}{\partial x} \quad \frac{\partial}{\partial y} \quad \frac{\partial}{\partial z} \right) \left[\begin{pmatrix} \sigma_x & \tau_{xy} & \tau_{xz} \\ \tau_{yx} & \sigma_y & \tau_{yz} \\ \tau_{zx} & \tau_{zy} & \sigma_z \end{pmatrix} \begin{pmatrix} \delta x \\ \delta y \\ \delta z \end{pmatrix} \right] \qquad (4\text{-}80)$$

Substituting Eq. 4-79 into Eq. 4-78 we have, upon combining terms,

$$\int_{\text{E.V.}} [\text{div}(T\delta) + \rho F^\star \delta] \, dV = 0 \qquad (4\text{-}81)$$

Since this holds everywhere in the body, it follows that

$$\text{div}(T\delta) + \rho F^\star \delta = 0 \qquad (4\text{-}82)$$

for any rigid virtual displacement corresponding to an equilibrium configuration of the body.

This relation will therefore hold for rigid virtual linear displacements and for rigid virtual angular displacements.

---

[7] Gauss' Theorem transforms a surface integral into a volume integral and for this reason the order of the matrix terms on the left-hand side of Equation 4-79 is immaterial. The resulting volume integral is as indicated on the right-hand side of this equation.

The first of the above will be true (as we saw) for

$$\delta x = \text{constant}$$

$$\delta y = \text{constant}$$

$$\delta z = \text{constant}$$

Thus, if we choose the following three sets of values we have a rigid virtual linear deformation, and if the body is in equilibrium, the expression Eq. 4-82 must hold.

$$\begin{Bmatrix} \delta x = 1 \\ \delta y = 0 \\ \delta z = 0 \end{Bmatrix}, \quad \begin{Bmatrix} \delta x = 0 \\ \delta y = 1 \\ \delta z = 0 \end{Bmatrix}, \quad \begin{Bmatrix} \delta x = 0 \\ \delta y = 0 \\ \delta z = 1 \end{Bmatrix} \tag{4-83}$$

If we substitute these values successively in Eq. 4-82 we obtain the following three equations

$$\left. \begin{aligned} \frac{\partial \sigma_x}{\partial x} + \frac{\partial \tau_{yx}}{\partial y} + \frac{\partial \tau_{zx}}{\partial z} + \rho F_x = 0 \\[2mm] \frac{\partial \tau_{xy}}{\partial x} + \frac{\partial \sigma_y}{\partial y} + \frac{\partial \tau_{yz}}{\partial z} + \rho F_y = 0 \\[2mm] \frac{\partial \tau_{xz}}{\partial x} + \frac{\partial \tau_{yz}}{\partial y} + \frac{\partial \sigma_z}{\partial z} + \rho F_z = 0 \end{aligned} \right\} \tag{4-84}$$

which, in tensor form, becomes

$$\text{div } T + \rho F^\star = 0 \tag{4-85}$$

and holds for any body in equilibrium. These are the *equations of equilibrium* in elasticity.

In Eq. 4-65, the expanded form of D is shown. As pointed out, D can also be zero when the body is subjected to a rigid body virtual rotation. In matrix terms, the connection between the rigid body virtual rotation and D may be expressed as follows:

Let

$$M = \begin{pmatrix} \dfrac{\partial(\delta x)}{\partial x} & \dfrac{\partial(\delta x)}{\partial y} & \dfrac{\partial(\delta x)}{\partial z} \\[3mm] \dfrac{\partial(\delta y)}{\partial x} & \dfrac{\partial(\delta y)}{\partial y} & \dfrac{\partial(\delta y)}{\partial z} \\[3mm] \dfrac{\partial(\delta z)}{\partial x} & \dfrac{\partial(\delta z)}{\partial y} & \dfrac{\partial(\delta z)}{\partial z} \end{pmatrix} \tag{4-86}$$

then (see Eq. 4-65) we have

$$D = \frac{M + M^\star}{2} \qquad (4\text{-}87)$$

and D may be zero, even though δx, δy, and δz are not constants, if

$$M = -M^\star \qquad (4\text{-}88)$$

This being so, the diagonal terms (top left to bottom right) of M must be zero, and the off-diagonal terms must be skew-symmetric or

$$\left.\begin{array}{c} \dfrac{\partial}{\partial y}(\delta x) = -\dfrac{\partial}{\partial x}(\delta y) \\[2ex] \dfrac{\partial}{\partial z}(\delta y) = -\dfrac{\partial}{\partial y}(\delta z) \\[2ex] \dfrac{\partial}{\partial x}(\delta z) = -\dfrac{\partial}{\partial z}(\delta x) \end{array}\right\} \qquad (4\text{-}89)$$

These relations are just the matrix expressions for the rigid body virtual rotation.

Going back to Eq. 4-82, which holds for a body in equilibrium, and expanding this, we obtain

$$\frac{\partial}{\partial x}(\delta x \sigma_x) + \frac{\partial}{\partial y}(\delta x \tau_{yx}) + \frac{\partial}{\partial z}(\delta x \tau_{zx}) + \rho \delta x F_x + \frac{\partial}{\partial x}(\delta y \tau_{xy}) + \frac{\partial}{\partial y}(\delta y \sigma_y)$$
$$+ \frac{\partial}{\partial z}(\delta y \tau_{zy}) + \rho \delta y F_y + \frac{\partial}{\partial x}(\delta z \tau_{xz}) + \frac{\partial}{\partial y}(\delta z \tau_{yz}) + \frac{\partial}{\partial z}(\delta z \sigma_z) + \rho \delta z F_z = 0 \qquad (4\text{-}90)$$

Now using the relations given in Eq. 4-89 above for the rigid virtual rotation condition, we find that *for a body in equilibrium*

$$\frac{\partial}{\partial x}(\delta y)(\tau_{xy} - \tau_{yx}) + \frac{\partial}{\partial x}(\delta z)(\tau_{xz} - \tau_{zx}) + \frac{\partial}{\partial y}(\delta z)(\tau_{yz} - \tau_{zy}) = 0 \quad (4\text{-}91)$$

and for this to be equal to zero as required, for any values of δx, δy and δz, it follows that

$$\left.\begin{array}{c} \tau_{xy} = \tau_{yz} \\[1ex] \tau_{xz} = \tau_{zx} \\[1ex] \tau_{yz} = \tau_{zy} \end{array}\right\} \qquad (4\text{-}92)$$

which proves the equality of shear stresses on right angle faces, and therefore proves that the stress tensor is a symmetric tensor. This was already proven, in a direct application of the moment balance form of Newton's Laws in Art. 4-3. However, the above demonstration ties in more directly with our matrix-tensor approach.[8]

Summarizing the derivation of the equilibrium equations:

1. We used the Principle of Virtual Deformations to derive the equilibrium equations. To this end, we obtained first the matrix D, the variational deformation matrix, given by

$$D = \frac{1}{2}\left\{\left[\frac{\partial(\delta_x)}{\partial x}\right]^\star + \left[\frac{\partial(\delta_x)}{\partial x}\right]\right\} \qquad (4\text{-}93)$$

2. It was pointed out that D will be zero when the body is subjected to a rigid virtual linear deformation and also when subjected to a rigid virtual rotation.

3. For a body in equilibrium, the virtual work done by the forces acting on the body will be zero when this body is subjected to a rigid virtual linear deformation and a rigid virtual rotation.

4. A consideration of the forces required that we introduce body forces and surface forces (which are related to the stress tensor), where T, the stress tensor, is given by

$$T = \begin{pmatrix} \sigma_x & \tau_{xy} & \tau_{xz} \\ \tau_{yx} & \sigma_y & \tau_{yz} \\ \tau_{zx} & \tau_{zy} & \sigma_z \end{pmatrix} \qquad (4\text{-}94)$$

5. The boundary condition which the stresses (i.e., stress tensor) must satisfy is $\bar{x} = NT$, in which $\bar{x}$ represents the boundary stresses and N is the unit normal to the boundary.

6. By considering both body and surface forces on a body in equilibrium we obtained

or

$$\left.\begin{aligned} \operatorname{div} T + \rho F^\star &= 0 \\[4pt] \frac{\partial \sigma_x}{\partial x} + \frac{\partial \tau_{yx}}{\partial y} + \frac{\partial \tau_{zx}}{\partial z} + \rho F_x &= 0 \\[4pt] \frac{\partial \tau_{xy}}{\partial x} + \frac{\partial \sigma_y}{\partial y} + \frac{\partial \tau_{zy}}{\partial z} + \rho F_y &= 0 \\[4pt] \frac{\partial \tau_{xz}}{\partial x} + \frac{\partial \tau_{yz}}{\partial y} + \frac{\partial \sigma_z}{\partial z} + \rho F_z &= 0 \end{aligned}\right\} \qquad (4\text{-}95)$$

the three equations of equilibrium.

7. Finally, by considering a particular case of $D = 0$, and because

---

[8] Because of this symmetry of the stress tensor, it follows that the equilibrium equations, Equation 4-85, can also be stated as $\operatorname{div} T^\star + \rho F^\star = 0$ and this form may be used in certain later sections of the book.

the body is in equilibrium, we obtained the important equations concerning the equality of shear stresses on right angled faces,

$$\left.\begin{array}{c} \tau_{xy} = \tau_{yx} \\ \tau_{yz} = \tau_{zy} \\ \tau_{zx} = \tau_{xz} \end{array}\right\} \tag{4-96}$$

which shows that the stress tensor is a symmetric tensor.

4-5 Derivation of Hooke's Law. We now alter our method of reasoning as follows:

Given a body in equilibrium, we subject this body to *any* virtual deformation (not necessarily a rigid virtual deformation or one for which $D = 0$). Then, since the body is in equilibrium,

$$\text{div } T + \rho F^{\star} = 0 \tag{4-97}$$

and the virtual work per unit volume (see Eq. 4-82) is given, because of the symmetry of the matrices, by the sum of each D matrix element (Eq. 4-65) times the corresponding stress matrix element of (Eq. 4-38).

This is called "taking the trace," is shown as $[TD]$, and is equal to the sum of the diagonal elements of the matrix product.

$$TD \tag{4-98}$$

Note from Eqs. 4-79 and 4-80 that this represents the contribution from the *surface* stresses only.

We now apply the conservation of energy requirement in the following form:

"The external work done during a virtual deformation by the forces acting on the body must be equal to the work stored in the body in the form of strain energy."

As an example of this, consider again the beam shown in Fig. 4.11, and repeated in Fig. 4.14. The original (unloaded) position of the beam was ABC. Under the load of 100, the beam is in equilibrium and has the shape $AB'C$.

We apply a virtual deformation (not a rigid one) consistent with the support conditions. This moves the beam to $AB''C$. Then,

external change of work = internal change of work

or

$100(B'B'')$ = change in strain energy stored in beam

In the general case, the virtual work done by the surface forces is given by

$$\int_{\text{E.V.}} [TD] \, dV$$

and, neglecting frictional and other similar forces, this must equal the change in strain energy stored in the body.

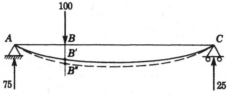

100

A   B

B'

B''

C

75   25

Fig. 4.14

If the total strain energy stored in the body in its equilibrium position is given by

$$\int_{\text{E.V.}} \phi(\eta)\rho \, dV$$

where $\phi(\eta)$ is a function of the strain elements and represents the strain energy per unit mass, then the above noted conservation of energy requirement leads to

$$\int_{\text{E.V.}} [TD] \, dV = \delta\left\{\int_{\text{E.V.}} \phi(\eta)\rho \, dV\right\} \tag{4-99}$$

or (stated in words)

change in external work = change in internal work

Now $\rho \, dV$ on the right-hand side is given with respect to x coordinates, but because of the constancy of mass we have

$$\rho_a \, dV_a = \rho_x \, dV_x \tag{4-100}$$

(the subscripts representing the variable at which points ρ and dV are given) and therefore the variation may be taken under the integral sign and applied to $\phi(\eta)$ only, as follows

$$\int_{\text{E.V.}} [TD] \, dV = \int_{\text{E.V.}} \rho\{\delta\phi(\eta)\} \, dV \tag{4-101}$$

Therefore

$$[TD] = \rho\{\delta\phi(\eta)\} \tag{4-102}$$

But, in view of the fact that (for our purposes) δ is equivalent to a differentiation, we have

$$\left.\begin{aligned}
\delta\phi(\eta) = {}&\frac{\partial\phi}{\partial\eta_{aa}}\delta\eta_{aa} + \frac{\partial\phi}{\partial\eta_{ab}}\delta\eta_{ab} + \frac{\partial\phi}{\partial\eta_{ac}}\delta\eta_{ac} + \frac{\partial\phi}{\partial\eta_{ba}}\delta\eta_{ba} + \frac{\partial\phi}{\partial\eta_{bb}}\delta\eta_{bb} \\
& + \frac{\partial\phi}{\partial\eta_{bc}}\delta\eta_{bc} + \frac{\partial\phi}{\partial\eta_{ca}}\delta\eta_{ca} + \frac{\partial\phi}{\partial\eta_{cb}}\delta\eta_{cb} + \frac{\partial\phi}{\partial\eta_{cc}}\delta\eta_{cc}
\end{aligned}\right\} \tag{4-103}$$

or, recalling the definition of the bracket [ ], and because η is a symmetric matrix, we have

$$\delta\phi(\eta) = \left[\frac{\partial\phi}{\partial\eta}\delta\eta\right] \tag{4-104}$$

where

$$\frac{\partial\phi}{\partial\eta} = \begin{pmatrix} \dfrac{\partial\phi}{\partial\eta_{aa}} & \dfrac{\partial\phi}{\partial\eta_{ab}} & \dfrac{\partial\phi}{\partial\eta_{ac}} \\[2ex] \dfrac{\partial\phi}{\partial\eta_{ba}} & \dfrac{\partial\phi}{\partial\eta_{bb}} & \dfrac{\partial\phi}{\partial\eta_{bc}} \\[2ex] \dfrac{\partial\phi}{\partial\eta_{ca}} & \dfrac{\partial\phi}{\partial\eta_{cb}} & \dfrac{\partial\phi}{\partial\eta_{cc}} \end{pmatrix} \tag{4-105}$$

and

$$\delta\eta = \begin{pmatrix} \delta\eta_{aa} & \delta\eta_{ab} & \delta\eta_{ac} \\ \delta\eta_{ba} & \delta\eta_{bb} & \delta\eta_{bc} \\ \delta\eta_{ca} & \delta\eta_{cb} & \delta\eta_{cc} \end{pmatrix} \tag{4-106}$$

From Eqs. 4-102 and 4-103, it follows that

$$[TD] = \rho\left[\frac{\partial\phi}{\partial\eta}\delta\eta\right] \tag{4-107}$$

and recalling (see Eq. 4-51) that

$$\delta\eta = J\star DJ \tag{4-108}$$

we have

$$\left.\begin{aligned} [TD] &= \rho\left[\frac{\partial\phi}{\partial\eta}(J\star DJ)\right] \\ &= \rho\left[J\frac{\partial\phi}{\partial\eta}J\star D\right] \end{aligned}\right\} \tag{4-109}$$

since $[AB] = [BA]$ as may easily be verified, and also because of the associative property of matrix multiplication.

Therefore, in order that Eq. 4-109 hold for all values of D,

$$T = \rho J\frac{\partial\phi}{\partial\eta}J\star \tag{4-110}$$

which is the fundamental relation between stresses and strains in the theory of elasticity. If $\phi(\eta)$ is known, this equation (which represents six equations) may be solved for stresses as functions of strains.

In the ordinary theory of elasticity, we use a degenerate form of this equation, obtained as follows:

We assume small displacements and neglect $\partial u/\partial a$ with respect to unity and $(\partial u/\partial a)^2$ with respect to $\partial u/\partial a$.

Then, to these approximations, $J = E$, the unit matrix, and we have

$$T = \rho \frac{\partial \phi}{\partial \eta} \tag{4-111}$$

Since the assumption $J = E$ is equivalent to assuming constant density, this becomes

$$T = \frac{\partial(\rho\phi)}{\partial \eta} \tag{4-112}$$

which states, in words, *the stress tensor is the gradient of the energy (per unit volume) of deformation with respect to the strain tensor.*

We point out once more that this is an approximate expression. However, for most engineering applications it is sufficiently accurate, and in any case the more exact relation is all but unusable.

We now require some assumption regarding the elastic material with which we are dealing In general, in the theory of elasticity, we are concerned with two types of materials: (1) isotropic or non-crystalline, (2) anisotropic or crystalline.

The first material is the most important for engineering applications. Many metals fall in this group. In fact, even those metals which have definite grain properties are very frequently assumed to be isotropic.

The second group contains materials such as wood, built-up materials, rolled metals, etc. For these materials the properties may vary considerably, depending upon grain direction.

We shall assume that we are dealing with an isotropic material. This means that the properties are independent of direction. Mathematically speaking, this means that the function $\phi(\eta)$ must be independent of the orientation of the axes.

We saw in Chapter 2 that there are only certain quantities which have this property of being independent of axial orientation, these being the invariants of the tensor. In other words, if $\phi(\eta)$ is to be independent of direction, then $\phi(\eta)$ must be given in terms of the invariants of the strain tensor. Or, if we assume terms in decreasing order of magnitude, and assume $\phi(\eta)$ can be given as a series expansion, we have as a consequence of the requirements of isotropy,

$$\rho\phi(\eta) = \rho\phi_0 + \alpha I_1 + \frac{\lambda + 2\mu}{2} I_1{}^2 - 2\mu I_2 + \cdots \tag{4-113}$$

where ϕ_0, α, λ, and μ are constants, $I_1 = \eta_{xx} + \eta_{yy} + \eta_{zz}$, an invariant, and $I_2 = \eta_{yy}\eta_{zz} - \eta_{yz}\eta_{zy} + \eta_{zz}\eta_{xx} - \eta_{zx}\eta_{xz} + \eta_{xx}\eta_{yy} - \eta_{xy}\eta_{yx}$, an invariant.

Now, recalling that, in its approximate form,

$$\frac{\partial(\rho\phi)}{\partial\eta} = \begin{pmatrix} \dfrac{\partial(\rho\phi)}{\partial\eta_{xx}} & \dfrac{\partial(\rho\phi)}{\partial\eta_{xy}} & \dfrac{\partial(\rho\phi)}{\partial\eta_{xz}} \\[2ex] \dfrac{\partial(\rho\phi)}{\partial\eta_{yx}} & \dfrac{\partial(\rho\phi)}{\partial\eta_{yy}} & \dfrac{\partial(\rho\phi)}{\partial\eta_{yz}} \\[2ex] \dfrac{\partial(\rho\phi)}{\partial\eta_{zx}} & \dfrac{\partial(\rho\phi)}{\partial\eta_{zy}} & \dfrac{\partial(\rho\phi)}{\partial\eta_{zz}} \end{pmatrix} \tag{4-114}$$

we find by substituting Eq. 4-13 in Eq. 4-112 that

$$T = \frac{\partial(\rho\phi)}{\partial\eta} = (\alpha + \lambda I_1)E_3 + 2\mu\eta \tag{4-115}$$

and if we assume that $T = 0$ when $\eta = 0$, this becomes

$$T = \lambda I_1 E_3 + 2\mu\eta \tag{4-116}$$

which is Hooke's Law, and which takes the expanded form (in the standard American notation for stresses and strains):

$$\begin{pmatrix} \sigma_x & \tau_{xy} & \tau_{xz} \\ \tau_{yx} & \sigma_y & \tau_{yz} \\ \tau_{zx} & \tau_{zy} & \sigma_z \end{pmatrix} = \lambda(e_x + e_y + e_z)\begin{pmatrix} 1 & 0 & 0 \\ 0 & 1 & 0 \\ 0 & 0 & 1 \end{pmatrix} + 2\mu\begin{pmatrix} e_x & \tfrac{1}{2}\gamma_{xy} & \tfrac{1}{2}\gamma_{xz} \\ \tfrac{1}{2}\gamma_{yx} & e_y & \tfrac{1}{2}\gamma_{yz} \\ \tfrac{1}{2}\gamma_{zx} & \tfrac{1}{2}\gamma_{zy} & e_z \end{pmatrix} \tag{4-117}$$

so that, for example,

$$\sigma_x = \lambda(e_x + e_y + e_z) + 2\mu e_x$$
$$= (\lambda + 2\mu)e_x + \lambda(e_y + e_z) \tag{4-118}$$

and

$$\tau_{xy} = 2\mu\frac{\gamma_{xy}}{2} = \mu\gamma_{xy} \tag{4-119}$$

Eq. 4-116, which is a *linear* relation in the strains, is called the linearized form of Hooke's law.

The constants μ and λ are connected with the tension and shear moduli of elasticity E and G and with Poisson's ratio, ν, as follows:

μ = shear modulus of elasticity, G

$$\lambda = \frac{\nu E}{(1+\nu)(1-2\nu)}$$

ν = Poisson's ratio

E = tension-compression modulus of elasticity

Compare this with p. 10 of Ref. (13).

We point out once more the assumptions and approximations that are inherent in the linearized form of Hooke's Law:

1. We assume $\partial/\partial a = \partial/\partial x$.

2. We neglect $\partial u/\partial a$ with respect to unity and $(\partial u/\partial a)^2$ with respect to $\partial u/\partial a$. That is, we assume small deformations.

3. We assume isotropy.

We also emphasize that the equation

$$T = \rho J \frac{\partial \phi}{\partial \eta} J^{\star} \tag{4-120}$$

is an exact expression, with none of the above limitations. If it were possible to handle mathematically the resulting expression and if $\phi(\eta)$ were known, then, theoretically Eq. 4-120 would give the exact relation between stresses and strains for large deformations, anisotropy and other conditions as well.

Problem 8 at end of this chapter is concerned with the higher-order (i.e., non-linear) form of Hooke's Law.

4-6   The Compatibility Conditions. The significance of the compatibility equations, their place in the general theory of elasticity, and the manner in which they are obtained is occasionally a matter of some confusion.

We may begin this discussion of the compatibility requirements by pointing out that these are relations or equations that must be satisfied in order to insure uniqueness, single-valuedness, and continuity of the deformations of the structure.

To illustrate in a simple, non-mathematical form what we mean by uniqueness, single-valuedness, and continuity of the deformations, let us consider the structure shown in Fig. 4.15a. In Fig. 4.15b this structure is shown with finite, single-valued deformations.

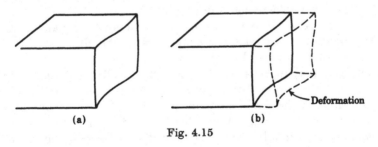

(a)   (b)

Deformation

Fig. 4.15

Now consider the same structure in Fig. 4.16a. In Fig. 4.16b we have a discontinuity in the deformation at the fiber ABC. We see that at the original point A there are two deformations, AB and AC. Thus, the deformation of this structure is *not* single-valued and continuous, and the compatibility conditions are the equations which—because they are not satisfied—will indicate that this is so.

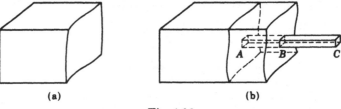

(a) (b)

Fig. 4.16

The usual procedure in deriving the compatibility conditions is to point out that the six different elements of the strain tensor, Eq. 4-121,

$$
\begin{pmatrix}
e_x & \tfrac{1}{2}\gamma xy & \tfrac{1}{2}\gamma xz \\
\tfrac{1}{2}\gamma yx & e_y & \tfrac{1}{2}\gamma yz \\
\tfrac{1}{2}\gamma zx & \tfrac{1}{2}\gamma zy & e_z
\end{pmatrix}
=
\begin{pmatrix}
\dfrac{\partial u}{\partial x} & \dfrac{1}{2}\left(\dfrac{\partial u}{\partial y}+\dfrac{\partial v}{\partial x}\right) & \dfrac{1}{2}\left(\dfrac{\partial u}{\partial z}+\dfrac{\partial w}{\partial x}\right) \\[3mm]
\dfrac{1}{2}\left(\dfrac{\partial v}{\partial x}+\dfrac{\partial u}{\partial y}\right) & \dfrac{\partial v}{\partial y} & \dfrac{1}{2}\left(\dfrac{\partial v}{\partial z}+\dfrac{\partial w}{\partial y}\right) \\[3mm]
\dfrac{1}{2}\left(\dfrac{\partial w}{\partial x}+\dfrac{\partial u}{\partial z}\right) & \dfrac{1}{2}\left(\dfrac{\partial w}{\partial y}+\dfrac{\partial v}{\partial z}\right) & \dfrac{\partial w}{\partial z}
\end{pmatrix}
$$

$$(4\text{-}121)$$

are given in terms of the three components of the displacements u, v, w, and hence the strains cannot be independent of each other. Following this, the various strain elements are partially differentiated twice with respect to the various coordinates, (x, y, z), certain of these added, some subtracted, and finally six equations, the compatibility equations, are obtained.

The above described procedure is given in Love (12), Timoshenko (13), and other books which follow these.

Just why one stops at two differentiations is not explained. One is led to wonder if other relations might perhaps be obtained with additional or different partial differentiations.

Southwell (14) derives the strain compatibility conditions by requiring that the total strain energy of a strained body shall vanish for all variations of the stress components which are compatible with the equations of equilibrium. While this is an interesting method, it does not seem to get directly to the matter of finiteness or single-valuedness of the deformations.

Murnaghan (16) obtains the equations by pointing out that the curvature tensor of the three-dimensional metric space is the zero tensor.

Using this fact, and the strain tensor, a fairly complicated analysis finally leads to the desired equations. This method does not clearly indicate the physical significance of the operations or the final result. It does, however, result in a non-linear (or higher order) form of the compatibility conditions.

Sokolnikoff (15), following Cesaro, obtains the equations by noting that continuity of deformations can be obtained by means of a line integral over a simple continuous curve, which in turn leads to the fact that certain integrands must be exact differentials. This finally determines the compatibility conditions.

The equations (or conditions) will now be derived in a different manner from all of the above.

Physically, the problem may be presented in the following form:

In many solutions of elasticity problems, we solve for the *unit strains* directly. In the solutions so obtained, the unit strain expressions may correspond to continuous, finite, and single valued *unit strains*, but (as pointed out in Fig. 4.16) it is necessary that the *deformations* be continuous, finite, and single-valued, and the fact that the *unit strains* are so is not, in itself, assurance that the *deformations* are also continuous, finite, and single-valued. The *compatibility conditions* are a set of conditions or equations which the *unit strains* must satisfy in order that we may be assured that a solution given in terms of the unit strains also corresponds to continuous, finite, and single-valued *deformations*.

For purposes of clarity, the development will be presented in detail for the two-dimensional case. The procedure is identical for three dimensions and will only be outlined for this case.

TWO-DIMENSIONAL CASE

In two dimensions, the strain tensor η is given by

$$\eta = \begin{pmatrix} e_x & \frac{1}{2}\gamma_{xy} \\ \frac{1}{2}\gamma_{yx} & e_y \end{pmatrix} = \begin{pmatrix} \dfrac{\partial u}{\partial x} & \dfrac{1}{2}\left(\dfrac{\partial u}{\partial y}+\dfrac{\partial v}{\partial x}\right) \\ \dfrac{1}{2}\left(\dfrac{\partial v}{\partial x}+\dfrac{\partial u}{\partial y}\right) & \dfrac{\partial v}{\partial y} \end{pmatrix} \qquad (4\text{-}122)$$

where $u(x, y)$ and $v(x, y)$ are the displacements. We ask, "What conditions must be placed upon the strains e_x, e_y, γ_{xy}, and γ_{yx} in order that the displacements be single-valued, continuous functions?"

In general, the requirements that a function $u(x, y)$ be continuous at a point (x_0, y_0) are that all partial derivatives be finite and continuous at (x_0, y_0) and that the remainder term of the Taylor series approach zero as the number of terms increases.

We assume that all of the above requirements are formally satisfied and look for any supplementary relations or conditions in order that this be so.

In two dimensions the Taylor series has the form,

$$u(x,y) = u(x_0,y_0) + (x-x_0)\left(\frac{\partial u}{\partial x}\right)_0 + (y-y_0)\left(\frac{\partial u}{\partial y}\right)_0 + \frac{1}{2!}(x-x_0)^2\left(\frac{\partial^2 u}{\partial x^2}\right)_0$$

$$+ \frac{1}{2!}(x-x_0)(y-y_0)\left(\frac{\partial^2 u}{\partial x \partial y}\right)_0 + \frac{1}{2!}(y-y_0)^2\left(\frac{\partial^2 u}{\partial y^2}\right)_0 + \cdots \quad (4\text{-}123)$$

There are four equations among the four partial derivatives—$\partial u/\partial x$, $\partial u/\partial y$, $\partial v/\partial x$, $\partial v/\partial y$—in terms of the strain components and the rotation, as follows (ω_z = rotation):

$$\left. \begin{aligned} \frac{\partial u}{\partial x} &= e_x \\[4pt] \frac{\partial v}{\partial y} &= e_y \\[4pt] \frac{\partial u}{\partial y}+\frac{\partial v}{\partial x} &= \gamma_{xy} \\[4pt] \frac{\partial u}{\partial y}-\frac{\partial v}{\partial x} &= 2\omega_z \end{aligned} \right\} \quad (4\text{-}124)$$

These are now treated as a set of linear non-homogeneous equations in terms of the variables $\partial u/\partial x$, $\partial u/\partial y$, $\partial v/\partial x$, $\partial v/\partial y$.

What requirement must be placed upon this set in order that $\partial u/\partial x$ [9] be regular?[10] This will then insure that $(\partial u/\partial x)_0$ will be regular.

We have

$$\frac{\partial u}{\partial x} = \frac{A}{B} \quad (4\text{-}125)$$

where

$$\left. \begin{aligned} A &= \begin{vmatrix} e_x & 0 & 0 & 0 \\ e_y & 0 & 0 & 1 \\ \gamma_{xy} & 1 & 1 & 0 \\ 2\omega_{xy} & 1 & -1 & 0 \end{vmatrix} \\[10pt] B &= \begin{vmatrix} 1 & 0 & 0 & 0 \\ 0 & 0 & 0 & 1 \\ 0 & 1 & 1 & 0 \\ 0 & 1 & -1 & 0 \end{vmatrix} \end{aligned} \right\} \quad (4\text{-}126)$$

[9] That is, we mean $\partial u/\partial x$, $\partial u/\partial y$, $\partial v/\partial x$, and $\partial v/\partial y$. We shall use this abbreviated notation throughout this section.

[10] By "regular" in the present sense, we mean that the partial derivatives are finite, continuous, and single-valued.

Since B is never zero, $\partial u/\partial x$ and the others are always regular, hence $(\partial u/\partial x)_0$ and the others are always regular, if the terms e, γ, and ω_z are finite and continuous.

Now check the terms $(\partial^2 u/\partial x^2)$ and the other similar second differentials for regularity. We do this by taking $\partial/\partial x$ and $\partial/\partial y$ of Eq. 4-124. This results in

$$
\begin{aligned}
\frac{\partial^2 u}{\partial x^2} &= \frac{\partial e_x}{\partial x} &&\text{(a)}\\[2mm]
\frac{\partial^2 u}{\partial x \partial y} &= \frac{\partial e_x}{\partial y} &&\text{(b)}\\[2mm]
\frac{\partial^2 v}{\partial x \partial y} &= \frac{\partial e_y}{\partial x} &&\text{(c)}\\[2mm]
\frac{\partial^2 v}{\partial y^2} &= \frac{\partial e_y}{\partial y} &&\text{(d)}\\[2mm]
\frac{\partial^2 u}{\partial x \partial y} + \frac{\partial^2 v}{\partial x^2} &= \frac{\partial \gamma_{xy}}{\partial x} &&\text{(e)}\\[2mm]
\frac{\partial^2 u}{\partial y^2} + \frac{\partial^2 v}{\partial x \partial y} &= \frac{\partial \gamma_{xy}}{\partial y} &&\text{(f)}\\[2mm]
\frac{\partial^2 u}{\partial x \partial y} - \frac{\partial^2 v}{\partial x^2} &= \frac{2 \partial \omega_z}{\partial x} &&\text{(g)}\\[2mm]
\frac{\partial^2 u}{\partial y^2} - \frac{\partial^2 v}{\partial x \partial y} &= \frac{2 \partial \omega_z}{\partial y} &&\text{(h)}
\end{aligned}
\right\} \quad (4\text{-}127)
$$

Equation 4-127 represents eight equations for six unknowns, $\partial^2 u/\partial x^2$. If these unknowns are to be single-valued, then we must have only six independent equations. By inspection we see

$$\frac{2 \partial e_x}{\partial y} - \frac{\partial \gamma_{xy}}{\partial x} = \frac{2 \partial \omega_z}{\partial x} \qquad (4\text{-}128)$$

and

$$\frac{\partial \gamma_{xy}}{\partial y} - 2\frac{\partial e_y}{\partial x} = \frac{2 \partial \omega_z}{\partial y} \qquad (4\text{-}129)$$

Thus, the six independent equations become Eqs. 4-127a through 4-127f and by inspection of the determinants it may be verified that these will give finite values for $\partial^2 u/\partial x^2$.[11]

[11] It should be noted that Eqs. 4-128 and 4-129 constitute a pair of compatibility conditions among the strains and *rotations* in order to insure that the deformations be finite, continuous, and single-valued. Otherwise stated, if, in a particular problem,

Footnote continued on page 124 .

We now check the third partial derivatives for regularity. To do this, we take $\partial/\partial x$ and $\partial/\partial y$ of Eqs. 4-127a through 4-127h, which leads to twelve equations in eight unknowns, as follows:

$$\frac{\partial^2 u}{\partial x^3} = \frac{\partial^2 e_x}{\partial x^2} \quad \text{(a)}$$

$$\frac{\partial^3 u}{\partial y\,\partial x^2} = \frac{\partial^2 e_x}{\partial x\,\partial y} \quad \text{(b)}$$

$$\frac{\partial^3 u}{\partial x\,\partial y^2} = \frac{\partial^2 e_x}{\partial y^2} \quad \text{(c)}$$

$$\frac{\partial^3 v}{\partial y\,\partial x^2} = \frac{\partial^2 e_y}{\partial x^2} \quad \text{(d)}$$

$$\frac{\partial^3 v}{\partial x\,\partial y^2} = \frac{\partial^2 e_y}{\partial x\,\partial y} \quad \text{(e)}$$

$$\frac{\partial^3 v}{\partial y^3} = \frac{\partial^2 e_y}{\partial y^2} \quad \text{(f)}$$

$$\frac{\partial^3 u}{\partial y\,\partial x^2} + \frac{\partial^3 v}{\partial x^3} = \frac{\partial^2 \gamma_{xy}}{\partial x^2} \quad \text{(g)}$$

$$\frac{\partial^3 u}{\partial x\,\partial y^2} + \frac{\partial^3 v}{\partial y\,\partial x^2} = \frac{\partial^2 \gamma_{xy}}{\partial x\,\partial y} \quad \text{(h)}$$

$$\frac{\partial^3 u}{\partial y^3} + \frac{\partial^3 v}{\partial x\,\partial y^2} = \frac{\partial^2 \gamma_{xy}}{\partial y^2} \quad \text{(i)}$$

$$\frac{\partial^3 u}{\partial y\,\partial x^2} - \frac{\partial^3 v}{\partial x^3} = 2\frac{\partial^2 \omega_z}{\partial x^2} \quad \text{(j)}$$

$$\frac{\partial^3 u}{\partial x\,\partial y^2} - \frac{\partial^3 v}{\partial y\,\partial x^2} = 2\frac{\partial^2 \omega_z}{\partial x\,\partial y} \quad \text{(k)}$$

$$\frac{\partial^3 y}{\partial y^3} - \frac{\partial^3 v}{\partial x\,\partial y^2} = 2\frac{\partial^2 \omega_z}{\partial y^2} \quad \text{(l)}$$

$$(4\text{-}130)$$

In order that $\partial^3 u/\partial x^3$ and the other third derivatives be single-valued,

----

Footnote continued from page 123.

the solution is obtained in terms of strains (e_x, etc.) and rotation (ω_z), then in order that the deformations u and v be compatible, it is necessary that the strains and rotation satisfy Eqs. 4-128 and 4-129. See the added requirement in Eqs. 4-131a.

TABLE 4.2

Set	Number of Terms	Number of Dependent Equations
$\dfrac{\partial u}{\partial x}$	4	Zero
e_x, ω_z	4	
$\dfrac{\partial^2 u}{\partial x^2}$	6	Two. Equations 4-128 and 4-129
$\dfrac{\partial e_x}{\partial x}$	8	
$\dfrac{\partial^3 u}{\partial x^3}$	8	Four. Equations 4-131a through 4-131d
$\dfrac{\partial^2 e_x}{\partial x^2}$	12	
$\dfrac{\partial^4 u}{\partial x^4}$	10	Six. $\partial/\partial x$ and $\partial/\partial y$ of Equations 4-131a through 4-131d
$\dfrac{\partial^3 e_x}{\partial x^3}$	16	
$\dfrac{\partial^n u}{\partial x^n}$	$2n+2$	$2n-2$. $\partial/\partial x$ and $\partial/\partial y$ of the equations from preceeding operation
$\dfrac{\partial^{n-1} e_x}{\partial x^{n-1}}$	$4n$	

it is necessary that these 12 equations be equivalent to only eight independent equations. By inspection it is seen that

$$
\begin{aligned}
\frac{\partial^2 e_x}{\partial y^2} + \frac{\partial^2 e_y}{\partial x^2} &= \frac{\partial^2 \gamma_{xy}}{\partial x \partial y} \quad &\text{(a)} \\[2mm]
\frac{\partial^2 \gamma_{xy}}{\partial x \partial y} - \frac{2 \partial^2 e_y}{\partial x^2} &= \frac{2 \partial^2 \omega_z}{\partial x \partial y} \quad &\text{(b)} \\[2mm]
\frac{\partial^2 \gamma_{xy}}{\partial y^2} - \frac{2 \partial^2 e_y}{\partial x \partial y} &= \frac{2 \partial^2 \omega_z}{\partial y^2} \quad &\text{(c)} \\[2mm]
\frac{2 \partial^2 e_x}{\partial x \partial y} - \frac{\partial^2 \gamma_{xy}}{\partial x^2} &= \frac{2 \partial^2 \omega_z}{\partial x^2} \quad &\text{(d)}
\end{aligned}
\right\} \quad (4\text{-}131)
$$

Therefore, the independent set of eight equations is Eq. 4-130a through Eqs. 4-130g and 4-130i, and it may be verified by examining the determinants that these will give finite values for the third derivatives.

Equation 4-131a is the ordinary strain compatibility condition as derived in elasticity. Equations 4-131b, 4-131c and 4-131d are obtained directly from Eqs. 4-128 and 4-129 by taking partials.[12]

We can show that no new additional relations are obtained (except those obtained by taking successive partial derivatives of Eq. 4-131). This is done as follows:

Consider the left-hand side of the set, Eqs. 4.130a to 4-130l. Each successive pair of partial differentiations adds two new terms (i.e., unknowns). Similarly, each successive pair of partial differentiations adds four new terms to the right-hand side. This means that two additional relations among the equations must be found for each successive pair of partial differentiations. But these are obtained from the set of Eqs. 4-131a through 4-131d. In addition, it may be seen that the determinants B are always different from zero. This may be summarized as in Table 4.2.

Thus, Eq. 4-131a is the required relation among the *strains* in order that u and v be finite and single-valued.

THREE-DIMENSIONAL CASE

For three dimensions the method follows along identical lines, except that now $\partial u / \partial x$ can be given in terms of six strain elements, Table 4.1, and three rotation elements (ω_x, ω_y, ω_z). The final results can be shown most concisely in tabular form, as in Table 4.3.

[12] See footnote on p. 123.

TABLE 4.3

Set	Number of Terms	Number of Dependent Equations
$\dfrac{\partial u}{\partial x}$	9	Zero
$e_x,\ \omega_x$	9	
$\dfrac{\partial^2 u}{\partial x^2}$	18	Nine, as follows:[13] $$2\frac{\partial \omega_z}{\partial x} = 2\frac{\partial e_x}{\partial y} - \frac{\partial \gamma_{xy}}{\partial x}$$
$\dfrac{\partial e_x}{\partial x}$	27	plus seven more similar
$\dfrac{\partial^3 u}{\partial x^3}$	30	18, obtained by differentiating above nine with respect to x, y, z, plus
$\dfrac{\partial^2 e_x}{\partial x^2}$	54	Six, in terms of strains, the compatibility conditions, as follows: $$\frac{\partial^2 e_y}{\partial z^2} + \frac{\partial^2 e_z}{\partial y^2} = \frac{\partial^2 \gamma_{yz}}{\partial y \partial z}$$ $$\frac{\partial^2 e_z}{\partial x^2} + \frac{\partial^2 e_x}{\partial z^2} = \frac{\partial^2 \gamma_{zx}}{\partial z \partial x}$$ $$\frac{\partial^2 e_x}{\partial y^2} + \frac{\partial^2 e_y}{\partial x^2} = \frac{\partial^2 \gamma_{xy}}{\partial x \partial y}$$ $$\frac{2\partial^2 e_z}{\partial x \partial y} = \frac{\partial}{\partial z}\left(-\frac{\partial \gamma_{yz}}{\partial x} + \frac{\partial \gamma_{zx}}{\partial y} - \frac{\partial \gamma_{xy}}{\partial z}\right)$$ $$\frac{2\partial^2 e_x}{\partial y \partial z} = \frac{\partial}{\partial x}\left(\frac{\partial \gamma_{zx}}{\partial y} + \frac{\partial \gamma_{xy}}{\partial z} - \frac{\partial \gamma_{yz}}{\partial x}\right)$$ $$\frac{2\partial^2 e_y}{\partial z \partial x} = \frac{\partial}{\partial y}\left(\frac{\partial \gamma_{xy}}{\partial z} + \frac{\partial \gamma_{yz}}{\partial x} - \frac{\partial \gamma_{zx}}{\partial y}\right)$$
$\dfrac{\partial^4 u}{\partial x^4}$	45	30, obtained by differentiating $\partial^2 \omega / \partial x^2$ with respect to x, y, z, plus
$\dfrac{\partial^3 e_x}{\partial x^3}$	90	15, from differentiating the compatibility relations (three interrelations out of 18)

[13] See once again the footnote on p. 123 regarding the strains and rotations.

Table 4.3—continued

Set	Number of Terms	Number of Dependent Equations
$\dfrac{\partial^n u}{\partial x^n}$	$\dfrac{3(n+1)(n+2)}{2}$	$q + r$, where $q = \frac{3}{2}n(n+1)$, $n \geqslant 2$ and q is obtained by differentiating the ω terms and $r = \frac{3}{2}(n-2)(n+1)$, $n \geqslant 3$ and r is obtained by differentiating the compatibility conditions.
$\dfrac{\partial^{n-1} e_x}{\partial x^{n-1}}$	$\dfrac{9n(n+1)}{2}$	

Summarizing the above:

If we obtain a solution to a given elasticity problem in terms of the *strain* elements, e_x, etc., then because the strain tensor contains *nine* elements which are, in turn, given in terms of *three* deformations, u, v, and w, it is necessary that the *strain* components satisfy certain relations in order to insure that the *deformations* will be continuous, finite, and single-valued.

By means of a Taylor expansion analysis, it was shown that the relations which must be satisfied by the elements of the strain tensor are (a) for the two-dimensional case,

$$\frac{\partial^2 e_x}{\partial y^2} + \frac{\partial^2 e_y}{\partial x^2} = \frac{\partial^2 \gamma_{xy}}{\partial x \partial y} \tag{4-131}$$

and (b) for the three-dimensional case,

$$\left.\begin{aligned}
\frac{\partial^2 e_y}{\partial z^2} + \frac{\partial^2 e_z}{\partial y^2} &= \frac{\partial^2 \gamma_{yz}}{\partial y \partial z} \\[2mm]
\frac{\partial^2 e_z}{\partial x^2} + \frac{\partial^2 e_x}{\partial z^2} &= \frac{\partial^2 \gamma_{zx}}{\partial z \partial x} \\[2mm]
\frac{\partial^2 e_x}{\partial y^2} + \frac{\partial^2 e_y}{\partial x^2} &= \frac{\partial^2 \gamma_{xy}}{\partial x \partial y} \\[2mm]
\frac{2\partial^2 e_z}{\partial x \partial y} &= \frac{\partial}{\partial z}\left(\frac{\partial \gamma_{yz}}{\partial x} + \frac{\partial \gamma_{zx}}{\partial y} - \frac{\partial \gamma_{xy}}{\partial z}\right) \\[2mm]
\frac{2\partial^2 e_x}{\partial y \partial z} &= \frac{\partial}{\partial x}\left(\frac{\partial \gamma_{zx}}{\partial y} + \frac{\partial \gamma_{xy}}{\partial z} - \frac{\partial \gamma_{yz}}{\partial x}\right) \\[2mm]
\frac{2\partial^2 e_y}{\partial z \partial x} &= \frac{\partial}{\partial y}\left(\frac{\partial \gamma_{xy}}{\partial z} + \frac{\partial \gamma_{yz}}{\partial x} - \frac{\partial \gamma_{zx}}{\partial y}\right)
\end{aligned}\right\} \tag{4-132}$$

The above represent the only independent relations among the strains and it was shown that these must be satisfied in order to insure uniqueness, finiteness, and continuity in the deformations u, v, and w.

Now it is possible to solve problems in elasticity in terms of the *stresses*. If this is done, then it is also necessary that these *stresses* satisfy certain compatibility conditions in order that the strains and stresses may be finite, single-valued, and continuous. These are the *stress compatibility conditions* and they may be obtained directly from the strain compatibility conditions (Eq. 4-132) by utilizing Hooke's Law (Eq. 4-116) and the equilibrium equations (Eq. 4-95). If this is done, then it will be found that the stresses must satisfy the following *six* stress compatibility conditions in order that u, v, and w be single valued, finite, and continuous,

$$\nabla^2 \sigma_x + \frac{1}{1+\nu} \frac{\partial^2 \Theta}{\partial x^2} = -\frac{\nu}{1-\nu} \operatorname{div} F - \frac{2\partial F_x}{\partial x}$$

$$\nabla^2 \sigma_y + \frac{1}{1+\nu} \frac{\partial^2 \Theta}{\partial y^2} = -\frac{\nu}{1-\nu} \operatorname{div} F - \frac{2\partial F_y}{\partial y}$$

$$\nabla^2 \sigma_z + \frac{1}{1+\nu} \frac{\partial^2 \Theta}{\partial z^2} = -\frac{\nu}{1-\nu} \operatorname{div} F - \frac{2\partial F_z}{\partial z}$$

$$\nabla^2 \tau_{yz} + \frac{1}{1+\nu} \frac{\partial^2 \Theta}{\partial y \partial z} = -\left(\frac{\partial F_y}{\partial z} + \frac{\partial F_z}{\partial y}\right) \qquad (4\text{-}133)$$

$$\nabla^2 \tau_{zx} + \frac{1}{1+\nu} \frac{\partial^2 \Theta}{\partial z \partial x} = -\left(\frac{\partial F_z}{\partial x} + \frac{\partial F_x}{\partial z}\right)$$

$$\nabla^2 \tau_{xy} + \frac{1}{1+\nu} \frac{\partial^2 \Theta}{\partial x \partial y} = -\left(\frac{\partial F_x}{\partial y} + \frac{\partial F_y}{\partial x}\right)$$

in which

$\Theta = \sigma_x + \sigma_y + \sigma_z$, the first invariant of the stress tensor

$F = (F_x \ F_y \ F_z)$, the body force

Finally, it is possible to solve the elasticity problem directly in term of u, v and w, the deformations. That is, we may obtain, directly, differential equations in terms of these three quantities. Then, obviously, the solution of these equations will themselves show whether u, v and w are single-valued, finite, and continuous. Hence no supplementary equations will be needed in this case.

4-7 The Compatibility Equations for Large Strains. For large strains, the strain tensor becomes (see Eq. 4-17),

$$\begin{pmatrix} \eta_{aa} & \eta_{ab} & \eta_{ac} \\ \eta_{ba} & \eta_{bb} & \eta_{bc} \\ \eta_{ca} & \eta_{cb} & \eta_{cc} \end{pmatrix} = \begin{pmatrix} \dfrac{\partial u}{\partial a} & \dfrac{1}{2}\left(\dfrac{\partial u}{\partial b}+\dfrac{\partial v}{\partial a}\right) & \dfrac{1}{2}\left(\dfrac{\partial u}{\partial c}+\dfrac{\partial w}{\partial a}\right) \\[2mm] \dfrac{1}{2}\left(\dfrac{\partial u}{\partial b}+\dfrac{\partial v}{\partial a}\right) & \dfrac{\partial v}{\partial b} & \dfrac{1}{2}\left(\dfrac{\partial v}{\partial c}+\dfrac{\partial w}{\partial b}\right) \\[2mm] \dfrac{1}{2}\left(\dfrac{\partial u}{\partial c}+\dfrac{\partial w}{\partial a}\right) & \dfrac{1}{2}\left(\dfrac{\partial v}{\partial c}+\dfrac{\partial w}{\partial b}\right) & \dfrac{\partial w}{\partial c} \end{pmatrix}$$

$$+\frac{1}{2}\begin{pmatrix} \left(\dfrac{\partial u}{\partial a}\right)^2+\left(\dfrac{\partial v}{\partial a}\right)^2+\left(\dfrac{\partial w}{\partial a}\right)^2 & \dfrac{\partial u}{\partial a}\dfrac{\partial u}{\partial b}+\dfrac{\partial v}{\partial a}\dfrac{\partial v}{\partial b}+\dfrac{\partial w}{\partial a}\dfrac{\partial w}{\partial b} \\[2mm] \dfrac{\partial u}{\partial b}\dfrac{\partial u}{\partial a}+\dfrac{\partial v}{\partial b}\dfrac{\partial v}{\partial a}+\dfrac{\partial w}{\partial b}\dfrac{\partial w}{\partial a} & \left(\dfrac{\partial u}{\partial b}\right)^2+\left(\dfrac{\partial v}{\partial b}\right)^2+\left(\dfrac{\partial w}{\partial b}\right)^2 \\[2mm] \dfrac{\partial u}{\partial c}\dfrac{\partial u}{\partial a}+\dfrac{\partial v}{\partial c}\dfrac{\partial v}{\partial a}+\dfrac{\partial w}{\partial c}\dfrac{\partial w}{\partial a} & \dfrac{\partial u}{\partial c}\dfrac{\partial u}{\partial b}+\dfrac{\partial v}{\partial c}\dfrac{\partial v}{\partial b}+\dfrac{\partial w}{\partial c}\dfrac{\partial w}{\partial b} \end{pmatrix}$$

$$\begin{pmatrix} \dfrac{\partial u}{\partial a}\dfrac{\partial u}{\partial c}+\dfrac{\partial v}{\partial a}\dfrac{\partial v}{\partial c}+\dfrac{\partial w}{\partial a}\dfrac{\partial w}{\partial c} \\[2mm] \dfrac{\partial u}{\partial b}\dfrac{\partial u}{\partial c}+\dfrac{\partial v}{\partial b}\dfrac{\partial v}{\partial c}+\dfrac{\partial w}{\partial b}\dfrac{\partial w}{\partial c} \\[2mm] \left(\dfrac{\partial u}{\partial c}\right)^2+\left(\dfrac{\partial v}{\partial c}\right)^2+\left(\dfrac{\partial w}{\partial c}\right)^2 \end{pmatrix}$$

$$(4\text{-}134)$$

In this case also, because there are *nine* elements and only *three* deformations, there are supplementary equations or conditions which must be satisfied by the strains in order to insure finiteness, uniqueness, and continuity of the deformations, u, v and w. These are the *compatibility conditions for higher-order strains*.

To obtain the compatibility conditions for higher-order strains, proceed as before. That is, we have

$$\frac{\partial u}{\partial a} = \eta_{aa}-\frac{1}{2}\left[\left(\frac{\partial u}{\partial a}\right)^2+\left(\frac{\partial v}{\partial a}\right)^2+\left(\frac{\partial w}{\partial a}\right)^2\right]$$

$$\frac{\partial v}{\partial b} = \eta_{bb}-\frac{1}{2}\left[\left(\frac{\partial u}{\partial b}\right)^2+\left(\frac{\partial v}{\partial b}\right)^2+\left(\frac{\partial w}{\partial b}\right)^2\right]$$

$$\frac{\partial w}{\partial c} = \eta_{cc}-\frac{1}{2}\left[\left(\frac{\partial u}{\partial c}\right)^2+\left(\frac{\partial v}{\partial c}\right)^2+\left(\frac{\partial w}{\partial c}\right)^2\right]$$

$$(4\text{-}135)$$

$$\frac{\partial u}{\partial b}+\frac{\partial v}{\partial a} = 2\eta_{ab}-\left[\frac{\partial u}{\partial a}\frac{\partial u}{\partial b}+\frac{\partial v}{\partial a}\frac{\partial v}{\partial b}+\frac{\partial w}{\partial a}\frac{\partial w}{\partial b}\right]$$

$$\frac{\partial u}{\partial c}+\frac{\partial w}{\partial a} = 2\eta_{ca}-\left[\frac{\partial u}{\partial c}\frac{\partial u}{\partial a}+\frac{\partial v}{\partial c}\frac{\partial v}{\partial a}+\frac{\partial w}{\partial c}\frac{\partial w}{\partial a}\right]$$

$$\frac{\partial v}{\partial c}+\frac{\partial w}{\partial b} = 2\eta_{bc}-\left[\frac{\partial u}{\partial b}\frac{\partial u}{\partial c}+\frac{\partial v}{\partial b}\frac{\partial v}{\partial c}+\frac{\partial w}{\partial b}\frac{\partial w}{\partial c}\right]$$

Then, for a first approximate higher-order form of the compatibility conditions—that is, squares of strains considered—we substitute the values of $\partial u/\partial a$ as given in the right-hand side of the above, retaining squares of strains only. To this approximation, we use for the right-hand-side terms,

$$\left.\begin{array}{l}\dfrac{\partial u}{\partial a} = \eta_{aa}, \quad \dfrac{\partial u}{\partial b} = \eta_{ab}+\omega_c, \quad \dfrac{\partial u}{\partial c} = \eta_{ca}-\omega_b, \quad \dfrac{\partial v}{\partial c} = \eta_{bc}+\omega_a \\[2mm] \dfrac{\partial v}{\partial a} = \eta_{ab}-\omega_c, \quad \dfrac{\partial w}{\partial a} = \eta_{ca}+\omega_b, \quad \dfrac{\partial w}{\partial b} = \eta_{bc}-\omega_a, \\[2mm] \qquad \dfrac{\partial v}{\partial b} = \eta_{bb}, \quad \dfrac{\partial w}{\partial c} = \eta_{cc}\end{array}\right\} \quad (4\text{-}136)$$

in which

$$\left.\begin{array}{l}\omega_a = -\dfrac{1}{2}\left(\dfrac{\partial w}{\partial b}-\dfrac{\partial v}{\partial c}\right) = \dfrac{1}{2}\left(\dfrac{\partial v}{\partial c}-\dfrac{\partial w}{\partial b}\right) \\[3mm] \omega_b = -\dfrac{1}{2}\left(\dfrac{\partial u}{\partial c}-\dfrac{\partial w}{\partial a}\right) = \dfrac{1}{2}\left(\dfrac{\partial w}{\partial a}-\dfrac{\partial u}{\partial c}\right) \\[3mm] \omega_c = -\dfrac{1}{2}\left(\dfrac{\partial v}{\partial a}-\dfrac{\partial u}{\partial b}\right) = \dfrac{1}{2}\left(\dfrac{\partial u}{\partial b}-\dfrac{\partial v}{\partial a}\right)\end{array}\right\} \quad (4\text{-}137)$$

To illustrate, we obtain the higher-order condition for the two-dimensional case. We have (neglecting terms higher than second order in η and ω)

$$\left.\begin{array}{ll}\dfrac{\partial u}{\partial a} = \eta_{aa}-\tfrac{1}{2}(\eta_{aa}{}^2+\eta_{ab}{}^2-2\eta_{ab}\omega+\omega^2) & \text{(a)} \\[3mm] \dfrac{\partial v}{\partial b} = \eta_{bb}-\tfrac{1}{2}(\eta_{bb}{}^2+\eta_{ab}{}^2+2\eta_{ab}\omega+\omega^2) & \text{(b)} \\[3mm] \dfrac{1}{2}\left(\dfrac{\partial u}{\partial b}+\dfrac{\partial v}{\partial a}\right) = \eta_{ab}-\tfrac{1}{2}[\eta_{aa}(\eta_{ab}+\omega)+\eta_{bb}(\eta_{ab}-\omega)] & \text{(c)} \\[3mm] \dfrac{1}{2}\left(\dfrac{\partial u}{\partial b}-\dfrac{\partial v}{\partial a}\right) = \omega & \text{(d)}\end{array}\right\} \quad (4\text{-}138)$$

These are

four equations
four unknowns

and the solution is regular for $\partial u/\partial a$.

Differentiating the above with respect to a and b, we get

eight unknowns
six unknowns

so that there must be two dependent relations. These are:

$$\left.\begin{array}{l}\dfrac{\partial(4\text{-}138\mathrm{d})}{\partial a} = \dfrac{\partial(4\text{-}138\mathrm{a})}{\partial b} - \dfrac{\partial(4\text{-}138\mathrm{c})}{\partial a} \\[2ex] \dfrac{\partial(4\text{-}138\mathrm{d})}{\partial b} = \dfrac{\partial(4\text{-}138\mathrm{c})}{\partial b} - \dfrac{\partial(4\text{-}138\mathrm{b})}{\partial a}\end{array}\right\} \qquad (4\text{-}139)$$

and represent higher-order compatibility conditions which must be satisfied by strains and rotations in order to insure uniqueness, finiteness, and continuity of the deformations u, v, and w.

Now differentiating these eight equations with respect to a and b, we get

twelve equations
eight unknowns

There are four dependent equations, three of which follow from Eq. 4-139 and one of which, the compatibility condition for the two-dimensional case, squares of strains considered, is

$$\frac{\partial^2 \eta_{aa}}{\partial b^2} + \frac{\partial^2 \eta_{bb}}{\partial a^2} = 2\frac{\partial^2 \eta_{ab}}{\partial a \partial b} - \left[-\frac{\partial \eta_{aa}}{\partial a}\frac{\partial \eta_{bb}}{\partial a} - \frac{\partial \eta_{aa}}{\partial b}\frac{\partial \eta_{aa}}{\partial b} + \frac{2\partial \eta_{aa}}{\partial a}\frac{\partial \eta_{ab}}{\partial b} \right.$$
$$\left. -\frac{\partial \eta_{bb}}{\partial a}\frac{\partial \eta_{bb}}{\partial a} - \frac{\partial \eta_{bb}}{\partial b}\frac{\partial \eta_{aa}}{\partial b} + \frac{2\partial \eta_{ab}}{\partial a}\frac{\partial \eta_{bb}}{\partial b} \right] \qquad (4\text{-}140)$$

This differs from the small deformation form of the compatibility equation in that it contains the terms in the brackets.

In a similar manner, one may obtain the three-dimensional forms of the higher-order compatibility conditions.

In the large deformation theory of plates, a higher-order compatibility equation is used (see Eq. 7-121a). However, as used, this equation is a special one which is based on the assumption that certain terms only of the non-linear strain tensor are important. It is not, therefore, the same equation as the one given above, Eq. 4-140.

4-8 Summary. We summarize the present discussion of the theory of elasticity.

Given an elastic body subject to forces and constraints. Then the stress and strain condition at any point in the body must be such that, if the body is in equilibrium—

1. The stresses satisfy the equilibrium equations:

or

$$
\begin{aligned}
\operatorname{div} T + \rho F^{\star} &= 0 \\
\frac{\partial \sigma_x}{\partial x} + \frac{\partial \tau_{yx}}{\partial y} + \frac{\partial \tau_{zx}}{\partial z} + \rho F_x &= 0 \\
\frac{\partial \tau_{xy}}{\partial x} + \frac{\partial \sigma_y}{\partial y} + \frac{\partial \tau_{zy}}{\partial z} + \rho F_y &= 0 \\
\frac{\partial \tau_{xz}}{\partial x} + \frac{\partial \tau_{yz}}{\partial y} + \frac{\partial \sigma_z}{\partial z} + \rho F_z &= 0
\end{aligned}
\right\} \qquad (4\text{-}141)
$$

2. The applied surface forces must satisfy the surface equilibrium relations, or, otherwise stated, the boundary conditions that must be satisfied by the stresses[14] are

or

$$
\begin{aligned}
\bar{x} &= NT \\
\bar{x} &= l\sigma_x + m\tau_{yx} + n\tau_{zx} \\
\bar{y} &= l\tau_{xy} + m\sigma_y + n\tau_{zy} \\
\bar{z} &= l\tau_{xz} + m\tau_{yz} + n\sigma_z
\end{aligned}
\right\} \qquad (4\text{-}142)
$$

3. In order to insure that an elasticity solution obtained in terms of the *strain* elements, e_x, etc., will correspond to one for which the *deformation* components $(u, v, w,)$ are finite, single-valued, and continuous, the strains must satisfy the strain compatibility conditions, which (a) in *small* deformation, two-dimensional theory is the single equation

$$
\frac{\partial^2 e_x}{\partial y^2} + \frac{\partial^2 e_y}{\partial x^2} = \frac{\partial^2 \gamma_{xy}}{\partial x \partial y} \qquad (4\text{-}143)
$$

(b) In *large* deformation, two-dimensional theory this becomes the following non-linear compatibility condition,

$$
\begin{aligned}
\frac{\partial^2 \eta_{aa}}{\partial b^2} + \frac{\partial^2 \eta_{bb}}{\partial a^2} = \frac{2\partial^2 \eta_{ab}}{\partial a \partial b} - \Bigg(&- \frac{\partial \eta_{aa}}{\partial a} \frac{\partial \eta_{bb}}{\partial a} - \frac{\partial \eta_{aa}}{\partial b} \frac{\partial \eta_{aa}}{\partial b} + \frac{2\partial \eta_{aa}}{\partial a} \frac{\partial \eta_{ab}}{\partial b} \\
&- \frac{\partial \eta_{bb}}{\partial a} \frac{\partial \eta_{bb}}{\partial a} - \frac{\partial \eta_{bb}}{\partial b} \frac{\partial \eta_{aa}}{\partial b} + \frac{2\partial \eta_{ab}}{\partial a} \frac{\partial \eta_{bb}}{\partial b} \Bigg)
\end{aligned}
$$

$$(4\text{-}144)$$

---

[14] In some cases the boundary conditions may be given in terms of the displacements, in which case these may be used instead of the stress boundary conditions.

(c) In *small* deformation, three-dimensional theory the compatibility conditions which must be satisfied by the strains are

$$
\left.
\begin{aligned}
\frac{\partial^2 e_y}{\partial z^2} + \frac{\partial^2 e_z}{\partial y^2} &= \frac{\partial^2 \gamma_{yz}}{\partial y \partial z} \\[2mm]
\frac{\partial^2 e_z}{\partial x^2} + \frac{\partial^2 e_x}{\partial z^2} &= \frac{\partial^2 \gamma_{zx}}{\partial z \partial x} \\[2mm]
\frac{\partial^2 e_x}{\partial y^2} + \frac{\partial^2 e_y}{\partial x^2} &= \frac{\partial^2 \gamma_{xy}}{\partial x \partial y} \\[2mm]
\frac{2 \partial^2 e_z}{\partial x \partial y} &= \frac{\partial}{\partial z}\left(\frac{\partial \gamma_{yz}}{\partial x} + \frac{\partial \gamma_{zx}}{\partial y} - \frac{\partial \gamma_{xy}}{\partial z} \right) \\[2mm]
\frac{2 \partial^2 e_x}{\partial y \partial z} &= \frac{\partial}{\partial x}\left(\frac{\partial \gamma_{zx}}{\partial y} + \frac{\partial \gamma_{xy}}{\partial z} - \frac{\partial \gamma_{yz}}{\partial x} \right) \\[2mm]
\frac{2 \partial^2 e_y}{\partial z \partial x} &= \frac{\partial}{\partial y}\left(\frac{\partial \gamma_{xy}}{\partial z} + \frac{\partial \gamma_{yz}}{\partial x} - \frac{\partial \gamma_{zx}}{\partial y} \right)
\end{aligned}
\right\}
\qquad (4\text{-}145)
$$

(d) Similar non-linear equations can be derived for the *large* deformation, three-dimensional theory.

(e) Because of the relations between stresses and strains, it is possible to transform the *strain* compatibility conditions into *stress* compatibility conditions. These will insure that a solution to a particular problem, given in terms of *stresses* is one which will give finite, single-valued, and continuous deformations. In the small deformation, three-dimensional theory, the *stress* compatibility conditions are

$$
\left.
\begin{aligned}
\nabla^2 \sigma_x + \frac{1}{1+\nu} \frac{\partial^2 \Theta}{\partial x^2} &= -\frac{\nu}{1-\nu} \operatorname{div} F - \frac{2 \partial F_x}{\partial x} \\[2mm]
\nabla^2 \sigma_y + \frac{1}{1+\nu} \frac{\partial^2 \Theta}{\partial y^2} &= -\frac{\nu}{1-\nu} \operatorname{div} F - \frac{2 \partial F_y}{\partial y} \\[2mm]
\nabla^2 \sigma_z + \frac{1}{1+\nu} \frac{\partial^2 \Theta}{\partial z^2} &= -\frac{\nu}{1-\nu} \operatorname{div} F - \frac{2 \partial F_z}{\partial z} \\[2mm]
\nabla^2 \tau_{yz} + \frac{1}{1+\nu} \frac{\partial^2 \Theta}{\partial y \partial z} &= -\left(\frac{\partial F_y}{\partial z} + \frac{\partial F_z}{\partial y} \right) \\[2mm]
\nabla^2 \tau_{zx} + \frac{1}{1+\nu} \frac{\partial^2 \Theta}{\partial z \partial x} &= -\left(\frac{\partial F_z}{\partial x} + \frac{\partial F_x}{\partial z} \right) \\[2mm]
\nabla^2 \tau_{xy} + \frac{1}{1+\nu} \frac{\partial^2 \Theta}{\partial x \partial y} &= -\left(\frac{\partial F_x}{\partial y} + \frac{\partial F_y}{\partial x} \right)
\end{aligned}
\right\}
\qquad (4\text{-}146)
$$

(f) Finally, it is possible to formulate problems in elasticity theory directly in terms of the deformations u, v and w. In these cases the solutions themselves will indicate finiteness, continuity, and single-valuedness and no additional compatibility conditions on strains are required.

4. The stresses and strains (for an assumed isotropic material) are related by Hooke's Law, which in its linearized form is given by (see Eq. 4-116)

$$T = \lambda I_1 E_3 + 2\mu\eta \tag{4-147}$$

In the next chapter we apply these results to the solution of several problems in the theory of elasticity.

Problems

1. (a) Obtain the linear and nonlinear forms of the strain matrix if

$$u = 6a^2 + 3bc + 2c^2$$

$$v = b^2 + 3ac^2$$

$$w = ab + 2bc + 3ac$$

   (b) Is this necessarily a strain compatible set of deformations in the linear system?

   (c) In the nonlinear system?

2. (a) Obtain the trace for

$$\begin{pmatrix} 2x^2 & 3t & 0 \\ -2yz & a & 3t \\ 2x & -y^3 & -z \end{pmatrix} \begin{pmatrix} 4 & 3xy & t^3 \\ e^{-z} & 7 & -20 \\ 3 & -2 & x^2 \end{pmatrix}$$

   (b) Verify that for this case

$$[AB] = [BA]$$

3. Given

$$A = \begin{pmatrix} 2 & -7 & 1 \\ -3 & 0 & 4 \\ 8 & 2 & 4 \end{pmatrix} \quad B = \begin{pmatrix} 3 & -5 & 2 \\ 8 & 3 & 1 \\ -3 & -2 & -1 \end{pmatrix} \quad C = \begin{pmatrix} 2 & 5 & -2 \\ 4 & -1 & 0 \\ 3 & -3 & 9 \end{pmatrix}$$

verify that

$$[ABC] = [CAB]$$

4. (a) Verify that Hooke's Law, in its linear form, as a relation between strains and stresses, is given by

$$e_x = \frac{\sigma_x}{E} - \frac{\nu}{E}(\sigma_y + \sigma_z)$$

$$\gamma_{xy} = \frac{\tau_{xy}}{G}$$

where

$$G = \frac{E}{2(1+\nu)}$$

(b) Show this relation in tensor form.

5. (a) Show that, to the first order, the change in volume of a strained cube of unit length each side is given by

$$\nabla V = e_x + e_y + e_z$$

(b) Hence, show that if a material is assumed to be *incompressible* this is equivalent to the assumption that $\nu = \frac{1}{2}$.

6. For the two-dimensional stress system σ_x, σ_y, τ_{xy}, τ_{yx}, prove by considering the equilibrium of forces on the faces of a plane element, that $T' = RTR^\star$ and that T, therefore, is a tensor of the second order.

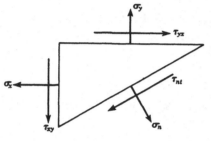

Prob. Fig. 4.6

7. Given a two-dimensional problem with

$$\sigma_x = x^2$$
$$\sigma_y = y^2$$
$$\tau_{xy} = -2xy$$

(a) Are the equilibrium equations satisfied everywhere in the body?

(b) Show that the linear strain compatibility equations are not satisfied. Hint: use Hooke's Law to obtain the strain components.

(c) By integrating the strains ($\partial u/\partial x$, $\partial v/\partial y$, etc.) directly show that the above stresses cannot represent a strain compatible solution to the problem.

8. One form of the nonlinear Hooke's Law is given by

$$T = \lambda I_1 E_3 + 2G\eta + (l I_1{}^2 - 2m I_2)E_3 + 2m I_1 \eta + n \operatorname{co} \eta$$

in which
l, m, n are new elastic constants
I_2 is the second strain invariant.
co η is the cofactor of the strain tensor.
Write out the six terms of this nonlinear form of Hooke's Law.

9. Given a nonlinear strain tensor

$$\eta = \frac{1}{2}\begin{pmatrix} 0 & K & 0 \\ K & K^2 & 0 \\ 0 & 0 & 0 \end{pmatrix}$$

determine Hooke's Law in its nonlinear form. Write out the six terms in their expanded form.

10. (a) Determine the equations for τ_{zx}, τ_{zy}, σ_z in order that the equilibrium equations be satisfied if

$$T = \begin{pmatrix} x^2+2yz & 3yz+xy & x \\ ? & 2z^3y & z^4y^4x \\ \tau_{zx} & \tau_{zy} & \sigma_z \end{pmatrix}$$

(b) What are the corresponding compatibility conditions?

11. The two-dimensional form of the equilibrium equations can be satisfied identically by introducing the Airy stress function ϕ, defined by

$$\sigma_x = \frac{\partial^2 \phi}{\partial y^2}$$

$$\sigma_y = \frac{\partial^2 \phi}{\partial x^2}$$

$$\tau_{xy} = -\frac{\partial^2 \phi}{\partial x \partial y}$$

(a) Show this.

(b) Using Hooke's Law and the strain compatibility condition, show that the Airy function satisfies the biharmonic equation,

$$\nabla^4 \phi = 0$$

where

$$\nabla^4 = \nabla^2 \nabla^2$$

$$= \frac{\partial^4}{\partial x^4} + \frac{2\partial^4}{\partial x^2 \partial y^2} + \frac{\partial^4}{\partial y^4}$$

(c) Show, therefore, that the real and imaginary parts of all $f(z)$, $z = x+iy$ are possible solutions to two-dimensional stress problems.

12. One form of nonisotropic (or aeolotropic) stress–strain relation is given by

$$\begin{pmatrix} e_x \\ e_y \\ e_z \\ \gamma_{yz} \\ \gamma_{zx} \\ \gamma_{xy} \end{pmatrix} = \begin{pmatrix} S_{11} & S_{12} & \cdots & S_{16} \\ S_{21} & \cdots & \cdots & S_{26} \\ \cdot & & & \\ \cdot & & & \\ \cdot & & & \\ S_{61} & \cdots & \cdots & S_{66} \end{pmatrix} \begin{pmatrix} \sigma_x \\ \sigma_y \\ \sigma_z \\ \tau_{yz} \\ \tau_{zx} \\ \tau_{xy} \end{pmatrix}$$

where

$$S_{rt} = S_{tr}, \qquad (r, t = 1, 2, \cdots 6)$$

(a) Show that this gives, in two dimensions,

$$e_x = S_{11}\sigma_x + S_{12}\sigma_y + S_{16}\tau_{xy}$$

$$e_y = S_{12}\sigma_x + S_{22}\sigma_y + S_{26}\tau_{xy}$$

$$\gamma_{xy} = S_{16}\sigma_x + S_{26}\sigma_y + S_{66}\tau_{xy}$$

(b) Show that the equilibrium equations are identically satisfied if we assume

$$\sigma_x = \frac{\partial^2 \phi}{\partial y^2}$$

$$\sigma_y = \frac{\partial^2 \phi}{\partial x^2}$$

$$\tau_{xy} = -\frac{\partial^2 \phi}{\partial x \partial y}$$

where ϕ is the Airy stress function.

(c) Show that compatibility requires

$$0 = S_{22}\frac{\partial^4 \phi}{\partial x^4} - 2S_{26}\frac{\partial^4 \phi}{\partial x^3 \partial y} + (2S_{12} + S_{66})\frac{\partial^4 \phi}{\partial x^2 \partial y^2} - 2S_{16}\frac{\partial^4 \phi}{\partial x \partial y^3} + S_{11}\frac{\partial^4 \phi}{\partial y^4}$$

(d) Finally, if we assume that tensile stresses do not cause shear strains and shear stresses do not cause tensile strains, then show that

$$S_{16} = S_{26} = 0$$

and the above equation becomes

$$\left(\frac{\partial^2}{\partial x^2} + \alpha_1 \frac{\partial^2}{\partial y^2}\right)\left(\frac{\partial^2}{\partial x^2} + \alpha_2 \frac{\partial^2}{\partial y^2}\right)\phi = 0$$

where

$$\alpha_1 \alpha_2 = \frac{S_{11}}{S_{22}}$$

and

$$\alpha_1 + \alpha_2 = \frac{S_{66} + 2S_{21}}{S_{22}}$$

13. Another form of anisotropic stress–strain theory can be developed assuming

$$e_x = \frac{\sigma_x}{E_x} - \frac{\nu_x \sigma_y}{E_y}$$

$$e_y = \frac{\sigma_y}{E_y} - \frac{\nu_y \sigma_x}{E_x}$$

$$\gamma_{xy} = \frac{\tau_{xy}}{G}$$

Prove that, for this case, the Airy stress function must satisfy the equation

$$\left(\frac{\partial^2}{\partial x^2} + K^2\frac{\partial^2}{\partial y^2}\right)\left(\frac{\partial^2}{\partial x^2} + \frac{\partial^2}{\partial y^2}\right)\phi = 0$$

with

$$K^2 = \frac{E_y}{E_x}$$

Determine the additional condition in order that this equation hold.

14. Using the indicial subscript notation and the summation convention show that

$$[AB] = [BA]$$

and that

$$[\widehat{ABC}] = [A\widehat{BC}] = [CAB]$$

15(a) Refer to Eqs. 3-84 through 3-88. Obtain the three stress equilibrium equations in general curvilinear coordinates.

(b) Consider a thick-walled circular cylinder subjected to a uniform internal pressure. Using cylindrical coordinates and noting the symmetry of the problem, determine the equilibrium equations using (a) above.

(c) Do the same for a thick-walled spherical container subjected to a uniform internal pressure.

16. Refer to Prob. 13. Show that

$$\phi = \int_0^\infty \frac{1}{\alpha^2} \left(A C^{-\alpha y} + B C^{-(\alpha y/K)} \right) \cos \alpha x \, d\alpha$$

satisfies the equation. A and B are constants.

Chapter 5

APPLICATIONS OF THE THEORY OF ELASTICITY

5-1 Introduction. In the previous chapter the fundamental equations between stresses, strains and deformations were derived. These were given in both a small deformation (linear) and a large deformation (nonlinear) form.

In this chapter the *linearized* results of the last chapter are applied to three classical problems in the mathematical theory of elasticity:

(a) the tension–compression bar problem,

(b) the pure bending problem,

(c) the torsion problem.

It will be seen that the solutions to these three problems follow naturally from the assumption of progressively more complicated forms to the stress tensor.

5-2 The Tension–Compression Bar in Elasticity. The equations of the linearized mathematical theory of elasticity in their three-dimensional forms are as follows:

1. Equilibrium (see Eq. 4-141), which for body forces neglected is given by

$$\left.\begin{aligned}
\frac{\partial \sigma_x}{\partial x}+\frac{\partial \tau_{yx}}{\partial y}+\frac{\partial \tau_{zx}}{\partial z} &= 0 \\[2mm]
\frac{\partial \tau_{xy}}{\partial x}+\frac{\partial \sigma_y}{\partial y}+\frac{\partial \tau_{zy}}{\partial z} &= 0 \\[2mm]
\frac{\partial \tau_{xz}}{\partial x}+\frac{\partial \tau_{yz}}{\partial y}+\frac{\partial \sigma_z}{\partial z} &= 0
\end{aligned}\right\} \tag{5-1}$$

2. Boundary conditions on stresses (see Eq. 4-142):

$$\left.\begin{aligned}
\bar{x} &= l\sigma_x+m\tau_{yx}+n\tau_{zx} \\
\bar{y} &= l\tau_{xy}+m\sigma_y+n\tau_{zy} \\
\bar{z} &= l\tau_{xz}+m\tau_{yz}+n\sigma_z
\end{aligned}\right\} \tag{5-2}$$

3. Compatibility (see Eq. 4-145):

$$\frac{\partial^2 e_y}{\partial z^2} + \frac{\partial^2 e_z}{\partial y^2} = \frac{\partial^2 \gamma_{yz}}{\partial y \partial z}$$

$$\frac{\partial^2 e_z}{\partial x^2} + \frac{\partial^2 e_x}{\partial z^2} = \frac{\partial^2 \gamma_{zx}}{\partial z \partial x}$$

$$\frac{\partial^2 e_x}{\partial y^2} + \frac{\partial^2 e_y}{\partial x^2} = \frac{\partial^2 \gamma_{xy}}{\partial x \partial y}$$

$$\frac{2\partial^2 e_z}{\partial x \partial y} = \frac{\partial}{\partial z}\left(\frac{\partial \gamma_{yz}}{\partial x} + \frac{\partial \gamma_{zx}}{\partial y} - \frac{\partial \gamma_{xy}}{\partial z}\right)$$

$$\frac{2\partial^2 e_x}{\partial y \partial z} = \frac{\partial}{\partial x}\left(\frac{\partial \gamma_{zx}}{\partial y} + \frac{\partial \gamma_{xy}}{\partial z} - \frac{\partial \gamma_{yz}}{\partial x}\right)$$

$$\frac{2\partial^2 e_y}{\partial z \partial x} = \frac{\partial}{\partial y}\left(\frac{\partial \gamma_{xy}}{\partial z} + \frac{\partial \gamma_{yz}}{\partial x} - \frac{\partial \gamma_{zx}}{\partial y}\right)$$

$$(5\text{-}3)$$

4. Hooke's Law (see Eq. 4-116 and Prob. 4, Chapter 4):

$$e_x = \frac{1}{E}[\sigma_x - \nu(\sigma_y + \sigma_z)]$$

$$e_y = \frac{1}{E}[\sigma_y - \nu(\sigma_z + \sigma_x)]$$

$$e_z = \frac{1}{E}[\sigma_z - \nu(\sigma_x + \sigma_y)]$$

$$\gamma_{xy} = \frac{\tau_{xy}}{G}$$

$$\gamma_{yz} = \frac{\tau_{yz}}{G}$$

$$\gamma_{zx} = \frac{\tau_{zx}}{G}$$

$$(5\text{-}4)$$

Let us assume that we are interested in investigating the behavior of a cylindrical (not necessarily circular) cross-section bar, as shown in Fig. 5.1. Let us assume further that we are interested in the behavior of this bar when it is subjected to a stress in the x direction only.

If we assume the stress tensor is given by

$$T = \begin{pmatrix} C & 0 & 0 \\ 0 & 0 & 0 \\ 0 & 0 & 0 \end{pmatrix} \qquad (5\text{-}5)$$

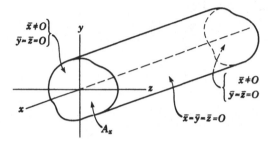

Fig. 5.1

in which $\sigma_x = C$, a constant, then obviously the equilibrium equations are satisfied (the student should verify this and the following statements), and we have from Hooke's Law

$$\left.\begin{aligned} e_x &= \frac{\sigma_x}{E} = \frac{C}{E} \\[2mm] e_y &= -\frac{\nu C}{E} \\[2mm] e_z &= -\frac{\nu C}{E} \\[2mm] \gamma_{xy} &= \gamma_{yz} = \gamma_{zx} = 0 \end{aligned}\right\} \qquad (5\text{-}6)$$

so that the strain compatibility conditions are also satisfied. For the cylindrical body shown in Fig. 5.2, the boundary conditions give, on the end faces which are perpendicular to the x axis (so that $l = \pm 1$, $m = n = 0$),

$$\bar{x} = l\sigma_x = \sigma_x \qquad (5\text{-}7)$$

on the near face and, similarly, $\bar{x} = \sigma_x$ on the far face.

The other equations that hold for the end faces and also for the longitudinal outside surface (for which $l = 0$, m and $n \neq 0$) are identically zero. This means that on the end planes normal to the x axis, the surface force per unit area must be equal to the constant stress, C. On all other boundary surfaces the stress is zero. This is in agreement with the conditions on the body which we wish to investigate.

In other words, the assumed form of the stress tensor, Eq. 5-5, corresponds to simple tension or compression in a cylindrical bar. The stress will be tension if C is positive and compression if C is negative.

Let us now determine the deformations of the structure. In order to do this we must use Hooke's Law. We have, from Eq. 5-4,

$$
\left.
\begin{aligned}
e_x &= \frac{C}{E} = \frac{F_x}{A_x E} = \frac{\partial u}{\partial x} \\[1em]
e_y &= -\frac{\nu F_x}{A_x E} = \frac{\partial v}{\partial y} \\[1em]
e_z &= -\frac{\nu F_x}{A_x E} = \frac{\partial w}{\partial z} \\[1em]
\gamma_{xy} &= \frac{\partial u}{\partial y} + \frac{\partial v}{\partial x} = 0 \\[1em]
\gamma_{yz} &= \frac{\partial v}{\partial z} + \frac{\partial w}{\partial y} = 0 \\[1em]
\gamma_{zx} &= \frac{\partial w}{\partial x} + \frac{\partial u}{\partial z} = 0
\end{aligned}
\right\} \qquad (5\text{-}8)
$$

in which F_x/A_x is the constant stress, caused by a force in the x direction, F_x, acting on a plane, A_x, which is normal to the x axis.[1]

It should be noted that our requirement that the stress distribution be a constant stress distribution means that the force F_x (which is, essentially, the *resultant* force due to the stresses on the end cross section) must be applied at the centroid. This is so since

$$
\left.
\begin{aligned}
F_x &= \int_{A_x} \sigma_x \, dA_x \\[0.5em]
&= C A_x
\end{aligned}
\right\} \qquad (5\text{-}9)
$$

and M_y = moment of stresses about the y axis $\left.\phantom{\rule{0pt}{1.5em}}\right\}$ (5-10)
$= F_x Z$

where Z is the distance between the point of application of F_x and the y axis. Now, we have also

$$
\left.
\begin{aligned}
M_y &= \int_{A_x} \sigma_x z \, dA_x \\[1em]
&= C \int_{A_x} z \, dA_x \\[1em]
&= 0, \text{ if } y \text{ is the centroidal axis}
\end{aligned}
\right\} \qquad (5\text{-}11)
$$

[1] Although the exact solution requires that the end face applied stress condition be the uniform stress, $\sigma_x = C$, in accordance with the well known St. Venant Principle, the stress condition at sections away from the ends will still be approximately uniform, of value $\sigma_x = C$, *no matter how the stress is applied on the end face*. The only requirement is that the resultant of this actual end stress condition be the same as the resultant due to $\sigma_x = C$ on the end face. By "sections away from the ends" we mean sections greater, roughly, than the beam depth away from the ends. See also Ref. (50).

Hence, since we require that $M_y = 0$,

and
$$\left. \begin{array}{c} F_x Z = 0 \\ Z = 0 \end{array} \right\} \tag{5-12}$$

or the force F_x is applied on the centroidal y axis. A similar argument may be applied to the centroidal z axis, so that finally, we have the requirement that F_x must act at the centroid of the cross section.

If we integrate the first of Eq. 5-8, using as the datum point or point of zero elongation the origin, $x = 0$, we get

$$u = \frac{F_x x}{A_x E} \tag{5-13}$$

and similarly, from the second and third equations,

$$\left. \begin{array}{c} v = -\dfrac{\nu F_x y}{A_x E} \\[2ex] w = -\dfrac{\nu F_x z}{A_x E} \end{array} \right\} \tag{5-14}$$

It may easily be verified, by substitution, that these values for u, v, and w satisfy the partial differential strain expressions given above.

The expression for u, the deformation in the x direction, will be recognized as the familiar elementary relation for the elongation (if F_x is tension) predicted by Hooke's Law.

The equations for v and w (the deformations in the y and z directions, respectively) represent the Poisson ratio effects corresponding to lateral contractions (or elongations) caused by longitudinal elongations (or contractions).

The above represents an exact solution of a problem in linearized theory of elasticity. We wish to emphasize that although the mathematical solution obtained above is a simple one, it is by no means a trivial one. Indeed, it represents an exact solution to a very important physical problem and as such is of considerable practical interest. And because it is an exact solution, it represents a possible point of departure for simplifications and approximations.[2]

---

[2] This pure normal stress solution is, in fact, the basis of techniques used in simple truss analysis. Trusses are generally assumed pin-connected at their ends and loaded only at the joints which, in turn, means that the forces acting on the bars or members of the truss must lie along the axes of the members. In other words, the members are subjected to simple tension—compression action, which is just the case considered above. It might be mentioned that when truss members are very deep and heavy, a more sophisticated analysis is required—but the simpler procedure applies in the majority of cases. See Ref. (10) for further discussion of trusses. See also Ref. (50).

5-3   The Bending Problem. This problem is also solved in an inverse manner—that is, a stress tensor is initially assumed; compatibility, boundary forces, and equilibrium equations are checked, and Hooke's Law is applied. It is then shown that the assumed stress tensor corresponds to a pure bending applied to the body.

Let us consider the linearized theory of elasticity solution for the stress tensor given by

$$T = \begin{pmatrix} \sigma_x & 0 & 0 \\ 0 & 0 & 0 \\ 0 & 0 & 0 \end{pmatrix} = \begin{pmatrix} Ky & 0 & 0 \\ 0 & 0 & 0 \\ 0 & 0 & 0 \end{pmatrix} \qquad (5\text{-}15)$$

for a cylindrical body as shown in the figure. It is clear that K is the stress at a unit distance, y, from the origin (see Fig. 5.2).

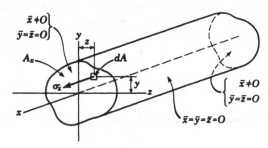

Fig. 5.2

The equilibrium equations (see Eq. 5-1) for body forces neglected,

$$\text{div } T = 0 \qquad (5\text{-}16)$$

are obviously identically satisfied by the assumed stress tensor, Eq. 5-15.

The boundary conditions become (since on the end face $l = \pm 1$, $m = 0$, $n = 0$, and on the side surface $l = 0$, $0 \leqslant m \leqslant \pm 1$, $0 \leqslant n \leqslant \pm 1$)

$$\bar{x} = \sigma_x \quad \text{on the end faces} \qquad (5\text{-}17)$$

and all other boundary conditions are identically satisfied, $0 = 0$.

Hooke's Law gives

$$e_x = \frac{1}{E}\sigma_x = \frac{Ky}{E} \qquad \text{(a)}$$

$$e_y = -\frac{\nu\sigma_x}{E} = -\frac{\nu Ky}{E} \qquad \text{(b)}$$

$$e_z = -\frac{\nu\sigma_x}{E} = -\frac{\nu Ky}{E} \qquad \text{(c)}$$

$$\gamma_{xy} = \gamma_{yz} = \gamma_{zx} = 0 \qquad \text{(d)}$$

$$(5\text{-}18)$$

so that the compatibility equations are also satisfied identically, as the reader may verify.

Thus, all formal requirements of the mathematical linearized theory of elasticity are satisfied by the assumed stress tensor.

We shall insist on the additional requirement that the *net* force be zero on the end faces (and hence all planes) which are perpendicular to the x axis—see Fig. 5-2. This means that

or

$$\left. \begin{array}{r} \displaystyle\int_{A_x} \sigma_x \, dA = 0 \\[2ex] \displaystyle K \int_{A_x} y \, dA = 0 \end{array} \right\} \tag{5-19}$$

and hence the z axis is a centroidal axis. The z axis, which is the line of zero stress, is called the neutral axis and is usually designated as N.A. We will require also that the moment about the y axis be zero, i.e.,

$$M_y = K \int_{A_x} yz \, dA = 0 \tag{5-20}$$

This means (see Eq. 2-71) that the y and z axes are principal axes of inertia.

Let us determine the moment of σ_x about the z axis. This is given by

which gives

$$\left. \begin{array}{r} \displaystyle\int_{A_x} \sigma_x y \, dA \\[2ex] \displaystyle K \int_{A_x} y^2 \, dA = KI_z \end{array} \right\} \tag{5-21}$$

and we call this quantity M_z, the bending moment about the z axis. Thus

$$M_z = KI_z \tag{5-22}$$

in which I_z is the area moment of inertia of the cross section about the z axis. Note that M_z is constant at all points, x, on the beam.

Hence, K, the stress per unit distance from the centroidal axis is given by

$$K = \frac{M_z}{I_z} \tag{5-23}$$

and the stress σ_x is given by

$$\sigma_x = Ky = \frac{M_z y}{I_z} \tag{5-24}$$

The above equation shows that at any section the stress varies linearly across the cross section from the neutral axis. This is as shown in Fig. 5.3, and it is clear that the maximum stress on the cross section occurs at the point furthest away from the neutral axis. This is of fundamental importance in the actual design of engineering beams.

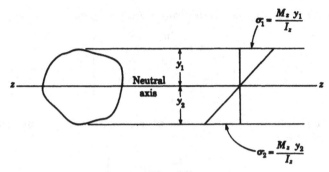

Fig. 5.3

To determine the deformation, we have from Hooke's Law, given in Eq. 5-4,

$$e_x = \frac{\partial u}{\partial x} = \frac{Ky}{E} \qquad (a)$$

$$e_y = \frac{\partial v}{\partial y} = -\frac{\nu Ky}{E} \qquad (b) \qquad \Bigg\} \qquad (5\text{-}25)$$

$$e_z = \frac{\partial w}{\partial z} = -\frac{\nu Ky}{E} \qquad (c)$$

From Eq. 5-25, by direct integration,

$$u = \frac{K}{E}yx + u_0(y, z) \qquad (5\text{-}26)$$

The first and third of Eq. 5-18d give

$$\frac{\partial v}{\partial x} = -\frac{\partial u}{\partial y} = -\frac{Kx}{E} - \frac{\partial u_0}{\partial y} \qquad (5\text{-}27)$$

and

$$\frac{\partial w}{\partial x} = -\frac{\partial u}{\partial z} = -\frac{\partial u_0}{\partial z} \qquad (5\text{-}28)$$

and these integrate at once to give

$$v = -\frac{K}{E}\frac{x^2}{2} - \frac{\partial u_0}{\partial y}x + v_0(y, z) \qquad (5\text{-}29)$$

and

$$w = -\frac{\partial u_0}{\partial z}x + w_0(y,\, z) \tag{5-30}$$

Substituting these in Eqs. 5-25b and 5-25c, we get

$$\left. \begin{aligned} \frac{\partial^2 u_0}{\partial y^2}x &= \frac{\partial v_0}{\partial y} + \frac{\nu K y}{E} \\[2mm] \frac{\partial^2 u_0}{\partial z^2}x &= \frac{\partial w_0}{\partial z} + \frac{\nu K y}{E} \end{aligned} \right\} \tag{5-31}$$

Note: the right-hand-side terms of Eq. 5-31 and are independent of x. Hence, we must have

$$\left. \begin{aligned} \frac{\partial^2 u_0}{\partial y^2} &= 0 \\[2mm] \frac{\partial^2 u_0}{\partial z^2} &= 0 \end{aligned} \right\} \tag{5-32}$$

and

$$\left. \begin{aligned} \frac{\partial v_0}{\partial y} + \frac{\nu K y}{E} &= 0 \\[2mm] \frac{\partial w_0}{\partial z} + \frac{\nu K y}{E} &= 0 \end{aligned} \right\} \tag{5-33}$$

The last two equations integrate to give

$$\left. \begin{aligned} v_0 &= -\frac{\nu K}{E}\frac{y^2}{2} + h_1(z) \\[2mm] w_0 &= -\frac{\nu K y z}{E} + h_2(y) \end{aligned} \right\} \tag{5-34}$$

Thus, Eqs. 5-29 and 5-30 now become

$$\left. \begin{aligned} v &= -\frac{K x^2}{2E} - \frac{\partial u_0}{\partial y}x - \frac{\nu K y^2}{2E} + h_1(z) \\[2mm] w &= -\frac{\partial u_0}{\partial z}x - \frac{\nu K y z}{E} + h_2(y) \end{aligned} \right\} \tag{5-35}$$

Substitute these in the second of Eq. 5-18d and find

$$\frac{\partial h_1}{\partial z} + \frac{\partial h_2}{\partial y} - \frac{\nu K z}{E} - 2x\frac{\partial^2 u_0}{\partial z \partial y} = 0 \tag{5-36}$$

The first three terms in this equation are independent of x. Therefore it follows that

$$\frac{\partial^2 u_0}{\partial z \partial y} = 0 \qquad (5\text{-}37)$$

and

$$\frac{\partial h_1}{\partial z} - \frac{\nu K z}{E} = -\frac{\partial h_2}{\partial y} \qquad (5\text{-}38)$$

In the last of these we see that the left-hand side is a function of z only and the right-hand side is a function of y only. This means that both sides must be equal to a constant, say C. Therefore,

$$\left. \begin{array}{l} \dfrac{\partial h_2}{\partial y} = -C \\[2ex] \dfrac{\partial h_1}{\partial z} - \dfrac{\nu K z}{E} = C \end{array} \right\} \qquad (5\text{-}39)$$

and from Eqs. 5-32 and 5-37 we also have

$$\left. \begin{array}{l} \dfrac{\partial^2 u_0}{\partial y^2} = 0 \\[2ex] \dfrac{\partial^2 u_0}{\partial z^2} = 0 \\[2ex] \dfrac{\partial^2 u_0}{\partial z \partial y} = 0 \end{array} \right\} \qquad (5\text{-}40)$$

and

The last three mean that

$$u_0 = C_1 y + C_2 z + C_3 \qquad (5\text{-}41)$$

and Eqs. 5-39 give

$$\left. \begin{array}{l} h_2 = -Cy + C_4 \\[2ex] h_1 = \dfrac{\nu K z^2}{2E} + Cz + C_5 \end{array} \right\} \qquad (5\text{-}42)$$

Thus, we have finally, for u, v, and w, the following expressions:

$$\left. \begin{array}{l} u = \dfrac{Kyx}{E} + C_1 y + C_2 z + C_3 \\[2ex] v = -\dfrac{Kx^2}{2E} - C_1 x - \dfrac{\nu K y^2}{2E} + \dfrac{\nu K z^2}{2E} + Cz + C_5 \\[2ex] w = -C_2 x - \dfrac{\nu K yz}{E} - Cy + C_4 \end{array} \right\} \qquad (5\text{-}43)$$

To determine the six constants we assume that at $x = y = z = 0$ (i.e. at the centroid),

$$\text{deformations} = 0 \qquad \text{or} \qquad u = v = w = 0 \qquad (5\text{-}44)$$

and distortion of a differential element is zero, or,

$$\frac{\partial u}{\partial y} = \frac{\partial v}{\partial x} = \frac{\partial v}{\partial z} = \frac{\partial w}{\partial y} = \frac{\partial w}{\partial x} = \frac{\partial u}{\partial z} = 0 \qquad (5\text{-}45)$$

These relations give

$$C = C_1 = C_2 = C_3 = C_4 = C_5 = 0 \qquad (5\text{-}46)$$

so that, finally,

$$\left. \begin{aligned} u &= \frac{Kyx}{E} \\ v &= -\frac{Kx^2}{2E} - \frac{\nu Ky^2}{2E} + \frac{\nu Kz^2}{2E} \\ w &= -\frac{\nu Kyz}{E} \end{aligned} \right\} \qquad (5\text{-}47)$$

These deformations, for a rectangular cross section, represent the anticlastic, or saddle, surface, as shown in Fig. 5.4. The reader can verify this configuration by grasping a soft rectangular rubber eraser at its ends and bending it into an arc.

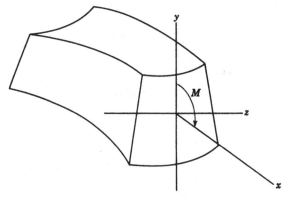

Fig. 5.4

The line $x = $ variable, $y = z = 0$, also deflects, as given by

$$\left. \begin{aligned} u &= 0 \\ v &= -\frac{Kx^2}{2E} \\ w &= 0 \end{aligned} \right\} \qquad (5\text{-}48)$$

In particular we may note that, since for this line

$$v = -\frac{Kx^2}{2E} \tag{5-49}$$

it follows that, for this curve or line (as well as for the entire beam),

$$\frac{d^2v}{dx^2} = \frac{M_z}{EI_z} \tag{5-50}$$

The right-hand side of Eq. 5-50 is a sort of *invariant* form in that it depends only on the applied moment, the cross section of the bar, and the material of the bar—but not on the coordinates.

The sign of the right-hand-side term in Eq. 5-50 depends upon the convention used for defining positive and negative moment. Thus, this equation can be given with either the plus or minus sign.

5-4 The Engineering Elasticity Solution for the Beam—The Bernoulli–Euler Solution.
In the previous section it was shown that the exact solution of the linearized equations of elasticity, corresponding to a pure moment M_z, constant along the axis x of a cylindrical bar, gives

$$\sigma_x = \frac{M_z y}{I_z} \tag{5-51}$$

and

$$\frac{d^2v}{dx^2} = \frac{M_z}{EI_z} \tag{5-52}$$

In these equations, σ_x is the stress at a distance y from the z axis, I_z is the moment of inertia of the cross section about the z axis, and E is the tension–compression modulus of elasticity; v is the deflection of the x axis in the y direction.

In the engineering application of the exact bending moment solution, the so-called Bernoulli–Euler solution, we take as a starting point the above two equations, which hold for a *constant* moment on the bar.

However, we make the assumption that these equations *hold at all points on the bar even if the moment varies from point to point on the beam.* In other words, the equations

$$\sigma_x = \frac{M_z y}{I_z} \tag{5-53}$$

and

$$\frac{d^2v}{dx^2} = \frac{M_z}{EI_z} \tag{5-54}$$

are assumed to be valid, *locally*, at all points on the beam.

In the engineering analysis, we say nothing about the cross-sectional deformations v and w; indeed, as a consequence of the above two relations, it follows that (for stresses σ_x proportional to strain e_x) a plane section normal to the x axis before bending is also a plane section after the bending deformation (which is a pure rotation of the cross section about the neutral axis) occurs. Furthermore, the deformation of this cross section (due to bending only) is limited to this rotation only—no distortion of the cross section in the y and z direction is assumed to occur. This means that the deflection of the line $y = 0$, $z = 0$, which is given by Eq. 5-48, is also, in the engineering beam, the deflection of the plane $z = 0$, the neutral plane, *and also the deflection of the beam.*

Finally, since the curvature of the deflected line $y = 0$, $z = 0$, is given by

$$\frac{d^2v/dx^2}{[1+(dv/dx)^2]^{3/2}} = \frac{1}{\rho_x} \tag{5-55}$$

it follows that, if we limit the beam to small deflections, so that

$$\frac{dv}{dx} \ll 1$$

we can neglect $(dv/dx)^2$ in comparison to unity, and we have

$$\frac{1}{\rho_x} \cong \frac{d^2v}{dx^2} = \frac{M_z}{EI_z} \tag{5-56}$$

Equations 5-53 and 5-56 are the fundamental equations for the deflected beam in the Bernoulli–Euler theory of beams. They can also be obtained starting with the assumption that plane sections of the beam before bending are plane after bending (see Ref. 10); however, for our purposes it was deemed preferable to arrive at them as a logical consequence of the exact solution of the linearized equations of elasticity— subject to the assumption noted above. See also Ref. (50).

5-5 Some Remarks on the Accuracy of the Engineering Form of the Flexure Formula.
As pointed out in the last section, in strength of materials courses the bending stress for a beam under any transverse loading is determined by assuming that Eqs. 5-53 and 5-56 hold for all types of transverse beam loading. This is the *Bernoulli–Euler theory of bending*, and a basic assumption in this theory is that plane sections before bending remain plane after bending. Also, in this theory anticlastic bending is neglected.

We will now show that plane sections cannot remain plane after bending if the bending moment is not either a constant or a linear function of x, the distance along the beam. This means that the transverse loading cannot be other than a concentrated force, and therefore the engineering form of the flexure formula for a beam under distributed loads is, at best, an approximation.

The development which follows is due to the English mathematician, Pearson.[3] He established an amount of approximation of the Bernoulli–Euler solution, as follows:

If Eq. 4-116 of the linearized theory of elasticity is solved for the stresses as functions of the strain, we would obtain

$$\sigma_x = A(e_x + e_y + e_z) + 2Ge_x \tag{5-57}$$

in which A is a constant, G is the shear modulus of elasticity, and $e_x + e_y + e_z = \mathscr{I}_1$, is the invariant of the strain tensor.

In the Bernoulli–Euler solution we assume that $e_y = e_z = 0$ and hence

$$\sigma_x = Ee_x = E\frac{\partial u}{\partial x} \tag{5-58}$$

in which E is the modulus of elasticity.

From the two equations given above we find

$$\mathscr{I}_1 = \frac{E - 2G}{A}\frac{\partial u}{\partial x} \tag{5-59}$$

and therefore

$$\frac{\partial \mathscr{I}_1}{\partial x} = \frac{E - 2G}{A}\frac{\partial^2 u}{\partial x^2} \tag{5-60}$$

Now, the equilibrium equation of the theory of elasticity is (see Eq. 5-1)

$$\frac{\partial \sigma_x}{\partial x} + \frac{\partial \tau_{yx}}{\partial y} + \frac{\partial \tau_{zx}}{\partial z} = 0 \tag{5-61}$$

and since

$$\left.\begin{aligned} \tau_{yx} &= G\gamma_{yx} = G\left(\frac{\partial u}{\partial y} + \frac{\partial v}{\partial x}\right) \\ \tau_{zx} &= G\gamma_{zx} = G\left(\frac{\partial u}{\partial z} + \frac{\partial w}{\partial x}\right) \\ \sigma_x &= A\mathscr{I}_1 + 2G\frac{\partial u}{\partial x} \end{aligned}\right\} \tag{5-62}$$

this gives

$$(A + G)\frac{\partial \mathscr{I}_1}{\partial x} + G\left(\frac{\partial^2 u}{\partial x^2} + \frac{\partial^2 u}{\partial y^2} + \frac{\partial^2 u}{\partial z^2}\right) = 0 \tag{5-63}$$

[3] In a paper "On the flexure of heavy beams subjected to continuous systems of loads," *Quart. J. Math.*, Vol. 24, p. 63–110, 1890.

and, using the above expression for $\partial \mathscr{I}_1 / \partial x$, this gives

$$K\frac{\partial^2 u}{\partial x^2} + K_1\left(\frac{\partial^2 u}{\partial y^2} + \frac{\partial^2 u}{\partial z^2}\right) = 0 \tag{5-64}$$

Now, the Bernoulli–Euler expression is also given by

$$\sigma_x = \frac{M(x)y}{I_z} = E\frac{\partial u}{\partial x} \tag{5-65}$$

in which $M(x)$ is a function of the x only. Hence, differentiating and integrating, we get this:

$$E\frac{\partial^2 u}{\partial x^2} = \frac{y}{I_z}\frac{dM(x)}{dx} \tag{5-66}$$

and

$$Eu = \frac{y}{I_z}\int M(x)\,dx + \psi(y,\,z) \tag{5-67}$$

in which $\psi(y,\,z)$ is a function of y and z only.

Putting Eqs. 5-66 and 5-67 in Eq. 5-64 above, we get

$$K_2 y\frac{dM(x)}{dx} + K_3\left[\frac{\partial^2 \psi(y,\,z)}{\partial y^2} + \frac{\partial^2 \psi(y,\,z)}{\partial z^2}\right] = 0 \tag{5-68}$$

and since the last bracket of this equation is independent of x, then so also is the first term. Hence

$$\frac{dM(x)}{dx} = \text{constant} \tag{5-69}$$

and

$$M(x) = K_5 x + K_6 \tag{5-70}$$

which means $M(x)$ can only be due to a concentrated load or a concentrated moment. In other words, for beams subjected to loading other than these, there is a conflict between the Bernoulli–Euler theory and the linearized theory of elasticity, indicating that the Bernoulli–Euler theory is in error. However, tests indicate that the approximate Bernoulli–Euler theory gives results which are sufficiently accurate for engineering purposes.

The interpretation to be placed on the Pearson result given above is as follows:

The Bernoulli–Euler assumptions—(1) that $\sigma_x = M_z y / I_z$ everywhere in the beam and (2) that $d^2 v / dx^2 = M_z / EI_z$ everywhere in the beam—are consistent with the exact equation of equlibrium and with the exact linearized form of Hooke's Law only if the moment is either a constant or varies linearly along the length of the beam. This result is, therefore, *one possible* measure of the inaccuracy of the Bernoulli–Euler theory within the framework of the more exact relations of the mathematical theory of elasticity. There are, however, other possible measures of this inaccuracy. For example, one may test when the failure of the equivalent Bernoulli–Euler assumption—that plane sections before bending remain plane after bending—occurs. Or one may consider the compatibility of strain equation and determine the inaccuracy introduced by neglecting this requirement.

5-6   Summary of the Bending Problem Solution. We discussed the exact linearized mathematical theory of elasticity solution for pure bending on a beam and pointed out the connection between this and the engineering solution—the so-called Bernoulli–Euler theory of beams.

Finally, we described Pearson's theoretical treatment which gives a measure of the approximation involved in the engineering treatment of beams.

5-7   The Shear Problem and the St. Venant Torsion Problem.
INTRODUCTION
In the preceding sections we studied in some detail problems which arise from a consideration of the stress tensor having the form

$$\begin{pmatrix} \sigma_x & 0 & 0 \\ 0 & 0 & 0 \\ 0 & 0 & 0 \end{pmatrix} \tag{5-71}$$

These were the direct load problem and the beam-bending problem.

In this section we shall consider the problems which occur when the stress tensor is assumed to contain only the *shear stress*. We shall see that, for our purposes, two problems are encountered:

1. the near-trivial two-dimensional pure shear structure,
2. the much more important pure torsion structure.

Problem 1 will be discussed first. Following this we shall present, in rather detailed form, the classical St. Venant solution to the torsion problem. We shall discuss the significance of warping in this problem and, having the exact general solution, we will develop the solutions for two cross sections of engineering interest—the circle and the ellipse. Following this we shall indicate very briefly how solutions for more complicated cross sections are obtained.

THE PURE SHEAR STRUCTURE
Consider the general two-dimensional stress tensor (see Eq. 4-39),

$$\begin{pmatrix} \sigma_x & \tau_{xy} \\ \tau_{yx} & \sigma_y \end{pmatrix} \tag{5-72}$$

and let us assume that all elements are zero except the two shear stresses $\tau_{xy} = \tau_{yx}$. Then the stress tensor becomes

$$\begin{pmatrix} 0 & \tau_{xy} \\ \tau_{yx} & 0 \end{pmatrix} \tag{5-73}$$

and conditions are the same at all sections $z = $ constant and are as shown in Fig. 5.5. The equilibrium equations (see Eq. 5-1) require

$$\left. \begin{array}{c} \dfrac{\partial \tau_{xy}}{\partial x} = 0 \\[2ex] \dfrac{\partial \tau_{yx}}{\partial y} = 0 \end{array} \right\} \tag{5-74}$$

or $\tau_{xy} = \tau_{yx} = $ constant everywhere in the body.

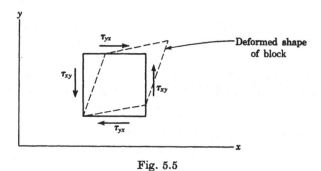

Fig. 5.5

The reader should verify that the boundary conditions are identically satisfied.

Hooke's Law gives

$$\left. \begin{array}{c} \gamma_{xy} = \dfrac{\tau_{xy}}{G} = \text{constant} \\[2ex] \gamma_{yx} = \dfrac{\tau_{yx}}{G} = \text{constant} \end{array} \right\} \tag{5-75}$$

and

and obviously the strain compatibility condition is identically satisfied, $0 = 0$.

Thus we have, essentially, the solution for a block of material subjected to constant, equal shear stresses acting on the four faces perpendicular to the x and y axes, and deforming as shown in Fig. 5.5. Although this represents an exact solution in the linearized mathematical theory of elasticity, it is really a near-trivial solution of rather limited practical interest. We shall not consider it further in this text.

THE CLASSICAL ST. VENANT TORSION SOLUTION

The solution of the torsion problem, as obtained by St. Venant, represents one of the truly great landmarks in the mathematical theory of elasticity. It accounted for many observed experimental phenomena which other, earlier theories could not predict. Moreover, this solution has been the starting point for many additional investigations and extensions of its theories in both elasticity and applied mathematics.

In the St. Venant Theory we start by stating that we are seeking a solution to the problem of a pure torsion acting on a cylindrical (not necessarily circular) cross section. This is shown in Fig. 5.6.

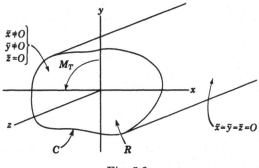

Fig. 5.6

The applied torque is M_T, the length of the bar is along the z axis, C is the boundary of any cross section, and R is the *region* or area of any cross section.

Experiments indicate that a torque applied as in Fig. 5.6 develops shear stresses on the cross section. Hence, we assume as our stress tensor

$$\begin{pmatrix} 0 & 0 & \tau_{xz} \\ 0 & 0 & \tau_{yz} \\ \tau_{zx} & \tau_{zy} & 0 \end{pmatrix} \tag{5-76}$$

which means we are assuming stresses acting as shown in Fig. 5.7.

Then the equilibrium equations (see Eq. 5-1) give, for everywhere in R,

$$\left(\frac{\partial}{\partial x} \ \frac{\partial}{\partial y} \ \frac{\partial}{\partial z} \right) \begin{pmatrix} 0 & 0 & \tau_{xz} \\ 0 & 0 & \tau_{yz} \\ \tau_{zx} & \tau_{zy} & 0 \end{pmatrix} = 0 \tag{5-77}$$

or

$$\frac{\partial \tau_{zx}}{\partial z} = 0 \tag{5-78}$$

$$\frac{\partial \tau_{zy}}{\partial z} = 0 \qquad (5\text{-}79)$$

$$\frac{\partial \tau_{xz}}{\partial x} + \frac{\partial \tau_{yz}}{\partial y} = 0 \qquad (5\text{-}80)$$

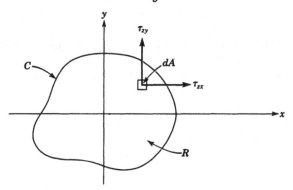

Fig. 5.7

The first two of the above imply that $\tau_{zx} = \tau_{xz} =$ independent of z, and that $\tau_{zy} = \tau_{yz} =$ independent of z.

Hooke's Law (see Eq. 5-4) is given by

$$\begin{pmatrix} e_x & \tfrac{1}{2}\gamma_{xy} & \tfrac{1}{2}\gamma_{xz} \\ \tfrac{1}{2}\gamma_{yx} & e_y & \tfrac{1}{2}\gamma_{yz} \\ \tfrac{1}{2}\gamma_{zx} & \tfrac{1}{2}\gamma_{zy} & e_z \end{pmatrix} = \frac{1+\nu}{E} \begin{pmatrix} \sigma_x & \tau_{xy} & \tau_{xz} \\ \tau_{yx} & \sigma_y & \tau_{yz} \\ \tau_{zx} & \tau_{zy} & \sigma_z \end{pmatrix}$$

$$- \frac{\nu}{E}(\sigma_x + \sigma_y + \sigma_z) \begin{pmatrix} 1 & 0 & 0 \\ 0 & 1 & 0 \\ 0 & 0 & 1 \end{pmatrix} \qquad (5\text{-}81)$$

and for our stress tensor this gives at once

$$\gamma_{xz} = \frac{2(1+\nu)}{E}\tau_{xz} = \frac{\tau_{xz}}{G} = \frac{\partial u}{\partial z} + \frac{\partial w}{\partial x} \qquad (5\text{-}82)$$

$$\gamma_{yz} = \frac{2(1+\nu)}{E}\tau_{yz} = \frac{\tau_{yz}'}{G} = \frac{\partial v}{\partial z} + \frac{\partial w}{\partial y} \qquad (5\text{-}83)$$

$$e_x = \frac{\partial u}{\partial x} = 0 \qquad (5\text{-}84)$$

$$\gamma_{yx} = \gamma_{xy} = \frac{\partial u}{\partial y} + \frac{\partial v}{\partial x} = 0 \qquad (5\text{-}85)$$

$$e_y = \frac{\partial v}{\partial y} = 0 \qquad (5\text{-}86)$$

$$e_z = \frac{\partial w}{\partial z} = 0 \qquad (5\text{-}87)$$

From the third, fourth, fifth and sixth of the above we have

$$\frac{\partial u}{\partial x} = \frac{\partial v}{\partial y} = \frac{\partial w}{\partial z} = \frac{\partial u}{\partial y} + \frac{\partial v}{\partial x} = 0 \qquad (5\text{-}88)$$

and therefore

$$\left. \begin{array}{l} u = u(y,\, z) \\ v = v(x,\, z) \\ w = w(x,\, y) \end{array} \right\} \qquad (5\text{-}89)$$

Now, when a torque such as shown in Fig. 5-6 is applied to a cylindrical bar, it is found that the cross section rotates about some point, such as O, Fig. 5.8, and the point P moves to P_1.

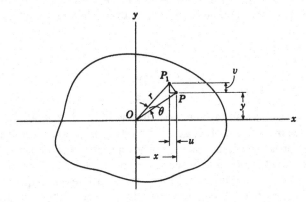

Fig. 5.8

Therefore, from the figure,

$$\overline{PP}_1 = r\theta$$

and

$$\frac{u}{\overline{PP}_1} = \frac{y}{r}$$

or

$$u = \frac{\overline{PP}_1 y}{r} = -y\theta \qquad (5\text{-}90)$$

Similarly

$$v = x\theta \qquad (5\text{-}91)$$

the negative sign being required since u is obviously in a negative direction.

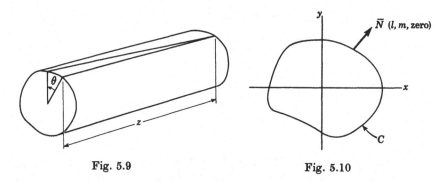

Fig. 5.9   Fig. 5.10

The bar twists through an angle θ that varies linearly from the end as shown in Fig. 5.9. Hence if we define an angle α as the twist per unit length, then

$$\theta = \alpha z \qquad (5\text{-}92)$$

and we have

$$u = -\alpha yz \qquad (5\text{-}93)$$

$$v = \alpha xz \qquad (5\text{-}94)$$

We must also consider the possibility of a deformation in the z direction, a *warping*, which is given by $w(x, y)$. Tests indicate that for non-circular cross sections, there is a warping of the face when a pure torque is applied, and this is considered in the St. Venant Theory. Indeed, it was the inclusion of this term and the subsequent solution obtained therewith that constituted St. Venant's great achievement, since none of the earlier torsion theories could properly account for this term.

Thus, we have

$$\left. \begin{aligned} u &= -\alpha yz \\ v &= \alpha xz \\ w &= w(x, y) \end{aligned} \right\} \qquad (5\text{-}95)$$

and we are in agreement with the requirements previously obtained; see Eq. 5-88 and 5-89.

Substituting these values for u, v, and w in Eqs. 5-82 and 5-83, we obtain

$$\gamma_{xz} = -\alpha y + \frac{\partial w}{\partial x} \qquad (5\text{-}96)$$

$$\gamma_{yz} = \alpha x + \frac{\partial w}{\partial y} \qquad (5\text{-}97)$$

and also

$$\tau_{xz} = G\gamma_{xz} = -G\alpha y + G\frac{\partial w}{\partial x} \qquad (5\text{-}98)$$

$$\tau_{yz} = G\gamma_{yz} = G\alpha x + G\frac{\partial w}{\partial y} \qquad (5\text{-}99)$$

It will simplify later expressions if we introduce a new function, the *warping function*, ϕ, defined by

$$\alpha\phi(x, y) = w(x, y) \qquad (5\text{-}100)$$

Then

$$\left.\begin{array}{l}\tau_{xz} = G\alpha\left(-y + \dfrac{\partial \phi}{\partial x}\right) \\[3mm] \tau_{yz} = G\alpha\left(x + \dfrac{\partial \phi}{\partial y}\right)\end{array}\right\} \qquad (5\text{-}101)$$

Using the above and substituting into Eq. 5-80, we have, for everywhere in R,

$$\frac{\partial}{\partial x}\left(-G\alpha y + G\alpha\frac{\partial \phi}{\partial x}\right) + \frac{\partial}{\partial y}\left(G\alpha x + G\alpha\frac{\partial \phi}{\partial y}\right) = 0 \qquad (5\text{-}102)$$

or

$$\nabla^2\phi = \frac{\partial^2\phi}{\partial x^2} + \frac{\partial^2\phi}{\partial y^2} = 0 \quad \text{in} \quad R \qquad (5\text{-}103)$$

Thus, the warping function, ϕ, is *harmonic* (or satisfies the Laplace equation), everywhere in R—i.e., throughout the bar.

Now let us consider the boundary conditions (see Fig. 5-10). Note that normals to the boundary, C, have components (l, m, zero) since the normal is always perpendicular to the direction of the z axis. Hence the boundary condition (see Eq. 5-2, and Fig. 5.6),

$$(l \ m \ n)\begin{pmatrix} \sigma_x & \tau_{xy} & \tau_{xz} \\ \tau_{yx} & \sigma_y & \tau_{yz} \\ \tau_{zx} & \tau_{zy} & \sigma_z \end{pmatrix} = (\bar{x} \ \bar{y} \ \bar{z}) \qquad (5\text{-}104)$$

becomes, since $\bar{x} = \bar{y} = \bar{z} = 0$ on C

$$\tau_{xz}l + \tau_{yz}m = 0 \qquad (5\text{-}105)$$

or, using Eq. 5-101,

$$\left(-y+\frac{\partial \phi}{\partial x}\right)l + \left(x+\frac{\partial \phi}{\partial y}\right)m = 0, \quad \text{everywhere on } C \quad [4] \qquad (5\text{-}106)$$

Thus, we have shown that the equations of equilibrium, Hooke's Law, and boundary conditions for the assumed stress tensor and deformation of Fig. 5.8 require that we find, for each bar, a function, ϕ, *the warping function*, such that

Everywhere in R, $\quad \nabla^2 \phi = 0 \qquad$ (a)

Everywhere on C, $\quad \left(-y+\dfrac{\partial \phi}{\partial x}\right)l + \left(x+\dfrac{\partial \phi}{\partial y}\right)m = 0 \qquad$ (b) $\qquad$ (5-107)

The boundary condition on C may be given in an alternate, simpler form, as follows; see Fig. 5.11 which shows a portion of the boundary with a unit normal, $\bar{n}$, and a unit tangent, $\bar{t}$, drawn at any point. The direction of these is as shown by the (n, s) coordinates, and these have components as indicated on the figure, as the student may verify.

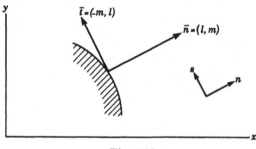

Fig. 5.11

Now, the vector

$$\nabla \phi = \text{grad}\, \phi = \left(\frac{\partial \phi}{\partial x} \quad \frac{\partial \phi}{\partial y}\right)$$

and the scalar product of this vector with a unit normal $(l\ m)$ is given by

$$\nabla \phi \cdot \bar{n} = l\frac{\partial \phi}{\partial x} + m\frac{\partial \phi}{\partial y} \qquad (5\text{-}108)$$

[4] The boundary conditions on the end faces, where the unit normals have components $(0\ 0\ \pm 1)$ simply states that on these end faces the shear stresses τ_{zx} and τ_{zy} exist—as we have assumed. However, see the footnote on p. 143 of this chapter, and Ref. (50).

By definition of scalar product, the above is the component of $\nabla\phi$ in the direction of n, and by definition of "gradient" this is just equal to $\partial\phi/\partial n$. Thus

$$\frac{\partial\phi}{\partial n} = l\frac{\partial\phi}{\partial x} + m\frac{\partial\phi}{\partial y} \qquad (5\text{-}109)$$

Therefore, from Eq. 5-107b,

$$\frac{\partial\phi}{\partial n} = ly - mx \qquad (5\text{-}110)$$

Now, it is obvious that the vector $(x\ y)$ is given by

$$(x\ y) = \nabla[\tfrac{1}{2}(x^2+y^2)] \qquad (5\text{-}111)$$

Furthermore, the scalar product

$$\nabla[\tfrac{1}{2}(x^2+y^2)] \cdot \hat{t} = (x\ y) \cdot (-m\ l) = ly - mx \qquad (5.112)$$

is just equal to the component of $\nabla[\tfrac{1}{2}(x^2+y^2)]$ in the direction of $\hat{t}$, i.e., in the s direction.

Hence, everywhere on C,

$$\frac{\partial\phi}{\partial n} = ly - mx = \frac{\partial}{\partial s}[\tfrac{1}{2}(x^2+y^2)] \qquad (5\text{-}113)$$

and this boundary condition may be used to replace the condition as given in Eq. 5-107b.

The next steps proceed most directly by application of elementary complex variable theory. Refer to Chapter 1, Art. 1-5.

Given

$$\phi + i\psi = f(z)$$

where

$$z = x + iy$$

then it was shown that ϕ and ψ satisfy Laplace's equation, and in addition, ϕ and ψ are related by the Cauchy–Riemann equation,

$$\frac{\partial\phi}{\partial n} = \frac{\partial\psi}{\partial s} \qquad (5\text{-}114)$$

Thus, we now look for a function ψ, such that

$$\frac{\partial^2\psi}{\partial x^2} + \frac{\partial^2\psi}{\partial y^2} = 0 \qquad (5\text{-}115)$$

everywhere in the cross section, and (see Eq. 5-113)

$$\frac{\partial \psi}{\partial s} = \frac{\partial}{\partial s}[\tfrac{1}{2}(x^2 + y^2)] \tag{5-116}$$

or

$$\psi = \tfrac{1}{2}(x^2 + y^2) + K \tag{5-117}$$

on the boundary. Equation 5-117 may alternatively be stated as

$$\psi - \tfrac{1}{2}(x^2 + y^2) = K \tag{5-118}$$

on the boundary or the boundary is a level (constant) surface of $\psi - \tfrac{1}{2}(x^2 + y^2)$.

Up to this point, only the equilibrium equations, boundary conditions, and Hooke's Law have been satisfied for an assumed stress distribution. It still remains to

1. prove compatibility is satisfied
2. show that the stress-system corresponds to a pure torque
3. obtain, if possible, functions ψ which will satisfy Eqs. 5-115 and 5-118 for given cross sections.

To check compatibility it will be recalled that

$$\left. \begin{aligned} e_x = e_y = e_z = \gamma_{xy} &= 0 \\[4pt] \gamma_{xz} &= \alpha\left(\frac{\partial \phi}{\partial x} - y\right) \\[4pt] \gamma_{yz} &= \alpha\left(\frac{\partial \phi}{\partial y} + x\right) \end{aligned} \right\} \tag{5-119}$$

If these values are substituted in the strain compatibility conditions it is found that these are satisfied, $0 = 0$.

To prove that the assumed stress distribution corresponds to a pure torque, we proceed as follows:

It must be shown that

$$\iint_R \tau_{zx}\, dxdy = 0 \tag{5-120}$$

$$\iint_R \tau_{zy}\, dxdy = 0 \tag{5-121}$$

and

$$\iint_R (\tau_{zy}x - \tau_{zx}y)\, dxdy = M_T \tag{5-122}$$

where M_T is the applied torque.

The solution of Eq. 5-120 will be shown in detail. Equation 5-121 may be shown similarly, and is left as an exercise to the student. First note

$$\iint_R \tau_{zx}\,dxdy = G\alpha \iint_R \left(\frac{\partial \phi}{\partial x} - y\right) dxdy \qquad (5\text{-}123)$$

and because $(\partial^2 \phi / \partial x^2) + (\partial^2 \phi / \partial y^2) = 0$, this may be replaced by

$$\iint_R \tau_{zx}\,dxdy = G\alpha \iint_R \left\{ \frac{\partial}{\partial x}\left[x\left(\frac{\partial \phi}{\partial x} - y\right)\right] + \frac{\partial}{\partial y}\left[x\left(\frac{\partial \phi}{\partial y} + x\right)\right]\right\} dxdy$$

$$(5\text{-}124)$$

This area integral is transformed to a line integral by means of Green's Theorem

$$\iint_{area} \left(\frac{\partial P}{\partial x} + \frac{\partial Q}{\partial y}\right) dxdy = \int_{line} P\,dy - Q\,dx \qquad (5\text{-}125)$$

Then Eq. 5-124 becomes

$$\left. \begin{aligned} \iint_R \tau_{zx}\,dxdy &= G\alpha\left[\int_C x\left(\frac{\partial \phi}{\partial x} - y\right) dy - \int_C x\left(\frac{\partial \phi}{\partial y} + x\right) dx\right] \\ &= G\alpha \int_C x\left[\left(\frac{\partial \phi}{\partial x} - y\right)\frac{dy}{ds} - \left(\frac{\partial \phi}{\partial y} + x\right)\frac{dx}{ds}\right] ds \end{aligned} \right\} \quad (5\text{-}126)$$

Now the vector coordinates of a unit normal to the boundary C (see Fig. 5.12) are

$$\bar{n} = (l\ \ m) = \left(\frac{dy}{ds}\quad -\frac{dx}{ds}\right) \qquad (5\text{-}127)$$

or Eq. 5-126 becomes

$$\iint_R \tau_{zx}\,dxdy = G\alpha \int_C x\left[\left(\frac{\partial \phi}{\partial x} - y\right)l + \left(\frac{\partial \phi}{\partial y} + x\right)m\right] ds \quad (5\text{-}128)$$

and because of the boundary condition Eq. 5-107b, the right-hand side is zero. This was to be shown.

The value of

$$\iint_R (\tau_{zy}x - \tau_{zx}y)\,dxdy = M_T \qquad (5\text{-}129)$$

will now be determined.

This is also given by

$$G\alpha \iint_R \left[\left(\frac{\partial \phi}{\partial y} + x\right)x - \left(\frac{\partial \phi}{\partial x} - y\right)y\right] dxdy = M_T \qquad (5\text{-}130)$$

or

$$G\alpha \iint_R \left(\frac{\partial \phi}{\partial y}x - \frac{\partial \phi}{\partial x}y + x^2 + y^2 \right) dxdy = M_T \qquad (5\text{-}131)$$

and if we call

$$D = G \iint_R \left(x^2 + y^2 + x\frac{\partial \phi}{\partial y} - y\frac{\partial \phi}{\partial x} \right) dxdy \qquad (5\text{-}132)$$

the "torsional rigidity" of the cross section, then

$$M_T = D\alpha \qquad (5\text{-}133)$$

The solution may be obtained as follows in terms of another quantity, Ψ, defined by

$$\Psi = \psi - \tfrac{1}{2}(x^2 + y^2) \qquad (5\text{-}134)$$

From Eqs. 5-134, 5-101, and 5-114,

$$\frac{\partial \Psi}{\partial x} = \frac{\partial \psi}{\partial x} - x = -\frac{\partial \phi}{\partial y} - x = -\frac{\tau_{yz}}{G\alpha} \qquad (5\text{-}135)$$

$$\frac{\partial \Psi}{\partial y} = \frac{\partial \psi}{\partial y} - y = \frac{\partial \phi}{\partial x} - y = \frac{\tau_{xz}}{G\alpha} \qquad (5\text{-}136)$$

$$D = -G \iint_R \left(x\frac{\partial \Psi}{\partial x} + y\frac{\partial \Psi}{\partial y} \right) dxdy \qquad (5\text{-}137)$$

We repeat here, for convenience, the requirements for an exact solution to the torsion problem as given above.

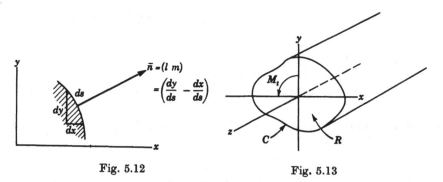

Fig. 5.12   Fig. 5.13

Given any cylindrical bar, having a cross section of any shape, Fig. 5.13, then, the solution will be obtained if we can find a function $\psi(x, y)$ such that

$$\nabla^2 \psi = 0 \quad \text{everywhere in } R \qquad (5\text{-}138)$$

$$\psi - \tfrac{1}{2}(x^2 + y^2) = \text{constant, on } C \qquad (5\text{-}139)$$

Having this function $\psi(x, y)$, then

$$\left.\begin{aligned} \tau_{zx} &= G\alpha\left(\frac{\partial\psi}{\partial y} - y\right) \\[2mm] \tau_{zy} &= -G\alpha\left(\frac{\partial\psi}{\partial x} - x\right) \end{aligned}\right\} \tag{5-140}$$

and

$$\alpha = \text{twist per unit length} = \frac{M_T}{G\iint_R\left(x^2 + y^2 - x\dfrac{\partial\psi}{\partial x} - y\dfrac{\partial\psi}{\partial y}\right)dA} \tag{5-141}$$

Or, if we introduce a new variable, Ψ, defined by

$$\Psi = \psi - \tfrac{1}{2}(x^2 + y^2) \tag{5-142}$$

the solution will be obtained if we determine a function Ψ such that

$$\nabla^2\Psi = -2, \quad \text{everywhere in } R \tag{5-143}$$

$$\Psi = 0, \text{ on } C \tag{5-144}$$

and having this function Ψ, then (see Prob. 13, p. 177)

$$\left.\begin{aligned} \tau_{zx} &= G\alpha\frac{\partial\Psi}{\partial y} \\[2mm] \tau_{zy} &= -G\alpha\frac{\partial\Psi}{\partial x} \end{aligned}\right\} \tag{5-145}$$

$$\alpha = \text{twist per unit length} = \frac{M_T}{2G\iint_R \Psi\, dx\,dy} \tag{5-146}$$

5-8 The Circular Cross Section. To indicate an application of this theory to a problem of practical interest, let us consider the bar with a circular cross section, Fig. 5.14, of radius r. The given applied torque is M_T. Determine the stresses, deformation, and warping for this bar.

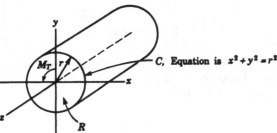

Fig. 5.14

SOLUTION

Let us try for ψ, the constant quantity K.[5] Then

$$\nabla^2\psi = \nabla^2 K = 0, \quad \text{as required} \tag{5-147}$$

On C, we have

$$\psi - \tfrac{1}{2}(x^2 + y^2) = d, \quad \text{a constant} \tag{5-148}$$

so that for $\psi = K$,

$$(K - d)2 = x^2 + y^2 \tag{5-149}$$

Hence if

$$2(K - d) = r^2 \tag{5-150}$$

we satisfy the boundary condition on C. We choose K and d so that this is so. Then

$$\left. \begin{aligned} \tau_{zx} &= G\alpha\left(\frac{\partial\psi}{\partial y} - y\right) = -yG\alpha \\[2mm] \tau_{zy} &= -G\alpha\left(\frac{\partial\psi}{\partial x} - x\right) = xG\alpha \end{aligned} \right\} \tag{5-151}$$

and we see that the stresses, τ_{zx} and τ_{zy} correspond to a *resultant* stress, τ_{res} (see Fig. 5.15), such that

1. τ_{res} varies linearly with distance, ρ from the center of the cross section O.
2. τ_{res} is perpendicular to a radial line from the center O to the point in question.

Also, α, the twist per unit length, is given by

$$\left. \begin{aligned} \alpha &= \frac{M_T}{G\displaystyle\iint_R \left(x^2 + y^2 - x\frac{\partial\psi}{\partial x} - y\frac{\partial\psi}{\partial y}\right) dA} \\[4mm] &= \frac{M_T}{G\displaystyle\iint_R (x^2 + y^2)\, dA} = \frac{M_T}{GI_\rho} \end{aligned} \right\} \tag{5-152}$$

where I_ρ is the polar moment of inertia of the cross section with respect to the origin O (see Eq. 2-59).

Also (see Fig. 5.15), since

$$\tau_{\text{res}} = \alpha G\rho \tag{5-153}$$

---

[5] See Art. 5-11 in regard to uniqueness of solutions of torsion problems.

it follows from the above that

$$\tau_{res} = \frac{M_T \rho}{I_\rho} \qquad (5\text{-}154)$$

which permits us to find the resultant shear stress on a circular cross section due to an applied torque, M_T. Note: this value will be a maximum at the outside fiber, where ρ is a maximum.

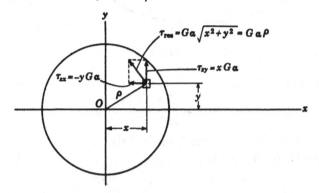

Fig. 5.15

Finally, since ψ is a constant, it follows that ϕ is a constant (see Eq. 1-166) and therefore the *warping* is zero, since this constant may be taken equal to zero (any other constant value for ϕ or w simply represents a rigid body movement of the entire bar and not a distortion of the cross section).

Summarizing the solution for the circular bar, we have

$$M_T = \text{given, applied torque}$$

then

$$\tau_{res} = M_T \rho / I_\rho = \text{resultant shear stress at } \rho \text{ from origin}$$
$$\alpha = M_T / G I_\rho = \text{twist per unit length}$$
$$w = 0 = \text{warping}$$

5-9 The Elliptical Cross Section. We now wish to determine the stresses, angle of twist and warping for an elliptical cross-section bar (Fig. 5.16) subjected to a constant torque M_T.

Solution

Without indicating how this function is obtained,[6] let us try for ψ, the function

$$\psi = e(x^2 - y^2) \qquad (5\text{-}155)$$

where e is some constant.

[6] See Art. 5-11 regarding uniqueness of solutions of torsion problems.

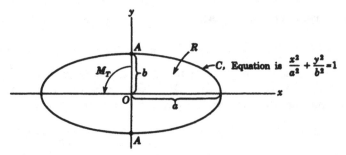

Fig. 5.16

Then certainly

$$\nabla^2\psi = \frac{\partial^2\psi}{\partial x^2} + \frac{\partial^2\psi}{\partial y^2} = 0 \qquad (5\text{-}156)$$

as required. The student should verify this.

Also, on C we require that

$$\psi - \tfrac{1}{2}(x^2 + y^2) = d, \quad \text{a constant} \qquad (5\text{-}157)$$

or that, on C, using the assumed value for ψ,

$$e(x^2 - y^2) - \tfrac{1}{2}(x^2 + y^2) = d \qquad (5\text{-}158)$$

This is equivalent to

$$\frac{x^2}{-2d/(1-2e)} + \frac{y^2}{-2d/(1+2e)} = 1 \qquad (5\text{-}159)$$

and for this to be true, i.e., for this to be given by

$$\frac{x^2}{a^2} + \frac{y^2}{b^2} = 1 \qquad (5\text{-}160)$$

requires that

$$-\frac{2d}{1-2e} = a^2 \qquad (5\text{-}161)$$

and

$$-\frac{2d}{1+2e} = b^2 \qquad (5\text{-}162)$$

which gives

$$e = \frac{a^2 - b^2}{2(a^2 + b^2)} \qquad (5\text{-}163)$$

Hence, the function ψ which satisfies all requirements in R and on the boundary C is given by

$$\psi = \frac{a^2 - b^2}{2(a^2 + b^2)}(x^2 - y^2) \qquad (5\text{-}164)$$

The stresses are then given by

$$\tau_{zx} = G\alpha\left(\frac{\partial\psi}{\partial y} - y\right) = -\frac{2G\alpha a^2}{a^2+b^2}y \atop \tau_{zy} = -G\alpha\left(\frac{\partial\psi}{\partial x} - x\right) = \frac{2G\alpha b^2}{a^2+b^2}x \right\} \qquad (5\text{-}165)$$

and

$$\alpha = \frac{M_T}{G\iint_R\left(x^2+y^2-x\frac{\partial\psi}{\partial x}-y\frac{\partial\psi}{\partial y}\right)dA} \atop = \frac{a^2+b^2}{\pi Ga^3b^3}M_T \right\} \qquad (5\text{-}166)$$

Finally, the warping function ϕ is given by

$$\phi = -\frac{a^2-b^2}{a^2+b^2}xy \qquad (5\text{-}167)$$

The above represents a complete solution to the problem of an elliptical bar subjected to a constant torque M_T. The student should note the following:

1. The maximum shear stress occurs on the boundary at the *minor* diameter at points A, Fig. 5.16. This is an unexpected result but one which is borne out by experiment.
2. This cross section *warps*. Indeed, the only cross section which will not warp when subjected to a pure torsion is the circular cross section solved earlier in this section.

5-10 Torsion Solutions for Other Cross-Sectional Shapes. It is possible to obtain exact solutions to the torsion problem for many cross sections of practical interest. However, there are many more cross sections for which exact solutions can not be obtained.

In these cases we must resort to some approximate method. There are various approximate procedures and techniques which have been developed for solving these problems. Some of these are:

1. The membrane analogy. In 1903, Ludwig Prandtl described his torsion-membrane analogy. He pointed out that the differential equation of the torsion problem is the same as the differential equation for small lateral displacements of the surface of a thin membrane subjected to lateral pressures. By properly correlating the two sets of boundary conditions it is possible to draw an analogy between the torsion and the membrane problems. Soap films have been used in place of membranes

and many irregular shapes have been investigated for torsion in this manner. It is necessary that the membrane be stretched across a hole having a boundary the same shape as that of the bar under investigation and that it be distorted using a slight lateral pressure. See Ref. (10) for further discussion of this.

2. There are various energy conservation theorems which can be utilized to give approximate solutions. The student is referred to Ref. (15) for a more detailed discussion of this point.

3. It is possible to solve the problem using finite difference methods. Further discussion of this will be found in Ref. (17) and Ref. (50).

5-11   Uniqueness of Solution of the Torsion Problem. In this section it will be proved that the solution to the torsion problem is unique. That is, it will be proved that if we have a solution to the torsion problem for a given cross section (no matter how obtained), then this solution is the *only* one for the given cross section.

Uniqueness proofs are extremely important in applied mechanics. This is so because in most problems there is not a set procedure for obtaining a solution. The solution, if one has been obtained, is very frequently obtained by some complicated, involved line of reasoning and one is naturally led to speculate if perhaps some simpler or other procedure might not give a different result. The uniqueness proof, if it exists, is absolute assurance that the given solution, no matter how obtained, is the only solution to the problem. Thus, in the previous sections in which solutions were obtained for circular and elliptical cross sections, the ψ functions used in the solutions were, apparently, rather arbitrarily chosen. This is, in reality, exactly so—that is, they were initially chosen arbitrarily. But, having them, it was shown that they correspond to solutions for circular and elliptical cross sections. Furthermore, the uniqueness proof of this section shows that it really does not matter how one obtains the ψ function for a particular problem. Any other method for obtaining the function *must* lead to the same ψ. In this connection, it may be well to point out, not all problems in applied mechanics have unique solutions. The simple Euler column buckling problem, for example, has an infinite number of solutions.

For the solution to the torsion problem, we require a function $\psi(x, y)$ which satisfies

$$\frac{\partial^2 \psi}{\partial x^2} + \frac{\partial^2 \psi}{\partial y^2} = 0 \qquad (5\text{-}168)$$

everywhere in the body R, and

$$\psi - \tfrac{1}{2}(x^2 + y^2) = \text{constant} \qquad (5\text{-}169)$$

at all points on the boundary, C.

To prove that this function ψ represents a unique solution, let us assume that there are two functions ψ and ψ_1, both of which satisfy Eqs. 5-168 and 5-169 above. We shall now show that these two ψ's can, at most, differ only by a constant value which means (see Eqs. 5-140 and 5-141) that stresses and deformations are the same for both values of ψ, i.e., the solutions are the same for both values of ψ.

Let

$$\psi - \psi_1 = \beta \qquad (5\text{-}170)$$

Then, it follows that

$$\frac{\partial^2 \beta}{\partial x^2} + \frac{\partial^2 \beta}{\partial y^2} = 0 \qquad (5\text{-}171)$$

at all points in the body R, and

$$\frac{\partial \beta}{\partial n} = 0, \quad \beta = \text{constant} \qquad (5\text{-}172)$$

on the boundary C.

Now let us consider the following identity

$$\iint_R \left[\frac{\partial}{\partial x}\left(\beta \frac{\partial \beta}{\partial x}\right) + \frac{\partial}{\partial y}\left(\beta \frac{\partial \beta}{\partial y}\right) \right] dxdy$$
$$= \iint_R \left[\beta\left(\frac{\partial^2 \beta}{\partial x^2} + \frac{\partial^2 \beta}{\partial y^2}\right) + \left(\frac{\partial \beta}{\partial x}\right)^2 + \left(\frac{\partial \beta}{\partial y}\right)^2 \right] dxdy \qquad (5\text{-}173)$$

By Green's Theorem, the left-hand side is also equal (see Fig. 5.12) to

$$\int_C \beta\left(\frac{\partial \beta}{\partial y}\frac{dx}{ds} - \frac{\partial \beta}{\partial x}\frac{dy}{ds}\right) ds = -\beta \int_C \left(l\frac{\partial \beta}{\partial x} + m\frac{\partial \beta}{\partial y}\right) ds \qquad (5\text{-}174)$$

and since C is a curve of constant β

$$\left(l\frac{\partial \beta}{\partial x} + m\frac{\partial \beta}{\partial y}\right) = \nabla\beta \cdot \bar{n} = \frac{\partial \beta}{\partial n} \qquad (5\text{-}175)$$

this integral is equivalent to

$$-\int_C \frac{\partial \beta}{\partial n} ds \qquad (5\text{-}176)$$

which is equal to zero, because of Eq. 5-172. Also, because of Eq. 5-171, the first term on the right-hand side of Eq. 5-173 is zero, so that, finally,

$$\iint_R \left[\left(\frac{\partial \beta}{\partial x}\right)^2 + \left(\frac{\partial \beta}{\partial y}\right)^2 \right] dxdy = 0 \qquad (5\text{-}177)$$

Since the square of a term is always positive (or zero), Eq. 5-177 can hold only if

$$\frac{\partial \beta}{\partial x} = \frac{\partial \beta}{\partial y} = 0 \qquad (5\text{-}178)$$

everywhere in R. Therefore β must be constant everywhere in R. This means that ψ and ψ_1 differ by at most, a constant, and hence the stresses and deformations are the same for ψ and ψ_1—or, the solution is unique.

5-12 Summary of the Shear and Torsion Discussion. It was pointed out in this chapter that by considering a single pair of shear forces only in the stress tensor, the near-trivial shear block solution is obtained.

By considering *two* pairs of shear stresses we obtain the classical St. Venant solution to the pure torsion problem for a cylindrical bar.

We indicated in some detail the steps in this exact solution of the linearized equations of elasticity. Having this general solution, we applied the theory to two cross sections of practical interest: (a) the circular cross section and (b) the elliptical cross section. Complete solutions were obtained for these two cases. It was then noted that there are various approximate techniques available for solving the torsion problem and some of these were briefly described.

Finally, uniqueness of torsion solutions was proved. That is, it was shown that the function ψ (no matter how obtained) which is a solution to a particular cross section is the only possible solution (except for an additive constant which does not effect the stresses and deformations).

Problems

1. Assume the stress tensor is given by

$$T = \begin{pmatrix} Ky^2 & 0 & 0 \\ 0 & 0 & 0 \\ 0 & 0 & 0 \end{pmatrix}$$

for a cylindrical bar with x axis along the length of the bar. Discuss the resulting solution.

2. Assume the stress tensor is given by

$$T = \begin{pmatrix} Ky^3 & 0 & 0 \\ 0 & 0 & 0 \\ 0 & 0 & 0 \end{pmatrix}$$

for a cylindrical bar with z axis along the length of the bar. Discuss the resulting solution.

3. If for the stress tensor we assume $\sigma_x = K_1 y + K_2 z$, then show that if y and z are centroidal but not principal axes, the stress is given by

$$\sigma_x = \frac{M_z I_{yy} - M_y I_{yz}}{I_{yy} I_{zz} - I_{yz}{}^2} y + \frac{M_y I_{zz} - M_z I_{yz}}{I_{yy} I_{zz} - I_{yz}{}^2} z$$

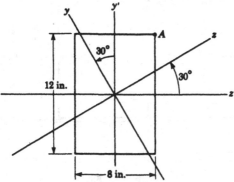

Prob. Fig. 5.4

4. Given a rectangular cross section as shown with an applied moment of 100 in.-lb about the z axis. Determine the stress using the equation given in Prob. 3 above, and verify the result for point A by comparing the result with the simple Bernoulli–Euler value, after resolving the moment into two components about the principal axes.

5. Given a circular steel bar, 3 in. dia., 8 ft long, with a torque M_T applied at its ends. If the allowable stress is 15,000 psi, determine the allowable moment.

6. Consider a thin hollow cylindrical (not necessarily circular) tube with $r \gg h$. The tube is twisted by a moment M_T.

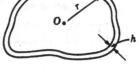

Prob. Fig. 5.6a

   (a) The boundary condition requires that at all points the shear stress be in the direction of the center line of the thin tube. Prove this.

   (b) The shear stress, τ, multiplied by the thickness h (not necessarily constant) at each point gives a "shear flow," q, in pounds per inch. By balancing the torque (about any point, such as O) due to this shear flow and the applied torque M_T, show that $q = M_T/2A$, where A is the area included within the center line of the tube.

   (c) Verify this result for the thin *circular tube* using the exact solution for a circular section.

   (d) Determine the shear flows and shear stresses in the walls of the tube shown if the applied torque is 10,000 in. lb.

7. Using the Bernoulli–Euler form of the beam deflection expression, determine, by integration, the end deflection and end slope for the beams and loadings shown.

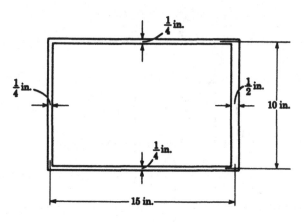

Prob. Fig. 5.6d

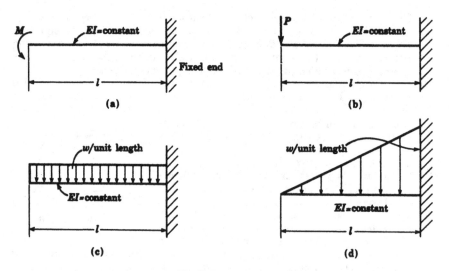

Prob. Fig. 5.7

8. Using the Bernoulli–Euler form of the beam stress equation, determine the maximum bending stress for the beams shown in Prob. 7 if the beam cross section is as shown in the figure and if $l = 10$ ft, $E = 1,500,000$ lb/in.^{2}, $M = 2000$ in. lb, $P = 350$ lb, $w = 50$ lb/ft.

9. For the bar of elliptical cross section prove that the maximum torque shearing stress occurs on the boundary at the minimum radius.

10. Prove that for a circular cross section bar, the warping, w, is equal to zero.

11. The torsion formulas for a circular cross section are

$$\tau = \frac{M_T \rho}{I_\rho} \qquad \alpha = \frac{M_T}{G I_\rho}$$

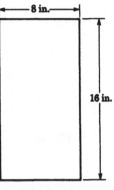

Prob. Fig. 5.8

(a) Consider a bar of elliptical cross section, with $a = 2b$. What would be the errors in τ and α if they are determined using the circular cross-section formulas?

12. Verify equations (5-165), (5-166) and (5-167) for the elliptical cross-section torsion solution.

13. Refer to Eq. 5-137. Show that if we assume the constant value of Ψ on the boundary is zero, then

$$D = 2G \iint_R \Psi \, dx \, dy$$

Hint: Use Green's Theorem and the zero boundary value for Ψ.

14(a) Refer to Eqs. 5-115 and 5-118. Show that the function

$$\psi = \frac{1}{2a}(x^3 - 3xy^2) + b$$

where a and b are constants, satisfies the equations. What must the equation of the boundary be?

(b) If $b = -(2/27)a^2$, show that the equation of the boundary is

$$(x - \sqrt{3}y - \tfrac{2}{3}a)(x + \sqrt{3}y - \tfrac{2}{3}a)(x + \tfrac{1}{3}a) = 0$$

and, therefore, the solution holds for an equilateral triangle.

(c) Show that τ_{max} occurs at the middle of the sides of the triangle, and is given by

$$\tau_{max} = \frac{\alpha G a}{2}$$

(d) Show that

$$\alpha = \frac{5}{3} \frac{M_T}{G I_\rho}$$

Chapter 6

THE DEFLECTION TENSOR IN THE THEORY OF STRUCTURES

6-1 Introduction. The *engineering* applications of the results obtained in the mathematical theory of elasticity are generally presented in courses of strength of materials or structural analysis and design. The exact solutions of the linearized theory (as given in Chapters 4 and 5) are generally approximated in some manner suitable for engineering use. An example of this was given in Chapter 5 in connection with the Bernoulli–Euler Theory of beam bending—and, in fact, it is this approximate bending theory which we consider in the present chapter.

We consider the problem of the static behavior of the thin structure in a plane. By a "thin structure" we mean one whose depth is less than, say, one-tenth of its length. The majority of structural components, such as beams, columns, plates, etc., fall in this category.

Following a short introductory discussion concerning beam deflections and the application of energy methods in determining deflections, a fundamental *deflection tensor* of structural analysis is derived.

It is shown that various techniques and results in the general theory of structural analysis follow directly because of the tensoral nature and properties of this deflection tensor.

6-2 The Deflection of a Plane, Thin Beam Element.[1] Consider a typical plane structure, the curved beam AB shown in Fig. 6.1. End A is free and end B is built-in. P_1 and P_2 are loads on the beam and the dotted figure is the deflected beam.

We note that the theory which follows is independent of the end conditions. It will hold for all thin, plane structures subjected to stresses within the proportional limit of the material, and subjected to

---

[1] See also the paper "The tensor in structural theory" by the author, *Annals of the New York Academy of Science*, 1962.

"small" deformations. By a small deformation we mean a deformation such that the resulting change in slope of the beam, $d\delta/dS$ is much

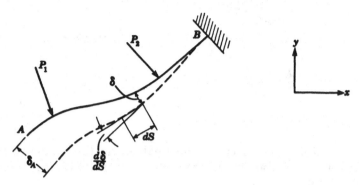

Fig. 6.1

less than unity. For a beam subjected to these conditions it can be shown (see Ref. (10) and Chapter 5) that

$$\frac{d^2\delta}{dS^2} = \frac{M(S)}{EI} \tag{6-1}$$

or

$$\frac{d\theta}{dS} = \frac{M(S)}{EI} \tag{6-2}$$

In the above equations,

δ = deflection of the beam at any point, S.

S = distance along the beam.

$M(S)$ = bending moment at any point S, on the beam due to the given external loading.

E = modulus of elasticity in tension or compression.

I = moment of inertia of the beam about the neutral axis, i.e., the unstrained axis.

$\theta = d\delta/dS$ = change in slope between the deflected and undeflected beam at any point on the beam. Hence $d\theta$ = the angle of rotation between cross sections a distance dS apart, at any point on the beam.

One way in which the deflection of such a structure can be obtained is by means of the so-called *unit load* or *dummy unit load method*. We shall describe this method by considering the structure of Fig. 6.1. Assume we want to find δ_A. Obviously if we find the x and y components of δ_A we shall then have δ_A, since (see Fig. 6.2)

$$\delta_A = (\delta_{A_x}^2 + \delta_{A_y}^2)^{1/2} \tag{6-3}$$

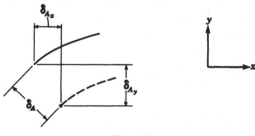

Fig. 6.2

Let us find δ_{A_y}, which is a typical value, and the technique to be described will then also apply, when altered in the obvious manner, to the determination of δ_{A_x}.

In our static analysis of structures (such as the present case) we assume all loads are "gradually applied." This, physically, means that the loads are applied so slowly that the beam does not vibrate under the application of loads. Otherwise stated, this means that the load starts at zero value and gradually increases to its final value—all the while remaining in essential equilibrium with the deflection of the beam. This means, for example, that if the final value of a particular load is P and its final deflection in the direction of its line of action is δ then the work done by this load, gradually applied, is given by

$$\text{work} = \frac{P\delta}{2} \tag{6-4}$$

In determining beam deflections we shall utilize a fundamental law of nature, the *conservation of energy*, used in a form appropriate to the given structure. When applied to a structure such as the beam (this also applies to the arch, plate, and other fundamental structural units), this law, simply stated, tells us that for a beam in static equilibrium the work done by a gradually applied external load is just equal to the work stored in the beam in the form of "strain energy," i.e., energy introduced by the deformation of the beam.

This strain energy can be due to bending strain, normal (tension or compression) strains, shear strains, and torsion strains. We shall consider only the first of these, bending strains, since for the thin beam structure this effect is most important.[2]

---

[2] If the beam is short and deep, then shearing effects are important and the shear strain should be considered. If the structure is a pure membrane, then tensile effects are significant and tension strain energy must be considered. Similar statements hold for other special structures.

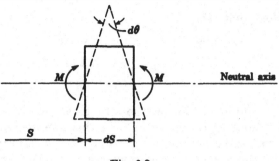

Fig. 6.3

Bending strain energy is caused by an internal moment deflecting through an angle—thus, at any point, S, on the beam (see Fig. 6.3) we have, for the gradually applied load,

$$\text{internal strain energy in the portion } dS = \frac{M\,d\theta}{2} = \frac{M^2\,dS}{2EI} \qquad (6\text{-}5)$$

and over the entire beam,

$$\text{internal strain energy} = \int_{\text{E. L.}} \frac{M^2\,dS}{2EI} \qquad (6\text{-}6)$$

(E.L. = entire length)

For our structure we proceed as follows:
1. Assume that before the given external loading is applied, we gradually apply a unit load at A in the direction of desired deflection, i.e., in the y direction. See Fig. 6.4.
2. Due to this gradually applied load, the end A deflects an amount δ_{AA} in the direction of the unit load, and the

Fig. 6.4

moment at any point is m, shown positive in Fig. 6.4.[3]
3. The external work done by the unit load is

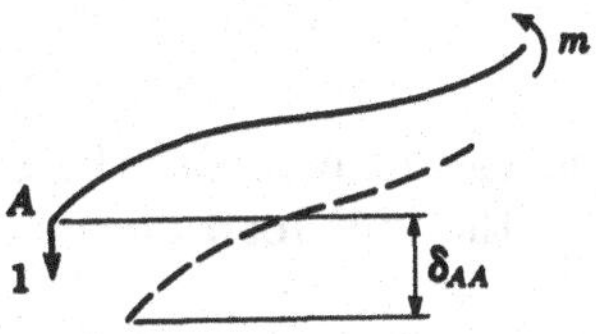

$$\frac{(1)\delta_{AA}}{2}$$

and the internal strain energy due to this load is

$$\int_{\text{E. L.}} \frac{m^2\,dS}{2EI} \qquad (6\text{-}7)$$

[3] Many sign conventions appear in the literature. We assume in this figure that a positive moment bends a beam as ⌣ and a negative moment bends the beam as ⌢ .

4. Up to this point then

$$
\text{external work} = \text{internal work} \\
\left.\frac{(1)\delta_{AA}}{2} = \int_{\text{E.L.}} \frac{m^2\, dS}{2EI}\right\}
$$

(6-8)

5. With the unit load still on the structure, now assume the given external P loading is applied. Due to this, the structure will deform and δ_{A_y} will be the *additional* deflection of point A, in the y direction, caused by this loading. Conditions are as shown in Fig. 6.5.

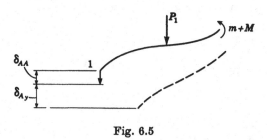

Fig. 6.5

6. Let us consider the effect of the P loading on the unit load. First, the unit load does external work (due to the P loading) given by

$$(1)\delta_{A_y}$$

(6-9)

7. Due to the P loading, the internal moments caused by the unit load do work given by

$$\int_{\text{E.L.}} m\left(\frac{M\, dS}{EI}\right)$$

(6-10)

8. Equating once more, *the unit load response only*, due to the P loading,

$$
\text{external work} = \text{internal work} \\
\left.(1)(\delta_{A_y}) = \int_{\text{E.L.}} \frac{Mm\, dS}{EI}\right\}
$$

(6-11)

and this is the general expression for the deflection of a point in a particular direction—due to bending effects only.

For emphasis, we redefine the terms in the above expression,

δ_{A_y} = deflection of point A in the y direction, due to the gradually applied P loading.

m = moment at any point on the beam due to a gradually applied unit load at A in the y direction.

M = moment at any point on the beam due to the given external P loading, gradually applied.

E, I, dS = elastic constants for the beam.

In view of Eq. 6-11, it follows that

$$\delta_{A_x} = \int_{E.L.} \frac{Mm_x \, dS}{EI} \tag{6-12}$$

and

$$\delta_{A_\theta} = \int_{E.L.} \frac{Mm_\theta \, dS}{EI} \tag{6-13}$$

where

δ_{A_x} = deflection of the point A in the x direction due to the given loading,

m_x = moment at any point due to a unit load at A in the x direction,

δ_{A_θ} = rotation of point A due to the given loading,

m_θ = moment at any point due to a unit *couple* at A.

In order to illustrate the application of Eqs. 6-11 to 6-13, we shall solve several simple deflection problems using them.

EXAMPLE 1

For the beam and loading shown, determine δ_A and θ_A.

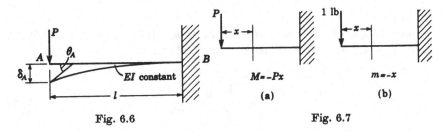

Fig. 6.6   Fig. 6.7

Solution

(a) To determine δ_A, we have

$$\delta_A = \int_0^l \frac{Mm \, dx}{EI} \tag{6-14}$$

in which M and m are obtained from Fig. 6.7 a and b. Then

$$\delta_A = \int_0^l \frac{(-Px)(-x)\,dx}{EI} = +\frac{Pl^3}{3EI} \qquad (6\text{-}15)$$

and the plus sign means the deflection is in the direction of the unit load, i.e., downward.

(b) To obtain θ_A we have $M = -Px$, as before, and m is equal to -1, as shown in Fig. 6.8.

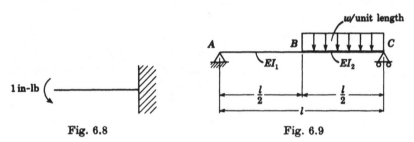

1 in-lb

Fig. 6.8   Fig. 6.9

Note: in this case the unit load is a unit *couple*, since we are interested in an *angular* deflection.

Then

$$\theta_A = \int_0^l \frac{(-Px)(-1)\,dx}{EI} = +\frac{Pl^2}{2EI} \qquad (6\text{-}16)$$

The plus sign means the deflection is in the direction of the unit couple, i.e., counterclockwise.

EXAMPLE 2

Determine δ_B for the beam and loading shown in Fig. 6.9.

Solution

For this case, we can write continuous M/EI expressions only in portions AB and CB. Hence our integrations will be as follows:

$$\int_{\text{E.L.}} = \int_{A,B} + \int_{C,B}$$
$$= \int_0^{l/2} + \int_0^{l/2} \qquad (6\text{-}17)$$

M is given (see Fig. 6.10), by

For AB, $M = \frac{1}{8}wlx$

For CB, $M = \frac{3}{8}wlx_1 - \frac{wx_1^2}{2}$

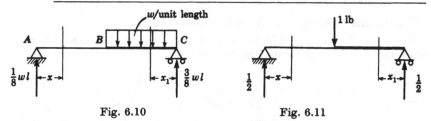

Fig. 6.10   Fig. 6.11

and similarly, m is given (see Fig. 6.11) by

For AB, $m = \frac{1}{2}x$

For CB, $m = \frac{1}{2}x_1$

so that

$$
\left.
\begin{aligned}
\delta_A &= \int_0^{l/2} \frac{(\frac{1}{8}wlx)\left(\dfrac{x}{2}\right)dx}{EI_1} + \int_0^{l/2} \frac{\left[(\frac{3}{8}wlx_1) - \dfrac{wx_1^2}{2}\right]\left(\dfrac{x_1}{2}\right)dx}{EI_2} \\[2mm]
&= \frac{wl^4}{384EI_1} + \frac{wl^4}{256EI_2}
\end{aligned}
\right\}
\tag{6-18}
$$

6-3 The Deflection Tensor. We shall now demonstrate that the following matrix is a *tensor* and, in fact, is the fundamental tensor of the theory of structures.

$$
\begin{pmatrix}
m_x m_x & m_x m_y & m_x m_\theta \\
m_y m_x & m_y m_y & m_y m_\theta \\
m_\theta m_x & m_\theta m_y & m_\theta m_\theta
\end{pmatrix}
\tag{6-19}
$$

To do this note first that the above is given by

$$
\begin{pmatrix} m_x \\ m_y \\ m_\theta \end{pmatrix}
(m_x \quad m_y \quad m_\theta)
\tag{6-20}
$$

In view of Eq. 2-29, therefore, it will be sufficient to show that

$$
\begin{pmatrix} m_x \\ m_y \\ m_\theta \end{pmatrix}
\tag{6-21}
$$

is a vector, i.e., satisfies the transformation law

$$
V' = RV
\tag{6-22}
$$

We do this directly as follows (see Fig. 6.12):

Assume the origin is at the end A. Actually, the origin can be taken anywhere—the result would be the same.

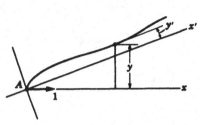

Fig. 6.12

Consider m_x, due to a unit load in the x direction. Its value is given by

$$m_x = y$$

and similarly, m'_x, the moment due to a unit load in the x' direction is given by

$$m'_x = y'$$

In the same way we have

$$m_y = -x$$
$$m'_y = -x'$$

(the minus sign is due to the moment sign convention[4]) and

$$m_\theta = 1$$
$$m'_\theta = 1$$

Thus,

and

$$\left.\begin{matrix} \begin{pmatrix} m_x \\ m_y \\ m_\theta \end{pmatrix} = \begin{pmatrix} y \\ -x \\ 1 \end{pmatrix} \\[2em] \begin{pmatrix} m'_x \\ m'_y \\ m'_\theta \end{pmatrix} = \begin{pmatrix} y' \\ -x' \\ 1 \end{pmatrix} \end{matrix}\right\}$$ (6-23)

Now, note first that the rotation of axes involved in this transformation is one in which z is not changed, that is, we simply rotate the x

---

[4] As pointed out on page 181, there are many possible sign conventions in structural analysis. In the present analysis, for simplicity, we use the following conventions: if the moment is in a counterclockwise direction it is positive; if clockwise, it is negative.

and y axes to x' and y', *about the z axis* at A. Hence, for this case, the rotation matrix, Eqs. 2-16 and 2-17, becomes

$$R = \begin{pmatrix} \cos\theta & \sin\theta & 0 \\ -\sin\theta & \cos\theta & 0 \\ 0 & 0 & 1 \end{pmatrix} \qquad (6\text{-}24)$$

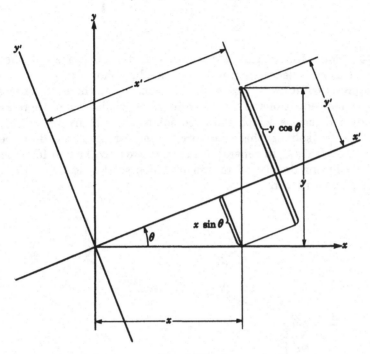

Fig. 6.13

Then it is certainly true (see Fig. 6-13) that

$$\begin{pmatrix} y' \\ -x' \\ 1 \end{pmatrix} = \begin{pmatrix} \cos\theta & \sin\theta & 0 \\ -\sin\theta & \cos\theta & 0 \\ 0 & 0 & 1 \end{pmatrix} \begin{pmatrix} y \\ -x \\ 1 \end{pmatrix} \qquad (6\text{-}25)$$

or

$$V' = RV \qquad (6\text{-}26)$$

which proves that

$$\begin{pmatrix} m_x m_x & m_x m_y & m_x m_\theta \\ m_y m_x & m_y m_y & m_y m_\theta \\ m_\theta m_x & m_\theta m_y & m_\theta m_\theta \end{pmatrix} \qquad (6\text{-}27)$$

is, in fact, a tensor.

Also, it is clear by inspection that Eq. 6-27 represents a symmetric tensor. Hence there is a set of axes, O, x', y', and z', for which this tensor can be given in diagonal (i.e., principal axes) form, in which case it becomes

$$\begin{pmatrix} m_{x'}m_{x'} & 0 & 0 \\ 0 & m_{y'}m_{y'} & 0 \\ 0 & 0 & m_{\theta'}m_{\theta'} \end{pmatrix} \tag{6-28}$$

6-4 The Redundant (Statically Indeterminate) Structure.
Now let us consider the structure and loading shown in Fig. 6.14. This structure is *triply indeterminate* in the sense that there are only three independent equations of static equilibrium available and there are six unknown reactions which must be solved for. Many techniques are available for solving this structure. Some of them are described in Ref. (10). We shall, in general terms, indicate how the unit load method and the tensor developed in the previous section can be utilized in solving this structure.

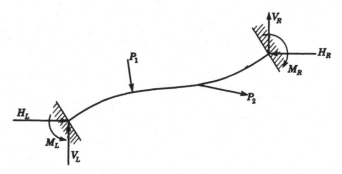

Fig. 6.14

Assume that H_L, V_L, and M_L are the three redundants, i.e., unknowns. Remove them. The structure (now determinate) and loading are shown in Fig. 6.15. Under the given loading, point A now has deflection components as shown, δ_{A_H}, δ_{A_V}, θ_A.

Physically, the reactions H_L, V_L, and M_L have such values that they just make these three deflections equal to zero—which is the requirement at the fixed end such as A.

Furthermore, the horizontal deflection due to H_L is (because of the linear nature of the problem) just H_L times the horizontal deflection due to the unit load in the direction of H_L. Similar statements can be made for the other deflections and reactions.

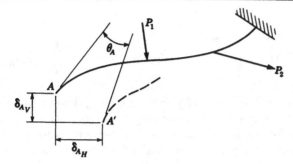

Fig. 6.15

All of the above is included in the following three equations which enable us to solve for the three redundants (H_L, V_L, and M_L) and hence to solve the problem (see Ref. (10)),

$$
\left.
\begin{aligned}
H_L \delta_{A_H A_H} + V_L \delta_{A_H A_V} + M_L \delta_{A_H A_\theta} + \delta_{A_H} &= 0 \\
H_L \delta_{A_V A_H} + V_L \delta_{A_V A_V} + M_L \delta_{A_V A_\theta} + \delta_{A_V} &= 0 \\
H_L \delta_{A_\theta A_H} + V_L \delta_{A_\theta A_V} + M_L \delta_{A_\theta A_\theta} + \theta_A &= 0
\end{aligned}
\right\}
\quad (6\text{-}29)
$$

in which typical terms of these equations are

$$
\left.
\begin{aligned}
\delta_{A_H A_H} &= \int_{\text{E.L.}} \frac{m_H m_H \, dS}{EI} \\
\delta_{A_H A_V} &= \int_{\text{E.L.}} \frac{m_H m_V \, dS}{EI} \\
\delta_{A_\theta A_V} &= \int_{\text{E.L.}} \frac{m_\theta m_V \, dS}{EI}
\end{aligned}
\right\}
\quad (6\text{-}30)
$$

so that the Eq. 6-29 can also be given in terms of our fundamental tensor as

$$
(H_L \ V_L \ M_L) \int_{\text{E.L.}}
\begin{pmatrix}
m_H m_H & m_H m_V & m_H m_\theta \\
m_V m_H & m_V m_V & m_V m_\theta \\
m_\theta m_H & m_\theta m_V & m_\theta m_\theta
\end{pmatrix}
\frac{dS}{EI} + (\delta_{A_H} \ \delta_{A_V} \ \theta_A) = 0
$$

$$(6\text{-}31)$$

Now Eqs. 6-29 and 6-31 represent three simultaneous equations in terms of three unknowns and they can usually be solved without difficulty. However, an essential simplification is introduced if our tensor is put in diagonal form, since then these equations become, respectively,

$$
\left.
\begin{aligned}
H'_L \delta_{A'_H A'_H} + \delta_{A'_H} &= 0 \\
V'_L \delta_{A'_V A'_V} + \delta_{A'_V} &= 0 \\
M'_L \delta_{A'_\theta A'_\theta} + \theta_{A'} &= 0
\end{aligned}
\right\}
\quad (6\text{-}32)
$$

or

$$(H'_L \; V'_L \; M'_L) \int_{\text{E.L.}} \begin{pmatrix} m'_H m'_H & 0 & 0 \\ 0 & m'_V m'_V & 0 \\ 0 & 0 & m'_\theta m'_\theta \end{pmatrix} \frac{dS}{EI} + (\delta_{A'_H} \delta_{A'_V} \theta_{A'}) = 0$$

(6-33)

and now we may solve for H'_L, V'_L, and M'_L, directly in each equation without requiring a simultaneous equation solution.

The transformation which makes our tensor diagonal is the basis of the so-called *elastic center method* of structural analysis. For the general structure, this method requires that the origin and orientation of axes be properly located, in order that the tensor be in diagonal form. Following this, the unknown reactions are obtained from equations similar to Eq. 6-33. These reactions, acting at the origin (or elastic center) must then be transferred by statics to the actual support location. See the paper cited in the footnote on p. 178 of this chapter.

6-5   Some Additional Diagonalization Examples. If we are concerned only with linear deflections, the deflection tensor becomes, (since $m_\theta = 0$)

$$\delta = \int_{\text{E.L.}} \begin{pmatrix} m_x m_x & m_x m_y \\ m_y m_x & m_y m_y \end{pmatrix} \frac{dS}{EI}$$

(6-34)

which may also be represented by

$$\delta = \begin{pmatrix} \delta_{xx} & \delta_{xy} \\ \delta_{yx} & \delta_{yy} \end{pmatrix}$$

(6-35)

in which δ_{xx} = deflection in the x direction due to a unit load in the x direction; the other terms are defined similarly.

Because δ is a symmetric tensor, it may be put in diagonal form. That is, there is a set of axes, $O-x'-y'$, such that

$$\delta_{x'y'} = \delta_{y'x'} = 0$$

(6-36)

Otherwise stated, *there is a set of axes, $O-x'-y'$, such that a load in the $\begin{Bmatrix} x' \\ y' \end{Bmatrix}$ direction will cause a deflection at A only in the $\begin{Bmatrix} x' \\ y' \end{Bmatrix}$ direction.* We may call these directions the principal directions.

As a simple example, consider the straight bar AB shown in Fig. 6.16. Obviously $\delta_{xy} = 0$ for the x and y axes as shown, since for a unit load in the x direction $m_x = 0$. It follows, therefore, that the x and y directions are the principal directions and therefore a unit load (or any load) in the y direction will cause no deflection in the x direction— which is the usual approximation in the elementary theory of beam deflections. We may note, however, that this assumption must follow

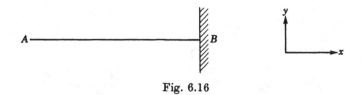

Fig. 6.16

as a direct consequence of the postulates which led to the derivation of the deflection tensor.

We now consider the quarter-circular bar, Fig. 6.17, free at A, built in at B, and loaded at A.

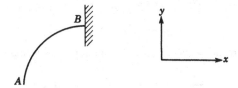

Fig. 6.17

Let us, for the above bar, determine the angle ϕ between the x and y axes and the x' and y' (principal) axes by use of the physical definition of principal directions, namely:

A unit load in the $\begin{Bmatrix} x' \\ y' \end{Bmatrix}$ direction will cause zero deflection in the $\begin{Bmatrix} y' \\ x' \end{Bmatrix}$ direction.

Assume a load applied in the direction x' such that its x component is K and its y component is 1. Then if $\delta_{y'y'} = 0$, $\delta_{x'x'} \neq 0$, δ_x/δ_y must be in the same ratio as $K/1$, where δ_x and δ_y are the x and y components, respectively, of the deflection due to this load. Thus, from Fig. 6.18 we see that the requirement for principal direction is that

$$\frac{K}{1} = \frac{K\delta_{xx} - \delta_{xy}}{\delta_{yy} - K\delta_{yx}} \qquad (6\text{-}37)$$

But

$$\tan \phi = \frac{1}{K} \qquad (6\text{-}38)$$

and therefore

$$\left. \begin{array}{c} \tan 2\phi = \dfrac{2 \tan \phi}{1 - \tan^2\phi} = \dfrac{2/K}{1 - (1/K^2)} \\[2ex] \tan 2\phi = \dfrac{2K}{K^2 - 1} \end{array} \right\} \qquad (6\text{-}39)$$

or

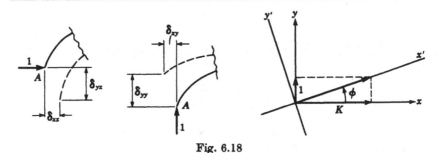

Fig. 6.18

Also, from Eq. 6-37,

or

$$K\delta_{yy} - K^2\delta_{yx} = K\delta_{xx} - \delta_{xy}$$
$$K(\delta_{yy} - \delta_{xx}) = \delta_{xy}(K^2 - 1)$$

$$(6\text{-}40)$$

from which the requirement for x' to be a principal axis is that

$$\frac{2K}{K^2 - 1} = \frac{2\delta_{xy}}{\delta_{yy} - \delta_{xx}} = \tan 2\phi \qquad (6\text{-}41)$$

which is (as it must be) the same relation as the principal axis location requirement previously derived from the inertia tensor (and hence the general symmetric tensor of the second order) in a different way.

6-6 Summary. Using an energy method, the deflection of a plane bar due to elastic bending effects was obtained.

It was shown that this led to a symmetric tensor of the second order—the deflection tensor, which is the fundamental tensor in the theory of elastic bending deflections.

This being a symmetric tensor, it follows that there are principal axes associated with it. The physical meaning of these principal axes for this tensor was discussed briefly.

It was pointed out that the *elastic center* method of structural analysis makes use of the symmetry property of the deflection tensor. In this method the axes are located so that the tensor is in diagonal form, and this simplifies the resulting algebraic solution.

As an additional application of basic tensor theory, the general equation for locating the principal axes of a symmetric second-order tensor was obtained directly from this deflection tensor by making use of the physical interpretation of principal axes.

Problems

1. In the structures shown, determine the deflections indicated, using the unit load method. Bending effects only are to be considered.
2. For the structures shown below, using the unit load method, determine

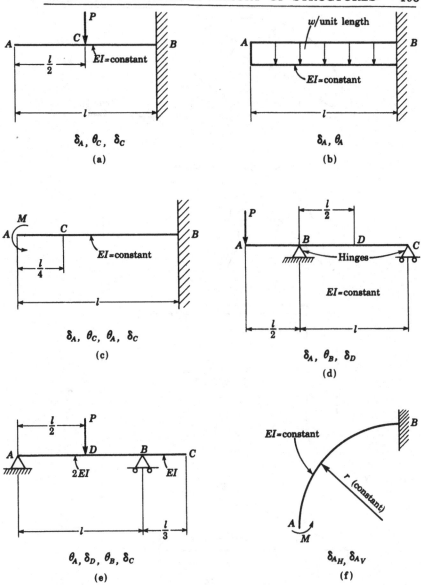

Fig. for Prob. 1

δ_{V_A}, δ_{H_A} and θ_A (bending only considered) due to (a) 1 lb horizontal at A, (b) 1 lb vertical at A, (c) 1 in. lb at A.

3. In the figures shown, AC is a rigid arm ($EI \to \infty$) of length l, making an angle θ with the horizontal. Using the physical definition of the elastic center, i.e., the elastic center is the point such that a vertical load acting on it causes it to deflect only in a vertical direction etc.—determine the

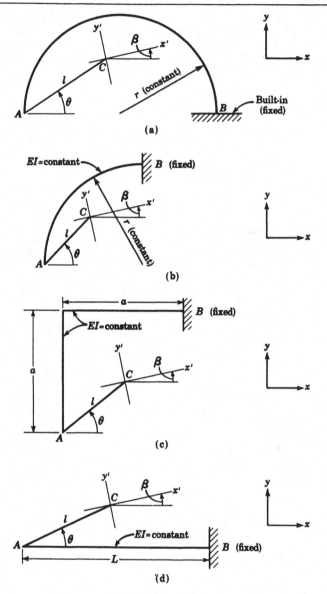

Fig. for Probs. 2, 3, 4

location of the elastic center, C, using the deflections obtained in Prob. 2. For a typical application of this technique see the reference cited in Footnote 1 of this chapter.

4. For the structures of Prob. 3 determine β, the angle between the x and y axes and the x' and y' axes at the elastic center. Use the same method as in Prob. 3.

Chapter 7

INTRODUCTION TO THE THEORY OF PLATES AND SHELLS

7-1 Introduction. In this chapter we derive by matrix methods the basic equations governing the behavior of thin plates. Some typical solutions are obtained for thin plates with small deflections. Based upon a tensoral generalization of thin plate equations, an anology is developed between thin plate theory and solutions of the mathematical theory of elasticity. Finally, a very brief discussion of Kármán's large deflection plate theory is given, and another nonlinear large deflection theory is derived that is based upon the elasticity–thin plate analogy and that is essentially an extension to thin plate theory of Murnaghan's finite deformation elasticity theory, Ref. (16).

The standard American reference on the subject in Ref. (19) and the notation and order of presentation will be based in broad outline on this text.

7-2 Basic Relations—Plates Subject to Pure Bending. We consider an O–x–y–z system directed as in Fig.7.1. Let

w = deflection, positive down

r = radius of curvature, positive up

n, t = mutually perpendicular directions making an angle α with x, y

e, γ = unit strain

σ = tensile or compressive stress

τ = shear stress

E = modulus of elasticity in tension and compression

G = modulus of elasticity in shear

ν = Poisson's ratio

h = plate thickness

$D = Eh^3/12(1-\nu^2)$ = plate flexural rigidity

M = bending moment per unit length

q = transverse loading per unit area

Q = shear force per unit length

First we consider the plates subjected to moments only, i.e., $q = 0$.

Referring to Fig. 7.1, for the deflection shown,

$$\frac{1}{r_{xx}} = -\frac{\partial}{\partial x}\left(\frac{\partial w}{\partial x}\right) = -\frac{\partial^2 w}{\partial x^2} \tag{7-1}$$

and in the same way

$$\frac{1}{r_{yy}} = -\frac{\partial}{\partial y}\left(\frac{\partial w}{\partial y}\right) = -\frac{\partial^2 w}{\partial y^2} \tag{7-2}$$

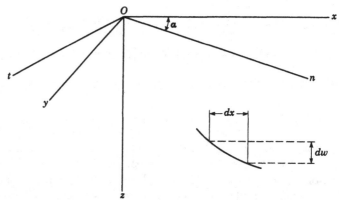

Fig. 7.1

Also, because $(\partial w/\partial x \quad \partial w/\partial y)$ is a vector, $\nabla w = \operatorname{grad} w$, we have (see Eq. 2-5 and note that we have taken the transpose of both sides)

$$\left(\frac{\partial w}{\partial n} \quad \frac{\partial w}{\partial t}\right) = \left(\frac{\partial w}{\partial x} \quad \frac{\partial w}{\partial y}\right)\begin{pmatrix} \cos\alpha & -\sin\alpha \\ \sin\alpha & \cos\alpha \end{pmatrix} \tag{7-3}$$

and in the same way, the vector ∇ becomes

$$\left(\frac{\partial}{\partial n} \quad \frac{\partial}{\partial t}\right) = \left(\frac{\partial}{\partial x} \quad \frac{\partial}{\partial y}\right)\begin{pmatrix} \cos\alpha & -\sin\alpha \\ \sin\alpha & \cos\alpha \end{pmatrix} \tag{7-4}$$

so that

$$\begin{pmatrix} \dfrac{\partial}{\partial x} \\ \dfrac{\partial}{\partial y} \end{pmatrix}\left(\frac{\partial w}{\partial x} \quad \frac{\partial w}{\partial y}\right) = \mathscr{R} \tag{7-5}$$

where $\mathscr{R}$ is a tensor of the second order, and from Eqs. 7-1 and 7-2 we may say

$$\mathscr{R} = \begin{pmatrix} -\dfrac{1}{r_{xx}} & \dfrac{1}{r_{xy}} \\ \dfrac{1}{r_{yx}} & -\dfrac{1}{r_{yy}} \end{pmatrix} = \begin{pmatrix} \dfrac{\partial^2 w}{\partial x^2} & \dfrac{\partial^2 w}{\partial x\,\partial y} \\ \dfrac{\partial^2 w}{\partial y\,\partial x} & \dfrac{\partial^2 w}{\partial y^2} \end{pmatrix} \tag{7-6}$$

Now $\mathscr{R}$ is obviously symmetric. Hence it can be put in diagonal form. In other words, there is a set of axes through O such that

$$\mathscr{R}' = \begin{pmatrix} -\dfrac{1}{r_{x'x'}} & 0 \\ 0 & -\dfrac{1}{r_{y'y'}} \end{pmatrix} \tag{7-7}$$

In addition, the previously derived transformations for the second-order tensor give at once (see Eq. 2-41)

$$\left.\begin{aligned} -\frac{1}{r_{nn}} &= \frac{\partial^2 w}{\partial n^2} = -\frac{1}{r_{xx}}\cos^2\alpha - \frac{1}{r_{yy}}\sin^2\alpha + \frac{2}{r_{xy}}\sin\alpha\cos\alpha \\ \frac{1}{r_{tt}} &= \frac{\partial^2 w}{\partial t^2} = -\frac{1}{r_{xx}}\sin^2\alpha - \frac{1}{r_{yy}}\cos^2\alpha - \frac{2}{r_{xy}}\sin\alpha\cos\alpha \\ \frac{1}{r_{nt}} &= \frac{\partial^2 w}{\partial n\partial t} = -\left(\frac{1}{r_{yy}}-\frac{1}{r_{xx}}\right)\sin\alpha\cos\alpha + \frac{1}{r_{xy}}(\cos^2\alpha - \sin^2\alpha) \end{aligned}\right\} \tag{7-8}$$

In the previous study of elasticity theory (see Chapter 4), it was shown that for an isotropic solid we have two fundamental symmetric tensors of the second order—the stress tensor and the strain tensor. For two dimensions, these take the respective forms

$$\left.\begin{aligned} T &= \begin{pmatrix} \sigma_x & \tau_{xy} \\ \tau_{yx} & \sigma_y \end{pmatrix} \quad (a) \\ \eta &= \begin{pmatrix} e_x & \tfrac{1}{2}\gamma_{xy} \\ \tfrac{1}{2}\gamma_{yx} & e_y \end{pmatrix} \quad (b) \end{aligned}\right\} \tag{7-9}$$

Also it was shown that the elements of Eq. 7-9, a and b, are connected by means of Hooke's Law, which in two dimensions is given by

$$\left.\begin{aligned} e_x &= \frac{1}{E}(\sigma_x - \nu\sigma_y) \quad (a) \\ e_y &= \frac{1}{E}(\sigma_y - \nu\sigma_x) \quad (b) \end{aligned}\right\} \tag{7-10}$$

Now, Eq. 7-10, a and b, are equivalent to

$$\left.\begin{aligned} e_x &= \frac{1+\nu}{E}\sigma_x - \frac{\nu}{E}(\sigma_x + \sigma_y) \quad (a) \\ e_y &= \frac{1+\nu}{E}\sigma_y - \frac{\nu}{E}(\sigma_x + \sigma_y) \quad (b) \end{aligned}\right\} \tag{7-11}$$

and because of the symmetric forms of Eq. 7-9, a and b, they may be put in diagonal form, from which, noting that $\sigma_x + \sigma_y$ is an invariant and using Eq. 7-11, a and b, we have the tensor equation

$$\begin{pmatrix} e_x & 0 \\ 0 & e_y \end{pmatrix} = \frac{1+\nu}{E} \begin{pmatrix} \sigma_x & 0 \\ 0 & \sigma_y \end{pmatrix} - \frac{\nu}{E}(\sigma_x + \sigma_y)\mathscr{E}_2 \qquad (7\text{-}12)$$

where $\mathscr{E}_2$ is the unit tensor

$$\mathscr{E}_2 = \begin{pmatrix} 1 & 0 \\ 0 & 1 \end{pmatrix}$$

Equation (7-12), as a tensor equation, holds for all orientation of axes. Hence, it becomes, in nondiagonal form

$$\begin{pmatrix} e_n & \tfrac{1}{2}\gamma_{nt} \\ \tfrac{1}{2}\gamma_{tn} & e_t \end{pmatrix} = \frac{1+\nu}{E} \begin{pmatrix} \sigma_n & \tau_{nt} \\ \tau_{tn} & \sigma_t \end{pmatrix} - \frac{\nu}{E}(\sigma_x + \sigma_y)\mathscr{E}_2 \qquad (7\text{-}13)$$

from which

$$\gamma_{nt} = \frac{2(1+\nu)}{E}\tau_{nt} \qquad (7\text{-}14)$$

Equation 7-14 introduces a new elastic constant, the modulus of elasticity in shear, defined by

or
$$\left.\begin{array}{c} \dfrac{2(1+\nu)}{E} = \dfrac{1}{G} \\[3mm] G = \dfrac{E}{2(1+\nu)} \end{array}\right\} \qquad (7\text{-}15)$$

Equation 7-13 is the general form of Hooke's Law, in two dimensions, for strains as functions of stresses. The results given in Eqs. 7-10 through 7-15 were all obtained earlier in Chapter 4. They were repeated because the technique utilized above in obtaining Hooke's Law will be used in our analysis of plate action which follows.

We now consider a differential rectangular element removed from the plate, Fig. 7.2. For small deflections there is a plane that, we can assume, remains unstressed—the neutral plane. This plane will be at mid-depth.

Then, at any plane a distance z below the neutral plane we have

and
$$\left.\begin{array}{c} e_x = \dfrac{z}{r_{xx}} \\[3mm] e_y = \dfrac{z}{r_{yy}} \end{array}\right\} \qquad (7\text{-}16)$$

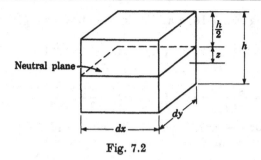

Fig. 7.2

Equation 7-11, a and b, solved for σ_x and σ_y, give

$$\left.\begin{aligned}
\sigma_x &= \frac{E}{1-\nu^2}(e_x+\nu e_y) \\[2mm]
\sigma_y &= \frac{E}{1-\nu^2}(e_y+\nu e_x)
\end{aligned}\right\} \tag{7-17}$$

or

$$\left.\begin{aligned}
\sigma_x &= \frac{Ez}{1-\nu^2}\left(\frac{1}{r_{xx}}+\nu\frac{1}{r_{yy}}\right) \quad \text{(a)} \\[2mm]
\sigma_y &= \frac{Ez}{1-\nu^2}\left(\frac{1}{r_{yy}}+\nu\frac{1}{r_{xx}}\right) \quad \text{(b)}
\end{aligned}\right\} \tag{7-18}$$

We may proceed with Eq. 7-18, a and b, just as with 7-11, a and b. Doing this, we obtain finally

$$\begin{pmatrix} \sigma_n & \tau_{nt} \\ \tau_{tn} & \sigma_t \end{pmatrix} = -\frac{Ez}{1+\nu}\begin{pmatrix} -\dfrac{1}{r_{nn}} & \dfrac{1}{r_{nt}} \\[2mm] \dfrac{1}{r_{tn}} & -\dfrac{1}{r_{tt}} \end{pmatrix} + \frac{Ez\nu}{1-\nu^2}\left(\frac{1}{r_{xx}}+\frac{1}{r_{yy}}\right)\mathscr{E}_2 \tag{7-19}$$

so that

$$\tau_{nt} = -2Gz\frac{1}{r_{nt}} = -2Gz\frac{\partial^2 w}{\partial n\partial t} \tag{7-20}$$

Going back to Eq. 7-18, a and b, and referring to Fig. 7-2, the requirement of static equilibrium becomes

$$\left.\begin{aligned}
M_{xx}\,dy &= \left\{\int_{-(h/2)}^{+(h/2)} \sigma_x z\,dz\right\}dy \quad \text{(a)} \\[2mm]
M_{yy}\,dx &= \left\{\int_{-(h/2)}^{+(h/2)} \sigma_y z\,dz\right\}dx \quad \text{(b)}
\end{aligned}\right\} \tag{7-21}$$

where M_{xx} = moment per unit length on a plane perpendicular to the x axis. M_{yy} is defined similarly.

Substituting Eq. 7-18, a and b, in the above, we get

$$M_{xx} = D\left(\frac{1}{r_{xx}} + \nu \frac{1}{r_{yy}}\right) = -D\left(\frac{\partial^2 w}{\partial x^2} + \nu \frac{\partial^2 w}{\partial y^2}\right) \qquad (a)$$

$$M_{yy} = D\left(\frac{1}{r_{yy}} + \nu \frac{1}{r_{xx}}\right) = -D\left(\frac{\partial^2 w}{\partial y^2} + \nu \frac{\partial^2 w}{\partial x^2}\right) \qquad (b)$$

(7-22)

and just as for the pairs Eqs. 7-11, a and b, and 7-18, a and b, we find from Eq. 7-22, a and b,

or

$$M_{nt} = D(1-\nu)\frac{1}{r_{nt}}$$

$$M_{nt} = D(1-\nu)\frac{\partial^2 w}{\partial n \partial t}$$

(7-23)

In the above M_{nt} is the *torque* per unit length on a plane perpendicular to n in a direction t.

An alternative derivation of Eq. 7-23 which points out the physical significance of M_{nt}, and in addition brings out an important fact concerning the signs of the moments, is the following (see Fig. 7-3):

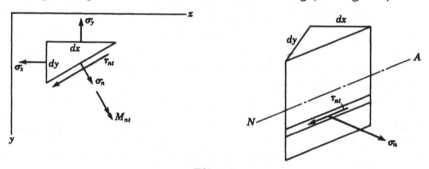

Fig. 7.3

τ is assumed positive if it tends to rotate the element clockwise. M_{nt} is a twisting moment and is positive, using the right-hand rule sign convention if it acts in the same direction as n. Hence, in Fig. 7.3,

$$M_{nt} = -\int_{-(h/2)}^{+(h/2)} \tau_{nt} z\, dz \qquad (7-24)$$

and, using Eq. 7-20,

$$M_{nt} = \frac{Gh^3}{6}\frac{\partial^2 w}{\partial n \partial t}$$

$$= D(1-\nu)\frac{\partial^2 w}{\partial n \partial t}$$

(7-25)

From the above stated sign convention it follows that

$$M_{nt} = -M_{tn} \qquad (7\text{-}26)$$

a relation which holds true independently of the subscripts and the sign convention used, that is,

$$M_{xy} = -M_{yx}, \text{ etc.} \qquad (7\text{-}27)$$

Because of Eqs. 7-21, a and b, and 7-24 it follows that M_{xx}, M_{yy}, and M_{xy} are components of a tensor, the *moment tensor*,

$$\mathcal{M} = \begin{pmatrix} M_{xx} & -M_{xy} \\ M_{yx} & M_{yy} \end{pmatrix} \qquad (7\text{-}28)$$

The following will summarize the theory of pure bending of thin plates up to this point:

(a) The curvature tensor,

$$\mathcal{R} = \begin{pmatrix} -\dfrac{1}{r_{xx}} & \dfrac{1}{r_{xy}} \\ \dfrac{1}{r_{yx}} & -\dfrac{1}{r_{yy}} \end{pmatrix} = \begin{pmatrix} \dfrac{\partial^2 w}{\partial x^2} & \dfrac{\partial^2 w}{\partial x \partial y} \\ \dfrac{\partial^2 w}{\partial y \partial x} & \dfrac{\partial^2 w}{\partial y^2} \end{pmatrix} \qquad (7\text{-}29)$$

was obtained and the rotation of axes equations governing this tensor were written at once.

(b) The stress–strain relations and Hooke's Law led to

$$\left.\begin{aligned}
M_{xx} &= D\left(\frac{1}{r_{xx}} + \nu \frac{1}{r_{yy}}\right) = -D\left(\frac{\partial^2 w}{\partial x^2} + \nu \frac{\partial^2 w}{\partial y^2}\right) \\
M_{yy} &= D\left(\frac{1}{r_{yy}} + \nu \frac{1}{r_{xx}}\right) = -D\left(\frac{\partial^2 w}{\partial y^2} + \nu \frac{\partial^2 w}{\partial x^2}\right) \\
M_{xy} &= -M_{yx} = D(1-\nu)\frac{1}{r_{xy}} = D(1-\nu)\frac{\partial^2 w}{\partial x \partial y}
\end{aligned}\right\} \qquad (7\text{-}30)$$

(c) We saw that in the theory of thin plates, the following fundamental tensors of the second order occur:

$$\mathcal{R} = \text{curvature tensor} = \begin{pmatrix} -\dfrac{1}{r_{xx}} & \dfrac{1}{r_{xy}} \\ \dfrac{1}{r_{yx}} & -\dfrac{1}{r_{yy}} \end{pmatrix} = \begin{pmatrix} \dfrac{\partial^2 w}{\partial x^2} & \dfrac{\partial^2 w}{\partial x \partial y} \\ \dfrac{\partial^2 w}{\partial y \partial x} & \dfrac{\partial^2 w}{\partial y^2} \end{pmatrix}$$

$$T = \text{stress tensor} = \begin{pmatrix} \sigma_x & \tau_{xy} \\ \tau_{yx} & \sigma_y \end{pmatrix}$$

$$\eta = \text{strain tensor} = \begin{pmatrix} e_x & \tfrac{1}{2}\gamma_{xy} \\ \tfrac{1}{2}\gamma_{yx} & e_y \end{pmatrix}$$

$$\mathcal{M} = \text{moment tensor} = \begin{pmatrix} M_{xx} & -M_{xy} \\ M_{yx} & M_{yy} \end{pmatrix}$$

Each of the above is a symmetric tensor and can therefore be put in diagonal form (that is, each has principal axes) and each tensor follows the rotation of axes transformation equations for the second-order tensor.

7-3 A Simple Exact Solution for the Thin Plate.

As an illustrative solution to the equations derived in the previous article, we consider the following problem:

Assume a rectangular plate with edge moments

$$\left.\begin{aligned} M_{xx} &= M_1 \\ M_{yy} &= M_2 \\ M_{xy} &= -M_{yx} = 0 \end{aligned}\right\} \tag{7-31}$$

Then Eqs. 7-22 and 7-23 become

$$\left.\begin{aligned} M_1 &= -D\left(\frac{\partial^2 w}{\partial x^2} + \nu\frac{\partial^2 w}{\partial y^2}\right) & \text{(a)} \\[2mm] M_2 &= -D\left(\frac{\partial^2 w}{\partial y^2} + \nu\frac{\partial^2 w}{\partial x^2}\right) & \text{(b)} \\[2mm] 0 &= D(1-\nu)\frac{\partial^2 w}{\partial x \partial y} & \text{(c)} \end{aligned}\right\} \tag{7-32}$$

and from the first pair of the above

$$\left.\begin{aligned} \frac{\partial^2 w}{\partial x^2} &= -\frac{M_1 - \nu M_2}{D(1-\nu^2)} \\[2mm] \frac{\partial^2 w}{\partial y^2} &= -\frac{M_2 - \nu M_1}{D(1-\nu^2)} \end{aligned}\right\} \tag{7-33}$$

We may integrate Eqs. 7-32c, 7-33 at once and obtain

$$C_1 f(x) + C_2 g(y) = w \tag{7-34}$$

$$-\frac{M_1 - \nu M_2}{D(1-\nu^2)}\frac{x^2}{2} + C_3 x + C_4 = w \tag{7-35}$$

$$-\frac{M_2 - \nu M_1}{D(1-\nu^2)}\frac{y^2}{2} + C_5 y + C_6 = w \tag{7-36}$$

where C_4 may be $C_4(y)$ and C_6 may be $C_6(x)$. We satisfy all of these equations by taking

$$w = -\left(\frac{M_1 - \nu M_2}{2D(1-\nu^2)}\right)x^2 - \left(\frac{M_2 - \nu M_1}{2D(1-\nu^2)}\right)y^2 + K_1 x + K_2 y + K_3 \qquad (7\text{-}37)$$

Now if we assume that (1) at $x = 0$, $y = 0$, $w = 0$—that is, the deflection is zero; (2) the $z = 0$ plane of reference is tangent to the deflected plate at $x = 0$, that is, at the origin

$$\frac{\partial w}{\partial x} = 0, \qquad \frac{\partial w}{\partial y} = 0$$

then

$$K_1 = K_2 = K_3 = 0$$

or

$$w = -\left(\frac{M_1 - \nu M_2}{2D(1-\nu^2)}\right)x^2 - \left(\frac{M_2 - \nu M_1}{2D(1-\nu^2)}\right)y^2 \qquad (7\text{-}38)$$

We now consider the following cases:

(a) $$M_2 = 0$$

Then

$$w = -\frac{M_1 x^2}{2D(1-\nu^2)} + \frac{\nu M_1 y^2}{2D(1-\nu^2)} \qquad (7\text{-}39)$$

which is a form of anticlastic surface (saddle surface).

(b) $$M_1 = M_2 = M$$

Then

$$w = -\frac{M(x^2 + y^2)}{2D(1+\nu)} \qquad (7\text{-}40)$$

which is a paraboloid.

(c) $$M_1 = -M_2$$

Then

$$w = -\frac{M_1}{2D(1+\nu)}(x^2 - y^2) \qquad (7\text{-}41)$$

which is an anticlastic surface.

7-4 Small Deflections of Laterally Loaded Plates. In this section the differential equations for the laterally loaded thin plate (with small deflections) will be derived and a "moment" matrix will be obtained. The boundary conditions will be discussed and certain simple solutions of the plate problem will be obtained.

We consider as a free body a differential piece of the transversely loaded plate, Fig. 7.4a. The moments acting on the element are shown schematically in Fig. 7.4b, and the shears in Fig. 7.4c.

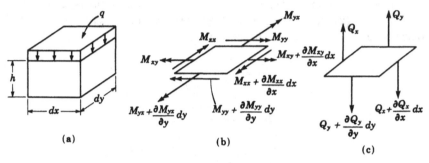

(a)   (b)   (c)

Fig. 7.4

Summing vertical forces, we require for static equilibrium,

$$q\,dxdy + \frac{\partial Q_x}{\partial x}\,dxdy + \frac{\partial Q_y}{\partial y}\,dxdy = 0 \qquad (7\text{-}42)$$

or

$$\frac{\partial Q_x}{\partial x} + \frac{\partial Q_y}{\partial y} + q = 0 \qquad (7\text{-}43)$$

Taking moments about the y axis and neglecting higher-order differentials, we require, for static equilibrium,

$$\frac{\partial M_{yx}}{\partial y}\,dxdy + \frac{\partial M_{xx}}{\partial x}\,dxdy - Q_x\,dxdy = 0 \qquad (7\text{-}44)$$

from which

$$\frac{\partial M_{yx}}{\partial y} + \frac{\partial M_{xx}}{\partial x} - Q_x = 0 \qquad (7\text{-}45)$$

Similarly, for moments about the x axis, we require for static equilibrium,

$$\frac{\partial M_{xy}}{\partial x}\,dxdy - \frac{\partial M_{yy}}{\partial y}\,dxdy + Q_y\,dxdy = 0 \qquad (7\text{-}46)$$

or

$$\frac{\partial M_{xy}}{\partial x} - \frac{\partial M_{yy}}{\partial y} + Q_y = 0 \qquad (7\text{-}47)$$

Now, in the chapter on theory of elasticity, it was shown that the equilibrium equations for an isotropic body with zero body forces are

$$\left.\begin{array}{l} \dfrac{\partial \sigma_x}{\partial x} + \dfrac{\partial \tau_{yx}}{\partial y} + \dfrac{\partial \tau_{zx}}{\partial z} = 0 \\[3mm] \dfrac{\partial \tau_{xy}}{\partial x} + \dfrac{\partial \sigma_y}{\partial y} + \dfrac{\partial \tau_{zy}}{\partial z} = 0 \\[3mm] \dfrac{\partial \tau_{xz}}{\partial x} + \dfrac{\partial \tau_{yz}}{\partial y} + \dfrac{\partial \sigma_z}{\partial z} = 0 \end{array}\right\} \qquad (7\text{-}48)$$

or, in matrix notation,

$$\operatorname{div} T = 0 \qquad (7\text{-}49)$$

where T is the symmetric stress tensor

$$T = \begin{pmatrix} \sigma_x & \tau_{xy} & \tau_{xz} \\ \tau_{yx} & \sigma_y & \tau_{yz} \\ \tau_{zx} & \tau_{zy} & \sigma_z \end{pmatrix} \qquad (7\text{-}50)$$

Equations 7-43, 7-45, and 7-47, which are the equilibrium equations for the laterally thin plate, can similarly be presented in matrix form by

$$\operatorname{div} \mathscr{M} = 0 \qquad (7\text{-}51)$$

where $\mathscr{M}$ is a symmetric moment tensor given by

$$\mathscr{M} = \begin{pmatrix} M_{xx} & -M_{xy} & -Q_x z \\ M_{yx} & M_{yy} & -Q_y z \\ -Q_x z & -Q_y z & -q\dfrac{z^2}{2} \end{pmatrix} \qquad (7\text{-}52)$$

In the next article an analogy between plate theory and linearized elasticity theory is developed. It is based upon the existence of this moment tensor.

Returning to our analysis of the loaded plate, if we substitute Q_x and Q_y from Eqs. 7-45 and 7-47 into Eq. 7-43, we obtain

$$\frac{\partial^2 M_{xx}}{\partial x^2} + \frac{\partial^2 M_{yx}}{\partial x \partial y} + \frac{\partial^2 M_{yy}}{\partial y^2} - \frac{\partial^2 M_{xy}}{\partial x \partial y} = -q \qquad (7\text{-}53)$$

and since $M_{yx} = -M_{xy}$ we have

$$\frac{\partial^2 M_{xx}}{\partial x^2} + \frac{\partial^2 M_{yy}}{\partial y^2} - 2\frac{\partial^2 M_{xy}}{\partial x \partial y} = -q \qquad (7\text{-}54)$$

Now we showed in Eqs. 7-22a, 7-22b, and 7-25 that

$$M_{xx} = -D\left(\frac{\partial^2 w}{\partial x^2} + \nu\frac{\partial^2 w}{\partial y^2}\right)$$

$$\left.\begin{array}{l} M_{yy} = -D\left(\dfrac{\partial^2 w}{\partial y^2} + \nu\dfrac{\partial^2 w}{\partial x^2}\right) \\[2ex] M_{xy} = D(1-\nu)\dfrac{\partial^2 w}{\partial x\,\partial y} \end{array}\right\} \tag{7-55}$$

Substituting these in Eq. 7-54, we obtain, finally,

$$\nabla^4 w = \frac{\partial^4 w}{\partial x^4} + 2\frac{\partial^4 w}{\partial x^2\,\partial y^2} + \frac{\partial^4 w}{\partial y^4} = \frac{q}{D} \tag{7-56}$$

Thus, the solution of the problem of the laterally loaded plate with small deflections requires the determination of a function $w(x, y)$ which satisfies Eq. 7-56 and also the boundary conditions for the given problem. We will now discuss the boundary conditions.

The edges of the plate may, in general, be either (a) built-in or clamped, (b) simply supported (that is, free to rotate in the plane of the plate but not free to deflect), or (c) free (that is, unsupported).

Condition (a), a built-in edge at say, $y = 0$ requires as boundary conditions

$$\left.\begin{array}{l} (w)_{y=0} = 0 \\[2ex] \left(\dfrac{\partial w}{\partial y}\right)_{y=0} = 0 \end{array}\right\} \tag{7-57}$$

For condition (b), a simply supported edge at, say, $y = 0$, the following boundary conditions are required,

$$\left.\begin{array}{l} (w)_{y=0} = 0 \\[1ex] (M_{yy})_{y=0} = 0 \end{array}\right\} \tag{7-58}$$

or (see Eq. 7.22b)

$$\left(\frac{\partial^2 w}{\partial y^2} + \nu\frac{\partial^2 w}{\partial x^2}\right)_{y=0} = 0 \tag{7-59}$$

and, since for this edge $\partial^2 w/\partial x^2 = 0$, we have as the second boundary condition

$$\left(\frac{\partial^2 w}{\partial y^2}\right)_{y=0} = 0$$

The condition (c), free edge, has been the cause of a most interesting controversy which extends even to this day. Poisson, in a famous memoir, originally formulated the boundary conditions for, say, a free edge at $x = a$ as

$$\left. \begin{array}{l} (M_{xx})_{x=a} = 0 \\ (M_{xy})_{x=a} = 0 \\ (Q_x)_{x=a} = 0 \end{array} \right\} \qquad (7\text{-}60)$$

Kirchhoff, however, by a variation of energy method, proved that it would, in general, be impossible to satisfy all three conditions—although in the special case of a circular plate (which Poisson considered) all three are satisfied because one of them is identically zero. Kirchhoff showed that only two boundary conditions are necessary and, in addition, he showed that the last two of Poisson's boundary conditions can be replaced by a single one.

It is interesting to note that Kirchhoff's conclusions were disputed by as eminent a mathematician as Mathieu, who derived his own boundary conditions for a free edge—which conditions, incidentally, appear to be in error. Even to this day we find discussions of the boundary conditions of the plate problem.[1]

The two boundary conditions for the free edge as derived by Kirchhoff are

$$\left. \begin{array}{l} (M_{xx})_{x=a} = 0 \\ \left(Q_x - \dfrac{\partial M_{xy}}{\partial y} \right)_{x=a} = 0 \end{array} \right\} \qquad (7\text{-}61)$$

or (see Eqs. 7-22a, 7-25, and 7-45),

$$\left. \begin{array}{l} \left(\dfrac{\partial^2 w}{\partial x^2} + \nu \dfrac{\partial^2 w}{\partial y^2} \right)_{x=a} = 0 \\ \left(\dfrac{\partial^3 w}{\partial x^3} + (2-\nu) \dfrac{\partial^3 w}{\partial x \partial y^2} \right)_{x=a} = 0 \end{array} \right\} \qquad (7\text{-}62)$$

The physical meaning of the reduction in the number of boundary conditions is discussed in some detail in Ref. (19). Another interesting discussion of the point is given in Ref. (20). At this time we only mention that as a consequence of this, a square plate will have concentrated reactions R of value

$$R = M_{xy} = -M_{yx} = 2D(1-\nu)\frac{\partial^2 w}{\partial x \partial y} \qquad (7\text{-}63)$$

acting at each corner.

[1] See, for example, a paper by Friedrichs in *Symposia of Applied Mathematics*, Vol 3, McGraw-Hill Book Co., 1951.

The following summarizes the discussion of the laterally loaded thin plate with small deflections:

(a) The equilibrium equations were obtained for the laterally loaded thin plate and it was shown that these may be given as

$$\text{div}\,\mathcal{M} = 0$$

where $\mathcal{M}$ is a moment matrix. This equation is similar in form to the equilibrium equations for the isotropic elastic body with zero body forces.

(b) The fundamental differential equation of the laterally loaded thin plate was obtained. This is

$$\frac{\partial^4 w}{\partial x^4} + 2\frac{\partial^4 w}{\partial x^2 \partial y^2} + \frac{\partial^4 w}{\partial y^4} = \frac{q}{D} \tag{7-64}$$

(c) The solution of the laterally loaded plate requires determining $w(x, y)$ which satisfies Eq. 7-64 and the boundary conditions. The boundary conditions for built-in, simply supported, and free edges were given. Having w from this solution, the moment and shear at any point are obtained from Eqs. 7-30, 7-45, and 7-47.

7-5 Some Solutions for the Laterally Loaded Thin Plate. As stated above, the solution to the laterally loaded plate with small deflections requires finding a function $w(x, y)$ which satisfies

$$\frac{\partial^4 w}{\partial x^4} + 2\frac{\partial^4 w}{\partial x^2 \partial y^2} + \frac{\partial^4 w}{\partial y^4} = \frac{q}{D} \tag{7-65}$$

and also the given boundary conditions.

The general solution to this problem has not been obtained. In fact, there are very few known exact solutions to this problem. Those that are known are either given in series form or they refer to an unusual mathematical loading (such as a sine form loading) or else have been obtained by what is essentially pure chance.

We shall discuss these types of solutions briefly by referring to specific problems.

PROBLEM 1

Given a rectangular plate, length of sides a and b, simply supported under a loading

$$q = A \sin\frac{m\pi x}{a} \sin\frac{n\pi y}{b} \tag{7-66}$$

Determine the deflection of this plate. The origin $x = y = o$ is at the corner of the plate.

Solution

Substituting this value of q in Eq. 7-65, we find

$$\nabla^4 w = \frac{A}{D} \sin\frac{m\pi x}{a} \sin\frac{n\pi y}{b} \qquad (7\text{-}67)$$

and the boundary conditions are

$$(1) \quad w = 0, \quad \frac{\partial^2 w}{\partial x^2} = 0 \quad \text{for} \quad x = 0, \quad x = a$$

$$(2) \quad w = 0, \quad \frac{\partial^2 w}{\partial y^2} = 0 \quad \text{for} \quad y = 0, \quad y = b$$

These boundary conditions are satisfied by assuming for w, a function

$$w = B \sin\frac{m\pi x}{a} \sin\frac{n\pi y}{b} \qquad (7\text{-}68)$$

If this is substituted in Eq. 7-67, we have

$$B\pi^4 \left(\frac{m^4}{a^4} + \frac{m^2 n^2}{a^2 b^2} + \frac{n^4}{b^4}\right) \sin\frac{m\pi x}{a} \sin\frac{n\pi y}{b} = \frac{A}{D} \sin\frac{m\pi x}{a} \sin\frac{n\pi y}{b} \quad (7\text{-}69)$$

from which we see that

$$B = \frac{A}{D}\frac{a^4 b^4}{\pi^4(m^2 b^2 + n^2 a^2)^2} \qquad (7\text{-}70)$$

Hence, the deflection at any point is given by

$$w = \frac{A a^4 b^4}{D\pi^4(m^2 b^2 + n^2 a^2)^2} \sin\frac{m\pi x}{a} \sin\frac{n\pi y}{b} \qquad (7\text{-}71)$$

Having w in this form, we can obtain the moment or shear at any point.

PROBLEM 2

Problem 1 corresponded to the case in which we have a double-sine type loading—not a loading usually met with in practice. A loading of more practical interest would be a uniform loading over the entire plate. To obtain a loading of this type we superpose loadings of the type of Eq. 7-66 to give a final net uniform load. This is done by means of a double Fourier series analysis as follows:

Solution

Assume the q loading is uniform and can be given by the double infinite series,

$$q = \sum_{m=1}^{\infty} \sum_{n=1}^{\infty} A_{mn} \sin\frac{m\pi x}{a} \sin\frac{n\pi y}{b} \qquad (7\text{-}72)$$

Then multiply both sides of Eq. 7-72 by $\sin(m'\pi x/a) \sin(n'\pi y/b)$ and integrate over the plate to give

$$
q \int_0^b \int_0^a \sin\frac{m'\pi x}{a} \sin\frac{n'\pi y}{b} \, dxdy
$$

$$
= \int_0^b \int_0^a \sum_{m=1}^\infty \sum_{n=1}^\infty A_{mn} \sin\frac{m\pi x}{a} \sin\frac{m'\pi x}{a} \sin\frac{n\pi y}{b} \sin\frac{n'\pi y}{b} \, dxdy
$$

(7-73)

or

$$
q \int_0^b \int_0^a \sin\frac{m'\pi x}{a} \sin\frac{n'\pi y}{b} \, dxdy = \frac{abA_{m'n'}}{4}
$$

(7-74)

as may easily be verified. From this,

$$
\begin{aligned}
A_{m'n'} &= \frac{4q}{ab} \int_0^b \int_0^a \sin\frac{m'\pi x}{a} \sin\frac{n'\pi y}{b} \, dxdy \\[2mm]
&= \begin{cases} \dfrac{16q}{m'n'\pi^2} & \text{if } m' \text{ and } n' \text{ are odd} \\[3mm] 0 & \text{if either } m' \text{ or } n' \text{ is even} \end{cases}
\end{aligned}
$$

(7-75)

so that

$$
q = \frac{16q}{\pi^2} \left\{ \begin{aligned} &\sin\frac{\pi x}{a}\left(\sin\frac{\pi y}{b} + \frac{1}{3}\sin\frac{3\pi y}{b} + \cdots\right) \\[2mm] &+ \frac{1}{3}\sin\frac{3\pi x}{a}\left(\sin\frac{\pi y}{b} + \frac{1}{3}\sin\frac{3\pi y}{b} + \cdots\right) \end{aligned} \right\}
$$

(7-76)

Now w, as given in Eq. 7-68, satisfies the required boundary conditions. Substituting w, Eq. 7-68, and q, Eq. 7-76, in Eq. 7-65 gives, finally,

$$
w = \frac{16a^4b^4q}{\pi^6 D} \left\{ \begin{aligned} &\frac{\sin\dfrac{\pi x}{a}\sin\dfrac{\pi y}{b}}{(a^2+b^2)^2} + \frac{1}{3}\frac{\sin\dfrac{\pi x}{a}\sin\dfrac{\pi y}{b}}{(3^2a^2+b^2)^2} + \cdots \\[4mm] &+ \frac{1}{3}\frac{\sin\dfrac{3\pi x}{a}\sin\dfrac{\pi y}{b}}{(a^2+3^2b^2)^2} + \cdots \end{aligned} \right\}
$$

(7-77)

from which the moments and shears can readily be obtained in series form. It must be emphasized that very important questions concerning convergence of the series used are here left unanswered. We merely showed that the assumed series does satisfy the formal manipulation processes used herein and hence the solution to the problem.

Also, it is interesting to note that the above solution is for the simply supported rectangular plate only. No such simple solution (not even in series form) exists for the clamped rectangular plate. That is, it has not as yet been possible to assume a simple function for the clamped plate. The clamped edge solutions which do exist are all lengthy numerical approximations. See in this connection Ref. (19).

Problem 3

We consider next a problem of physical interest for which an exact closed form solution exists—namely, the elliptic plate with clamped edges under a uniform load.

Solution

The equation of the boundary is

$$\frac{x^2}{a^2} + \frac{y^2}{b^2} = 1 \tag{7-78}$$

Let us assume

$$w = C\left(1 - \frac{x^2}{a^2} - \frac{y^2}{b^2}\right)^2 \tag{7-79}$$

then, substituting this in Eq. 7-65, we have

$$C\left(\frac{24}{a^4} + \frac{24}{b^4} + \frac{16}{a^2 b^2}\right) = \frac{q}{D} \tag{7-80}$$

so that q is constant and given by

$$q = DC\left(\frac{24}{a^4} + \frac{24}{b^4} + \frac{16}{a^2 b^2}\right) \tag{7-81}$$

Also, from Eq. 7-79 we see that along the boundary

$$\left.\begin{array}{c} w = 0 \\[2mm] \dfrac{\partial w}{\partial n} = 0 \end{array}\right\} \tag{7-82}$$

so that w as given in Eq. 7-79 represents the complete solution for the elliptic plate subject to the boundary conditions Eq. 7-82, i.e., built-in. Having w, the moment and shear at any point may be obtained as before.

We note again that the above solution was obtained by chance. It just happens that Eq. 7-79, which is a solution of the required differential equation, also satisfies boundary conditions of physical significance for this problem. Also, as previously pointed out, no such simple solution is known for the *simply supported* elliptic plate.

In concluding our discussion of small deflection theory, we mention that many plate problems not otherwise capable of solution may be solved approximately by the use of energy methods. Fundamentally, these methods make use of the fact that when a loaded plate is in equilibrium its potential energy has a stationary (minimum) value. This law is very useful in the general theory of structures; in plate and shell theory it very frequently is the only known method for obtaining even an approximate solution to a given problem. It is applied in the following manner:

The energy expressions for plates are known in terms of the deflections. It has been found that if deflections are assumed that satisfy the boundary conditions, then even though these are not the true deflections, the solutions for shears, stresses, moments, etc., obtained by making the potential energy a minimum will be quite close to the true values. We must emphasize that this has been shown by comparing certain approximate energy solutions with certain known exact solutions. That this will always be so is not at all evident. Indeed, it seems quite possible that in some cases so-called "pathological" deflection functions may be chosen (which satisfy the boundary conditions) that give values for the various functions that differ markedly from the exact or other approximate solutions. In addition, it may be worthwhile to point out that the approximate energy solution is one which satisfies the boundary conditions and the requirement for minimum potential energy only—it does not, in general, satisfy the equations of equilibrium or the equations of stress and strain compatibility.

Here is summarized the discussion of certain solutions to the thin plate equations:

(a) It was pointed out that in general there are four types of solutions available for the plate problem.

1. Exact closed-form solutions. As an example of this the solution was obtained for the clamped-edge elliptic plate.
2. Exact closed-form solutions with impractical or nonengineering-type loadings or boundary conditions. As an example of this, the simply supported rectangular plate loaded with sine-type loading was obtained.
3. Series type solutions. As an example of this, the simply supported rectangular plate under a uniform load was solved.
4. Approximate solutions based upon energy or similar methods.

In the next section we discuss an analogy between the linearized elasticity theory and the thin plate theory described herein. In order that the presentation be as coherent as possible we shall repeat, where necessary, different relations, equations, and statements that may have appeared in earlier sections of the text.

7-6   Two Analogies Between Elasticity Theory and Thin Plate Theory.[2]

It will first be shown that the equations governing the behavior of moment and torque loaded thin plates which have small deflections are analogous to the equations of the two-dimensional linearized mathematical theory of isotropic elasticity. Because of the analogy, all solutions of two-dimensional plane strain or plane stress problems may be taken as solutions of a particular thin plate problem, and vice versa.

Next a three-dimensional generalization of the above is developed. In this form, the theory draws an analogy between solutions in three-dimensional elasticity and those of thin plates subjected to lateral and shear loads as well as moments and torques.

The latter development is based on a tensoral generalization of the two-dimensional result to three dimensions. Because of this analogy all the known solutions for thin plates subjected to lateral loads represent *possible* solutions in a three-dimensional elastic structure, and vice versa.

Examples of both analogies will be given.

The Equations of the Two-Dimensional Linearized
Mathematical Theory of Isotropic Elasticity

The fundamental tensors of two-dimensional linearized elasticity are the stress tensor, T, given by

$$T = \begin{pmatrix} \sigma_x & \tau_{xy} \\ \tau_{yx} & \sigma_y \end{pmatrix} \tag{7-83}$$

and the strain tensor, η, given by

$$\eta = \begin{pmatrix} e_x & \tfrac{1}{2}\gamma_{xy} \\ \tfrac{1}{2}\gamma_{yx} & e_y \end{pmatrix} \tag{7-84}$$

Note that these tensors are symmetrical. Also they are tensors in the sense that they satisfy a particular rotation of axes law, which,

---

[2] See S. F. Borg, "Two Analogies in Solid Continuum Mechanics," *J. Franklin Inst.* Oct. 1961. H. Schaefer, "Die vollständige Analogie Scheibe-Platte" in the *Abhandlungen der Braunschweigischen Wissenschaftlichen Gesellschaft*, Braunschweig, 1956, Vol.VIII, develops an analogy similar to the two-dimensional one in this article but his is fundamentally different in that his analogous quantities differ from those of this discussion. See also R. V. Southwell, "On the Analogies Relating Flexure and Extension of Flat Plates," *Quart J. Mech. Appl. Math.*, Vol. 3, pp. 257-270 (1950), and Yi Yuan Yu, "On the Complex-Representation of the General Extensional and Flexural Problems of Thin Plates and Their Analogies," *J. Franklin Inst.*, Vol. 260, No. 4, Oct. 1955. See also Ref. (19) for another plate analogy.

typically (see Eq. 2-35), is

$$T' = RTR^\star \qquad (7\text{-}85)$$

The fundamental equations governing the behavior of a deformable body, in the two-dimensional linearized theory of elasticity, are

stress equilibrium, body forces neglected

or

$$\left.\begin{array}{c} \text{div } T = 0 \\[6pt] \left(\dfrac{\partial}{\partial x}\ \dfrac{\partial}{\partial y}\right)\begin{pmatrix} \sigma_x & \tau_{xy} \\ \tau_{yx} & \sigma_y \end{pmatrix} = 0 \end{array}\right\} \qquad (7\text{-}86)$$

boundary conditions on stresses

or

$$\left.\begin{array}{c} T\bar{N} = \bar{x} \\[6pt] \begin{pmatrix} \sigma_x & \tau_{xy} \\ \tau_{yx} & \sigma_y \end{pmatrix}\begin{pmatrix} l \\ m \end{pmatrix} = \begin{pmatrix} \bar{x} \\ \bar{y} \end{pmatrix} \end{array}\right\} \qquad (7\text{-}87)$$

in which (l, m) are the direction cosines of a normal to the boundary and $\bar{x}$, $\bar{y}$ are the boundary stress in the x, y directions.

Hooke's Law in its linearized isotropic form

$$\eta = \frac{1+\nu}{E}T - \frac{\nu}{E}\mathscr{I}_1\mathscr{E}_2 \qquad (7\text{-}88)$$

and a
strain compatibility equation, in its linearized form

$$\frac{\partial^2 e_x}{\partial y^2} + \frac{\partial^2 e_y}{\partial x^2} = \frac{\partial^2 \gamma_{xy}}{\partial x\,\partial y} \qquad (7\text{-}89)$$

In Eq. 7-88 ν is Poisson's ratio, E is the isotropic modulus of elasticity, $\mathscr{I}_1$ is the trace of the stress tensor, and $\mathscr{E}_2$ is the 2×2 unit matrix.

The Fundamental Equations of Small Deformation Thin Plate Theory

We may represent the loading, moments, and torques acting on a typical element of a thin plate as in Fig. 7.5. The moments, M_{xx} and M_{yy} and the torques, $M_{xy} = -M_{yx}$ are given per unit length of side. We may also define the curvatures, a typical value of which is

$$-\frac{1}{r_{xx}} = \frac{\partial}{\partial x}\left(\frac{\partial w}{\partial x}\right) = \frac{\partial^2 w}{\partial x^2} \qquad (7\text{-}90)$$

Then, as we saw, the following tensors occur in the theory of moment and torque loaded thin plates with small deflections:

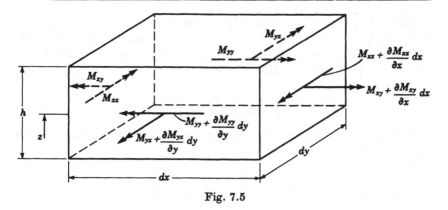

Fig. 7.5

moment tensor

$$\mathcal{M} = \begin{pmatrix} M_{xx} & -M_{xy} \\ M_{yx} & M_{yy} \end{pmatrix} \tag{7-91}$$

curvature tensor

$$\mathcal{R} = \begin{pmatrix} -\dfrac{1}{r_{xx}} & \dfrac{1}{r_{xy}} \\ \dfrac{1}{r_{yx}} & -\dfrac{1}{r_{yy}} \end{pmatrix}$$

$$= \begin{pmatrix} \dfrac{\partial}{\partial x}\left(\dfrac{\partial w}{\partial x}\right) & \dfrac{1}{2}\left[\dfrac{\partial}{\partial x}\left(\dfrac{\partial w}{\partial y}\right) + \dfrac{\partial}{\partial y}\left(\dfrac{\partial w}{\partial x}\right)\right] \\ \dfrac{1}{2}\left[\dfrac{\partial}{\partial y}\left(\dfrac{\partial w}{\partial x}\right) + \dfrac{\partial}{\partial x}\left(\dfrac{\partial w}{\partial y}\right)\right] & \dfrac{\partial}{\partial y}\left(\dfrac{\partial w}{\partial y}\right) \end{pmatrix} \tag{7-92}$$

We saw that the equations governing the action of thin, laterally loaded, small deflected plates are

equilibrium equations

or

$$\left. \begin{array}{c} \operatorname{div}\mathcal{M} = 0 \\ \left(\dfrac{\partial}{\partial x} \quad \dfrac{\partial}{\partial y}\right)\begin{pmatrix} M_{xx} & -M_{xy} \\ M_{yx} & M_{yy} \end{pmatrix} = 0 \end{array} \right\} \tag{7-93}$$

We may show also that the moments and torques satisfy

boundary conditions

or

$$\left. \begin{array}{c} \mathcal{M}\bar{N} = (\bar{M}) \\ \begin{pmatrix} M_{xx} & -M_{xy} \\ M_{yx} & M_{yy} \end{pmatrix}\begin{pmatrix} l \\ m \end{pmatrix} = \begin{pmatrix} \overline{M_{xx}} \\ \overline{M_{yy}} \end{pmatrix} \end{array} \right\} \tag{7-94}$$

in which $\overline{M_{xx}}$ and $\overline{M_{yy}}$ are the edge moments.

There is a *relation between curvatures and moments (Hooke's Law)*, since it may be shown that the following relation holds between R and $\mathscr{M}$ (from Eqs. 7-22 and 7-23):

$$\mathscr{R} = -\frac{1+\nu}{D(1-\nu^2)}\mathscr{M} + \frac{\nu}{D(1-\nu^2)}\mathscr{I}_1\mathscr{E}_2 \qquad (7\text{-}95)$$

In Eq. 7-95 $\mathscr{I}_1$ is the trace or first invariant of the moment tensor and D is the plate stiffness.

Finally, there is a

compatibility of curvature relation (as may be verified)

$$\frac{\partial^2\left(-\dfrac{1}{r_{xx}}\right)}{\partial y^2} + \frac{\partial^2\left(-\dfrac{1}{r_{yy}}\right)}{\partial x^2} = 2\frac{\partial^2\left(\dfrac{1}{r_{xy}}\right)}{\partial x\,\partial y} \qquad (7\text{-}96)$$

The Analogous Quantities and Relations for the Two-Dimensional Theory

In view of the foregoing, the following analogous quantities and relations may be set up for the two-dimensional isotropic elasticity—thin plate fields (the symbol $\sim$ means "is analogous to"):

the tensors

or

$$\left.\begin{aligned} T &\sim \frac{6\mathscr{M}}{h^2} \\[2mm] \text{Eq. 7-83} &\sim \frac{6}{h^2}\ \text{Eq. 7-91} \end{aligned}\right\} \qquad (7\text{-}97)$$

Also

or

$$\left.\begin{aligned} \eta &\sim -\frac{h}{2}\mathscr{R} \\[2mm] \text{Eq. 7-84} &\sim -\frac{h}{2}\ \text{Eq. 7-92} \end{aligned}\right\} \qquad (7\text{-}98)$$

equilibrium equations

or

$$\left.\begin{aligned} \operatorname{div} T = 0 &\sim \operatorname{div}\mathscr{M} = 0 \\ \text{Eq. 7-86} &\sim \text{Eq. 7-93} \end{aligned}\right\} \qquad (7\text{-}99)$$

boundary conditions

or

$$\left.\begin{aligned} T\bar{N} = \bar{x} &\sim \mathscr{M}\bar{N} = \bar{M} \\ \text{Eq. 7-87} &\sim \text{Eq. 7-94} \end{aligned}\right\} \qquad (7\text{-}100)$$

Hooke's Law

$$\text{Eq. 7-88} \sim \text{Eq. 7-95} \qquad (7\text{-}101)$$

compatibility condition

$$\text{Eq. 7-89} \sim \text{Eq. 7-96} \qquad (7\text{-}102)$$

We see, therefore, that there is a direct analogy between the two-dimensional equations of the linearized theory of isotropic two-dimensional elasticity and the governing equations for the thin, small deflection theory of moment and torque loaded plates.

Because of this analogy, it follows that solutions in the two-dimensional theory of isotropic elasticity may also be taken as analogous solutions in the theory of moment and torque loaded thin plates. In applying the analogy we compare a plate in the x-y plane (deflecting laterally in the z direction) to a two-dimensional deformable body loaded by stresses or forces in the x and y directions. In particular, the analogous quantities are (using the subscript E for elasticity solutions and P for plate solutions),

$$
\begin{array}{ccc}
\textit{Elasticity} & & \textit{Thin Plates} \\[1em]
\sigma_x & \sim & \dfrac{6M_{xx}}{h^2} \\[1em]
\sigma_y & \sim & \dfrac{6M_{yy}}{h^2} \\[1em]
\tau_{xy} = \tau_{yx} & \sim & -\dfrac{6M_{xy}}{h^2} = \dfrac{6M_{yx}}{h^2} \\[1em]
e_x & \sim & \dfrac{h}{2r_{xx}} \\[1em]
e_y & \sim & \dfrac{h}{2r_{yy}} \\[1em]
\tfrac{1}{2}\gamma_{xy} = \tfrac{1}{2}\gamma_{yx} & \sim & -\dfrac{h}{2r_{xy}} = -\dfrac{h}{2r_{yx}} \\[1em]
u_E & \sim & -\dfrac{h}{2}\left(\dfrac{\partial w}{\partial x}\right)_P \\[1em]
v_E & \sim & -\dfrac{h}{2}\left(\dfrac{\partial w}{\partial y}\right)_P
\end{array}
\qquad (7\text{-}103)
$$

ILLUSTRATIVE EXAMPLE

As an illustration of the application of the analogy, we consider the following problem in the theory of elasticity (see Ref. 13):

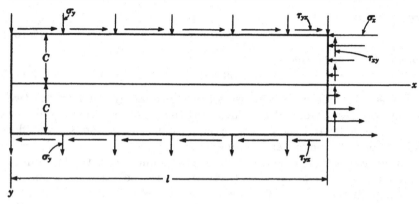

Fig. 7.6

The beam shown in Fig. 7.6, with stresses

$$\left. \begin{array}{c} \sigma_x = d_5(x^2y - \tfrac{2}{3}y^3) \\ \sigma_y = \tfrac{1}{3}d_5y^3 \\ \tau_{xy} = -d_5xy^2 \end{array} \right\} \qquad (7\text{-}104)$$

satisfies all the requirements of the two-dimensional linearized theory of elasticity. Along the top and bottom sides the normal stress σ_y is uniformly distributed. On the side $x = l$ the normal stresses follow a linear and cubic parabolic variation. The shear stress on the top and bottom is proportional to x and is parabolic on the face $x = l$.

Thus, the solution shown represents a thin plate solution, with

$$\frac{6M_{xx}}{h^2} \text{ in place of } \sigma_x$$

$$\frac{6M_{yy}}{h^2} \text{ in place of } \sigma_y$$

$$\text{and } -\frac{6M_{xy}}{h^2} \text{ in place of } \tau_{xy}$$

From Hooke's Law the elasticity strains may be obtained as

$$e_x = \frac{1}{E}(\sigma_x - \nu\sigma_y)$$

$$e_y = \frac{1}{E}(\sigma_y - \nu\sigma_x)$$

$$\gamma_{xy} = \frac{2(1+\nu)}{E}\tau_{xy}$$

and therefore the radii of curvature for the plate are known, using Eq. 7-103.

The Three-Dimensional Form of the Analogy

We consider next a thin plate having small deflections and subjected to transverse and shear loads in addition to the edge moments and torques; see Fig. 7.4, repeated again here.

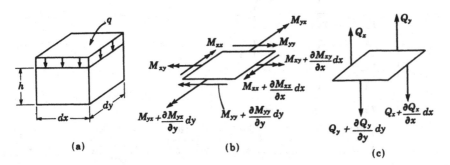

(a)   (b)   (c)

Then, as we saw in Eqs. 7-43, 7-45, and 7-47, the equilibrium equations for this structure are

$$\left.\begin{aligned}
\frac{\partial M_{xx}}{\partial x} + \frac{\partial M_{yx}}{\partial y} - Q_x &= 0 \\
-\frac{\partial M_{xy}}{\partial x} + \frac{\partial M_{yy}}{\partial y} - Q_y &= 0 \\
\frac{\partial Q_x}{\partial x} + \frac{\partial Q_x}{\partial y} + q &= 0
\end{aligned}\right\} \qquad (7\text{-}105)$$

which is also given by

$$\text{div}\,\mathcal{M} = 0$$

or

$$\left.\begin{pmatrix} \dfrac{\partial}{\partial x} & \dfrac{\partial}{\partial y} & \dfrac{\partial}{\partial z} \end{pmatrix} \begin{pmatrix} M_{xx} & -M_{xy} & -Q_x z \\ M_{yx} & M_{yy} & -Q_y z \\ -Q_x z & -Q_y z & -q\dfrac{z^2}{2} \end{pmatrix} = 0 \right\} \qquad (7\text{-}106)$$

In the above the symmetric tensor, $\mathcal{M}$, defined by

$$\mathcal{M} = \begin{pmatrix} M_{xx} & -M_{xy} & -Q_x z \\ M_{yx} & M_{yy} & -Q_y z \\ -Q_x z & -Q_y z & -\dfrac{q z^2}{2} \end{pmatrix} \qquad (7\text{-}107)$$

has been introduced. This tensor represents the three-dimensional generalization of the two-dimensional form given in Eq. 7-91 and includes the effect of transverse and shear loads.

We are therefore led to introduce a *generalized three-dimensional* form of $\mathscr{R}$, as given in Eq. 7-92. This becomes, in its three-dimensional form,

$$\mathscr{R} = \begin{pmatrix} -\dfrac{1}{r_{xx}} & \dfrac{1}{r_{xy}} & \dfrac{1}{r_{xz}} \\[2mm] \dfrac{1}{r_{yx}} & -\dfrac{1}{r_{yy}} & \dfrac{1}{r_{yz}} \\[2mm] \dfrac{1}{r_{zx}} & \dfrac{1}{r_{zy}} & -\dfrac{1}{r_{zz}} \end{pmatrix} =$$

$$\begin{pmatrix} \dfrac{\partial}{\partial x}\left(\dfrac{\partial w}{\partial x}\right) & \dfrac{1}{2}\left[\dfrac{\partial}{\partial x}\left(\dfrac{\partial w}{\partial y}\right)+\dfrac{\partial}{\partial y}\left(\dfrac{\partial w}{\partial x}\right)\right] & \dfrac{1}{2}\left[\dfrac{\partial}{\partial x}\left(\dfrac{\partial w}{\partial z}\right)+\dfrac{\partial}{\partial z}\left(\dfrac{\partial w}{\partial x}\right)\right] \\[3mm] \dfrac{1}{2}\left[\dfrac{\partial}{\partial x}\left(\dfrac{\partial w}{\partial y}\right)+\dfrac{\partial}{\partial y}\left(\dfrac{\partial w}{\partial x}\right)\right] & \dfrac{\partial}{\partial y}\left(\dfrac{\partial w}{\partial y}\right) & \dfrac{1}{2}\left[\dfrac{\partial}{\partial y}\left(\dfrac{\partial w}{\partial z}\right)+\dfrac{\partial}{\partial z}\left(\dfrac{\partial w}{\partial y}\right)\right] \\[3mm] \dfrac{1}{2}\left[\dfrac{\partial}{\partial x}\left(\dfrac{\partial w}{\partial z}\right)+\dfrac{\partial}{\partial z}\left(\dfrac{\partial w}{\partial x}\right)\right] & \dfrac{1}{2}\left[\dfrac{\partial}{\partial y}\left(\dfrac{\partial w}{\partial z}\right)+\dfrac{\partial}{\partial z}\left(\dfrac{\partial w}{\partial y}\right)\right] & \dfrac{\partial}{\partial z}\left(\dfrac{\partial w}{\partial z}\right) \end{pmatrix}$$

$$(7\text{-}108)$$

It should be noted that $\mathscr{M}$ and $\mathscr{R}$ as formulated above are tensors in the sense of Eq. 7-85.

Because of the generalized nature of these tensors one is led to postulate a three-dimensional Hooke's Law as a direct generalization of the two-dimensional case, Eq. 7-95. This is given at once by

$$\mathscr{R} = -\frac{1+\nu}{D(1-\nu^2)}\mathscr{M} + \frac{\nu}{D(1-\nu^2)}\mathscr{I}_1\mathscr{E}_3 \qquad (7\text{-}109)$$

In this equation, $\mathscr{R}$ and $\mathscr{M}$ are given by Eqs. 7-107 and 7-108. $\mathscr{I}_1$ is the trace of $\mathscr{M}$ and $\mathscr{E}_3$ is the 3×3 unit matrix.

The boundary or edge conditions on the moments, torque, and shears are

$$\mathscr{M}\bar{N} = \left(\frac{\bar{M}}{\overline{Q_{iz}}}\right)$$

or

$$\begin{pmatrix} M_{xx} & -M_{xy} & -Q_{x}z \\[2mm] M_{yx} & M_{yy} & -Q_{y}z \\[2mm] -Q_{x}z & -Q_{y}z & -\dfrac{qz^2}{2} \end{pmatrix} \begin{pmatrix} l \\[2mm] m \\[2mm] n \end{pmatrix} = \begin{pmatrix} \overline{M_{xx}} \\[2mm] \overline{M_{yy}} \\[2mm] \overline{Q_{iz}} \end{pmatrix} \qquad (7\text{-}110)$$

In Eq. 7-110, l, m and n are the direction cosines of a normal to the edge of the plate at the central plane $x = 0$, hence $n = 0$. The right-hand-side terms are the applied edge moments, and shear.

Also, the elements of the curvature tensor, $\mathscr{R}$, satisfy the following six compatibility conditions:

$$
\left.
\begin{array}{c}
\dfrac{\partial^2\left(-\dfrac{1}{r_{yy}}\right)}{\partial z^2} + \dfrac{\partial^2\left(-\dfrac{1}{r_{zz}}\right)}{\partial y^2} = 2\dfrac{\partial^2\left(\dfrac{1}{r_{yz}}\right)}{\partial y \partial z} \\[4ex]
\dfrac{\partial^2\left(-\dfrac{1}{r_{zz}}\right)}{\partial x^2} + \dfrac{\partial^2\left(-\dfrac{1}{r_{xx}}\right)}{\partial z^2} = 2\dfrac{\partial^2\left(\dfrac{1}{r_{zx}}\right)}{\partial z \partial x} \\[4ex]
\dfrac{\partial^2\left(-\dfrac{1}{r_{xx}}\right)}{\partial y^2} + \dfrac{\partial^2\left(-\dfrac{1}{r_{yy}}\right)}{\partial x^2} = 2\dfrac{\partial^2\left(\dfrac{1}{r_{xy}}\right)}{\partial x \partial y} \\[4ex]
\dfrac{\partial^2\left(-\dfrac{1}{r_{zz}}\right)}{\partial x \partial y} = \dfrac{\partial}{\partial z}\left[\dfrac{\partial\left(\dfrac{1}{r_{yz}}\right)}{\partial x} + \dfrac{\partial\left(\dfrac{1}{r_{zx}}\right)}{\partial y} - \dfrac{\partial\left(\dfrac{1}{r_{xy}}\right)}{\partial z}\right] \\[4ex]
\dfrac{\partial^2\left(-\dfrac{1}{r_{xx}}\right)}{\partial y \partial z} = \dfrac{\partial}{\partial x}\left[\dfrac{\partial\left(\dfrac{1}{r_{zx}}\right)}{\partial y} + \dfrac{\partial\left(\dfrac{1}{r_{xy}}\right)}{\partial z} - \dfrac{\partial\left(\dfrac{1}{r_{yz}}\right)}{\partial x}\right] \\[4ex]
\dfrac{\partial^2\left(-\dfrac{1}{r_{yy}}\right)}{\partial z \partial x} = \dfrac{\partial}{\partial y}\left[\dfrac{\partial\left(\dfrac{1}{r_{xy}}\right)}{\partial z} + \dfrac{\partial\left(\dfrac{1}{r_{yz}}\right)}{\partial x} - \dfrac{\partial\left(\dfrac{1}{r_{zx}}\right)}{\partial y}\right]
\end{array}
\right\} \quad (7\text{-}111)
$$

We now turn to the relations and equations of elasticity. In the three-dimensional theory of elasticity, the fundamental stress and strain tensors are,

$$
T = \begin{pmatrix} \sigma_x & \tau_{xy} & \tau_{xz} \\ \tau_{xy} & \sigma_y & \tau_{yz} \\ \tau_{zx} & \tau_{zy} & \sigma_z \end{pmatrix} \tag{7-112}
$$

and

$$
\eta = \begin{pmatrix} e_x & \tfrac{1}{2}\gamma_{xy} & \tfrac{1}{2}\gamma_{xz} \\ \tfrac{1}{2}\gamma_{yx} & e_y & \tfrac{1}{2}\gamma_{yz} \\ \tfrac{1}{2}\gamma_{zx} & \tfrac{1}{2}\gamma_{zy} & e_z \end{pmatrix}
$$

(Continued on p. 222)

$$
= \begin{pmatrix}
\dfrac{\partial u}{\partial x} & \dfrac{1}{2}\left(\dfrac{\partial u}{\partial y}+\dfrac{\partial v}{\partial x}\right) & \dfrac{1}{2}\left(\dfrac{\partial u}{\partial z}+\dfrac{\partial w}{\partial x}\right) \\[3mm]
\dfrac{1}{2}\left(\dfrac{\partial v}{\partial x}+\dfrac{\partial u}{\partial y}\right) & \dfrac{\partial v}{\partial y} & \dfrac{1}{2}\left(\dfrac{\partial v}{\partial z}+\dfrac{\partial w}{\partial y}\right) \\[3mm]
\dfrac{1}{2}\left(\dfrac{\partial w}{\partial x}+\dfrac{\partial w}{\partial z}\right) & \dfrac{1}{2}\left(\dfrac{\partial w}{\partial y}+\dfrac{\partial v}{\partial z}\right) & \dfrac{\partial w}{\partial z}
\end{pmatrix}
\tag{7-113}
$$

The equilibrium equations in three-dimensional elasticity are, for body forces reglected,

$$
\operatorname{div} T = 0
$$

or

$$
\left.
\begin{pmatrix} \dfrac{\partial}{\partial x} & \dfrac{\partial}{\partial y} & \dfrac{\partial}{\partial z} \end{pmatrix}
\begin{pmatrix}
\sigma_x & \tau_{xy} & \tau_{xz} \\
\tau_{yx} & \sigma_y & \tau_{yz} \\
\tau_{zx} & \tau_{zy} & \sigma_z
\end{pmatrix} = 0
\right\}
\tag{7-114}
$$

Hooke's Law, the relation between the strain and stress tensor, is

$$
\eta = \frac{1+\nu}{E}T - \frac{\nu}{E}\mathscr{I}_1\mathscr{E}_3
\tag{7-115}
$$

In Eq. 7-115, $\mathscr{I}_1$ is the trace of the stress tensor, and $\mathscr{E}_3$ is the 3×3 unit matrix.

The boundary conditions on stresses[3] are given by

$$
TN = \bar{x}
$$

or

$$
\left.
\begin{pmatrix}
\sigma_x & \tau_{xy} & \tau_{xz} \\
\tau_{yx} & \sigma_y & \tau_{yz} \\
\tau_{zx} & \tau_{zy} & \sigma_z
\end{pmatrix}
\begin{pmatrix} l \\ m \\ n \end{pmatrix}
=
\begin{pmatrix} \bar{x} \\ \bar{y} \\ \bar{z} \end{pmatrix}
\right\}
\tag{7-116}
$$

and finally, the strains must satisfy the following six compatibility conditions,

$$
\left.
\begin{aligned}
\frac{\partial^2 e_y}{\partial z^2}+\frac{\partial^2 e_z}{\partial y^2} &= \frac{\partial^2 \gamma_{yz}}{\partial y \partial z} \\[3mm]
\frac{\partial^2 e_z}{\partial x^2}+\frac{\partial^2 e_x}{\partial z^2} &= \frac{\partial^2 \gamma_{zx}}{\partial z \partial x} \\[3mm]
\frac{\partial^2 e_x}{\partial y^2}+\frac{\partial^2 e_y}{\partial x^2} &= \frac{\partial^2 \gamma_{xy}}{\partial x \partial y} \\[3mm]
\frac{2\partial^2 e_z}{\partial x \partial y} &= \frac{\partial}{\partial z}\left(\frac{\partial \gamma_{yz}}{\partial x}+\frac{\partial \gamma_{zx}}{\partial y}-\frac{\partial \gamma_{xy}}{\partial z}\right)
\end{aligned}
\right\}
\tag{7-117}
$$

[3] Although the boundary condition on stresses is given as $\bar{x} = NT$ in Art. 4-8, because the stress tensor is symmetric, this is equivalent to the expression given here.

$$\left.\begin{array}{l} \dfrac{2\partial^2 e_x}{\partial y \partial z} = \dfrac{\partial}{\partial x}\left(\dfrac{\partial \gamma_{zx}}{\partial y}+\dfrac{\partial \gamma_{xy}}{\partial z}-\dfrac{\partial \gamma_{yz}}{\partial x}\right) \\[4mm] \dfrac{2\partial^2 e_y}{\partial z \partial x} = \dfrac{\partial}{\partial y}\left(\dfrac{\partial \gamma_{xy}}{\partial z}+\dfrac{\partial \gamma_{yz}}{\partial x}-\dfrac{\partial \gamma_{zx}}{\partial y}\right) \end{array}\right\} \quad \text{(7-117)}_{\text{continued}}$$

By comparing the above, it becomes clear that an analogy may be drawn between the *postulated* thin plate relations, Eqs. 7-105–7-111, and the elasticity relations, Eqs. 7-112–7-117.

This means that solutions in plate theory *may* represent solutions in the three-dimensional elasticity theory and vice versa.

In Ref. (19) many different plate solutions are given. Each of these corresponds to a particular solution in a three-dimensional stresses structure *providing it satisfies all the relations* (Eqs. 7-105–7-111). The analogous quantities are the following;

$$\left.\begin{array}{ccc} Elasticity & & Plate \\[2mm] \sigma_x & \sim & \dfrac{6M_{xx}}{h^2} \\[4mm] \sigma_y & \sim & \dfrac{6M_{yy}}{h^2} \\[4mm] \tau_{xy}=\tau_{yx} & \sim & -\dfrac{6M_{xy}}{h^2}=\dfrac{6M_{yx}}{h^2} \\[4mm] \sigma_z & \sim & -\dfrac{3qz^2}{h^2} \\[4mm] \tau_{xz}=\tau_{zx} & \sim & -\dfrac{6Q_x z}{h^2} \\[4mm] \tau_{yz}=\tau_{zy} & \sim & -\dfrac{6Q_y z}{h^2} \\[4mm] u_E & \sim & -\dfrac{h}{2}\left(\dfrac{\partial w}{\partial x}\right)_P \\[4mm] v_E & \sim & -\dfrac{h}{2}\left(\dfrac{\partial w}{\partial y}\right)_P \\[4mm] w_E & \sim & -\dfrac{h}{2}\left(\dfrac{\partial w}{\partial z}\right)_P \end{array}\right\} \quad \text{(7-118}$$

The procedure may be summarized as follows:

1. Knowing $w(x, y)_P$, determine M_{xx}, M_{yy}, M_{xy}, and Q_x, Q_y, using the known plate relations. These are all functions of (x, y) only.

2. Using the values from (1), from Hooke's Law (Eq. 7-109) determine the $1/r$ terms. Note: these are now functions of (x, y, z).
3. Check the compatibility equations, Eq. 7-111. If these hold, then the analogous relations, Eq. 7-118, follow.

As an example, we saw (Eq. 7-79) that the exact solution for a uniformly loaded elliptical plate with built-in edges is given by

$$w(x, y) = C\left(1 - \frac{x^2}{a^2} - \frac{y^2}{b^2}\right)^2 \tag{7-119}$$

in which C is a constant and a and b are the major and minor elliptic radii respectively. Using the known plate relations, we obtain from Eq. 7-109 the various elements of $\mathscr{R}$, which, when substituted in Eq. 7-111, give six equations. The third, fourth, fifth, and sixth of these are satisfied identically. The first and second will be satisfied if $a = b$ (i.e., for a circle) and if $\nu = 1$.

Using these, we find the analogous stresses are given by

and

$$\left.\begin{aligned} \sigma_x &\sim \frac{6M_{xx}}{h^2} \\[2mm] \sigma_z &\sim -\frac{3qz^2}{h^2} \\[2mm] \tau_{xz} &\sim -\frac{6Q_xz}{h^2}, \quad \text{etc.} \end{aligned}\right\} \tag{7-120}$$

Then, using Eq. 7-115, we can obtain the strains, and from the strains we obtain the deformations u, v, and w—or these could be obtained using Eq. 7-118.[4]

7-7 Laterally Loaded Plate with Large Deflections.

The problem of the laterally loaded plate with large deflections is of some interest in connection with thin metal construction of airplanes and similar structures. By a "large deflection" in plate theory, we mean that the ratio w_0/h is in the neighborhood of two or ten or twenty. (w_0 is the plate center deflection and h is the plate thickness.) For w_0/h in this range, the ordinary small deflection plate theory cannot be

---

[4] The value $\nu = 1$ is physically unrealistic and introduces infinities (as in Eq. 7-22, for example), which further restricts the range of applicability in this example. However, the technique described above represents the procedure to be used in applying the analogy. It is clear that not all solutions of plate theory have analogous three-dimensional elasticity solutions—and the reverse is true as well. Whether or not a particular laterally loaded thin plate solution corresponds to a physically realistic three-dimensional elasticity solution (or vice versa) must be determined for each individual case by checking the procedure given above.

used. This theory does not apply, because we may have stresses and strains in the central plane and also because the higher-order strain terms of the strain tensor (see Chapter 4) may be important—both of which effects are neglected in the small deflection theory.

As might be expected, the large deflection theory becomes quite complicated. The equations are nonlinear, and there is no known closed form exact solution to them for any practical problem, though several series and numerical solutions have been obtained.

The fundamental equations were derived about 60 years ago by Kármán, Ref. (24). In the last 25 years, the problem has been studied by various American and German engineers. A fairly complete bibliography on the subject will be found in Ref. (25). In this brief descriptive discussion, we merely point out that the numerical solutions have been obtained up to about $w_0/h = 3$ for clamped and free edge rectangular plates under uniform loads.

Timoshenko, Ref. (19), describes several and mentions other methods of solution to this problem and Prescott, Ref. (20), discusses an approximate method of solution which he recommends. Wang, Ref. (25), describes a numerical finite–difference relaxation solution, but the labor involved in this procedure appears to prohibit wide application of this method.

Merely for the purpose of rounding out this discussion, we list the fundamental equations for the large deflection theory and describe briefly the methods for solving them.

The equations are (see Eqs. 10-36 and 10-37)

$$\frac{\partial^4 F}{\partial x^4}+\frac{2\partial^4 F}{\partial x^2 \partial y^2}+\frac{\partial^4 F}{\partial y^4}=E\left[\left(\frac{\partial^2 w}{\partial x \partial y}\right)^2-\frac{\partial^2 w}{\partial x^2}\frac{\partial^2 w}{\partial y^2}\right] \quad \text{(a)}$$

$$\frac{\partial^4 w}{\partial x^4}+\frac{2\partial^4 w}{\partial x^2 \partial y^2}+\frac{\partial^4 w}{\partial y^4}=\frac{q}{D}+\frac{h}{D}\left(\frac{\partial^2 F}{\partial x^2}\frac{\partial^2 w}{\partial y^2}+\frac{\partial^2 F}{\partial y^2}\frac{\partial^2 w}{\partial x^2}-\frac{2\partial^2 F}{\partial x \partial y}\frac{\partial^2 w}{\partial x \partial y}\right) \quad \text{(b)}$$

$$(7\text{-}121)$$

The first of these is essentially a compatibility of strain relation (see Ref. (19)), and the second is a balance of forces and effects lateral to the plate. In the above, all terms except F are as defined before, and F (a stress function) is defined by the median plane stress expressions—

$$\sigma'_x = \frac{\partial^2 F}{\partial y^2}$$

$$\sigma'_y = \frac{\partial^2 F}{\partial x^2} \qquad (7\text{-}122)$$

$$\tau'_{xy} = -\frac{\partial^2 F}{\partial x \partial y}$$

The solution of the problem requires finding w and F that satisfy Eqs. 7-121 and 7-122 and also the boundary conditions. Levy, Ref. (26), solved the problem by assuming q, w, and F as double Fourier series. Way, Ref. (27), solved the problem by an approximate energy method which utilized assumed functions, u, v, and w satisfying the boundary conditions. He then set the variation of energy equal to zero and solved for the various undetermined deflection term parameters. Hencky, Ref. (28), Kaiser, Ref. (29), and also Wang, Ref. (25), solved the equations in the form of difference equations. In conclusion, it may be noted that all three methods described above appear to check each other fairly well in the region where comparisons may be made.

7-8 Another Nonlinear Large Deflection Thin Plate Theory.[5]

In Art. 7-7 the Kármán large deflection theory was briefly discussed. It was noted that the Kármán relations assume essentially that the large deflection causes membrane stresses in addition to bending stresses, and that the membrane stresses add terms to the small-deflection equation of equilibrium-of-forces in the lateral direction. In addition, a first-order large deflection compatibility condition is utilized so that two nonlinear equations in terms of transverse deflection and a stress function are obtained.

We shall now describe a large deflection theory of a fundamentally different type. It will be assumed, essentially, that the large deflection is one which occurs *without the development of membrane stresses*. That is, a large deflection form of the bending deformation relation is obtained. The nonlinearity occurs in the second derivative of the deflection terms and arises as a consequence of the analogous behavior of a thin plate to that of an elastic body in general, as described in Art. 7-6. In a sense, therefore, the present analysis bears the same relation to linear, thin plate theory that the elastica (column) solution bears to the Euler column theory.

It should be noted that in the theory to be presented the equations are given in terms of curvilinear coordinates (a, b) which define the deformed shape of the neutral (unstressed or inextensional) plane. These correspond to the Lagrangian coordinates of the analogous large deflection deformation elasticity theory.

In Art. 7-6 it was shown (see Eq. 7-98) that an analogy may be made between the curvature tensor $\mathcal{R}$ of thin plate theory and the strain tensor η of elasticity theory, the analogy being

$$\eta \sim -\frac{h}{2}\mathcal{R} \tag{7-123}$$

[5] See the paper by the author "A specialized non-linear thin plate inextensional bending theory," *Journal of the Franklin Institute*, Aug. 1962. Also S. F. Borg, Kurt Hoppe, Gerald Kopchinski, "Extensional thin plates with large deflections," Rep. No. 23, Civil Engineering Department, Stevens Institute of Technology, Hoboken, N.J., Oct. 3, 1963.

In elasticity (see Art. 4.2) the strain tensor is obtained as follows:

We consider an element da, initially at (a, b, c) which in its deformed state becomes dx at (x, y, z) where

$$x = a+u \qquad y = b+v \qquad z = c+w \qquad (7\text{-}124)$$

Thus, u, v, and w are the *deformations*. Then, a definition of the "state of strain" is the difference in squared length of the final, deformed element from the initial, undeformed element (see Eq. 4-10) or (using $\mathscr{E}_3$ for the unit matrix).

$$\text{state of strain} = da{\star}(J{\star}J - \mathscr{E}_3)\,da \qquad (7\text{-}125)$$

and the strain tensor is defined as

$$\eta = \frac{J{\star}J - \mathscr{E}_3}{2} \qquad (7\text{-}126)$$

η is a nonlinear tensor and is shown in its expanded form in Eq. 4-17.

Now, as noted above, in elasticity theory the deformation of the elastic body is represented by movements of particles in the body—these being (at any point) movements (u, v, w) in the (x, y, z) directions.

In thin plate theory, conditions are fundamentally different in that the elastic body is a *thin plate* that deforms by *assumption* in accordance with the Bernoulli–Euler Theory—namely, a plane section of the plate before bending is assumed to be plane after bending. In view of these basic restrictions on the form of the structure, the deformations of the thin plate in the x and y directions are fully characterized by net movements of planes perpendicular to the x and y axes. That is, we may uniquely define at any point in the plate the deformation in the x direction as the quantity $(h/2)\,(\partial w/\partial a)$ (see Fig. 7.7). Note particularly how, in analogy to elasticity deformation and strain theory, it

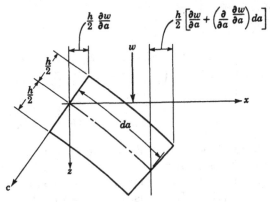

Fig. 7.7

follows that a measure of the *unit strain* in the x direction is just given by

$$\frac{h}{2}\left[\frac{\partial}{\partial a}\left(\frac{\partial w}{\partial a}\right)\right] \tag{7-127}$$

and, in a similar way, we may show the direct analogy and physical relation between all elements of the strain tensor η and the curvature tensor $\mathscr{R}$, in their 2×2 (i.e., x, y) forms.

Therefore the quantities $[(h/2)(\partial w/\partial a), (h/2)(\partial w/\partial b)]$ play the same role in thin plate theory that (u, v) play in general elasticity theory.

In other words, a "state of strain" expression for the thin plate can be developed just as was done for the elasticity problem, by noting that $[(h/2)(\partial w/\partial a), (h/2)(\partial w/\partial b)]$ is analogous to (u, v).

With this fundamental assumption we can proceed in an identical manner in thin plate theory in deriving a state-of-strain equation as was done in elasticity theory. If this is done we obtain, finally,

$$\text{state of strain} = da^\star(\mathscr{J}^\star\mathscr{J} - \mathscr{E}_2)\,da \tag{7-128}$$

where $\mathscr{J}$ and $\mathscr{J}^\star$ are similar to J and $J^\star$ except that $(h/2)(\partial w/\partial a)$ is used instead of u and $(h/2)(\partial w/\partial b)$ is used instead of v.

Then, as was done in elasticity we define the *curvature* tensor by

$$\frac{h}{2}\mathscr{R} = \frac{\mathscr{J}^\star\mathscr{J} - \mathscr{E}_2}{2} \tag{7-129}$$

which in its expanded, *nonlinear* form is given by

$$\frac{h}{2}\mathscr{R} = \frac{h}{2}\begin{pmatrix} \dfrac{\partial^2 w}{\partial a^2} & \dfrac{\partial^2 w}{\partial a\partial b} \\[2ex] \dfrac{\partial^2 w}{\partial b\partial a} & \dfrac{\partial^2 w}{\partial b^2} \end{pmatrix}$$

$$+\frac{h^2}{8}\begin{pmatrix} \left(\dfrac{\partial^2 w}{\partial a^2}\right)^2 + \left(\dfrac{\partial^2 w}{\partial a\partial b}\right)^2 & \dfrac{\partial^2 w}{\partial b^2}\dfrac{\partial^2 w}{\partial a\partial b} + \dfrac{\partial^2 w}{\partial a^2}\dfrac{\partial^2 w}{\partial a\partial b} \\[2ex] \dfrac{\partial^2 w}{\partial a^2}\dfrac{\partial^2 w}{\partial a\partial b} + \dfrac{\partial^2 w}{\partial a\partial b}\dfrac{\partial^2 w}{\partial b^2} & \left(\dfrac{\partial^2 w}{\partial b^2}\right)^2 + \left(\dfrac{\partial^2 w}{\partial a\partial b}\right)^2 \end{pmatrix} \tag{7-130}$$

In view of the above and utilizing the postulated analogous three-dimensional forms as shown in Art. 7-6, one may derive a *nonlinear thin plate theory* entirely analogous to the Murnaghan nonlinear elasticity theory (Ref. (16)). If this is done it is found that the governing relations and equations of this nonlinear plate theory are the following:

(a) The non-linear curvature tensor—

$$\mathscr{R} = \begin{pmatrix} \dfrac{\partial^2 w}{\partial a^2} & \dfrac{\partial^2 w}{\partial a \partial b} & \dfrac{\partial^2 w}{\partial a \partial c} \\[2mm] \dfrac{\partial^2 w}{\partial b \partial a} & \dfrac{\partial^2 w}{\partial b^2} & \dfrac{\partial^2 w}{\partial b \partial c} \\[2mm] \dfrac{\partial^2 w}{\partial c \partial a} & \dfrac{\partial^2 w}{\partial c \partial b} & \dfrac{\partial^2 w}{\partial c^2} \end{pmatrix}$$

$$+ \frac{1}{2} \left(\begin{array}{cc} \left(\dfrac{\partial^2 w}{\partial a^2}\right)^2 + \left(\dfrac{\partial^2 w}{\partial a \partial b}\right)^2 + \left(\dfrac{\partial^2 w}{\partial a \partial c}\right)^2 & \dfrac{\partial^2 w}{\partial a \partial a}\dfrac{\partial^2 w}{\partial a \partial b} + \dfrac{\partial^2 w}{\partial a \partial b}\dfrac{\partial^2 w}{\partial b \partial b} + \dfrac{\partial^2 w}{\partial a \partial c}\dfrac{\partial^2 w}{\partial b \partial c} \\[3mm] \left(\dfrac{\partial^2 w}{\partial a \partial a}\right)\dfrac{\partial^2 w}{\partial a \partial b} + \dfrac{\partial^2 w}{\partial a \partial b}\dfrac{\partial^2 w}{\partial b \partial b} + \dfrac{\partial^2 w}{\partial a \partial c}\dfrac{\partial^2 w}{\partial b \partial c} & \left(\dfrac{\partial^2 w}{\partial b^2}\right)^2 + \left(\dfrac{\partial^2 w}{\partial a \partial b}\right)^2 + \left(\dfrac{\partial^2 w}{\partial b \partial c}\right)^2 \\[3mm] \dfrac{\partial^2 w}{\partial a \partial a}\dfrac{\partial^2 w}{\partial a \partial c} + \dfrac{\partial^2 w}{\partial a \partial b}\dfrac{\partial^2 w}{\partial b \partial c} + \dfrac{\partial^2 w}{\partial a \partial c}\dfrac{\partial^2 w}{\partial c \partial c} & \dfrac{\partial^2 w}{\partial b \partial c}\dfrac{\partial^2 w}{\partial c \partial a} + \dfrac{\partial^2 w}{\partial b \partial b}\dfrac{\partial^2 w}{\partial c \partial b} + \dfrac{\partial^2 w}{\partial b \partial c}\dfrac{\partial^2 w}{\partial c \partial c} \end{array} \right.$$

$$\left. \begin{array}{c} \dfrac{\partial^2 w}{\partial a \partial a}\dfrac{\partial^2 w}{\partial a \partial c} + \dfrac{\partial^2 w}{\partial a \partial b}\dfrac{\partial^2 w}{\partial b \partial c} + \dfrac{\partial^2 w}{\partial a \partial c}\dfrac{\partial^2 w}{\partial c^2} \\[3mm] \dfrac{\partial^2 w}{\partial b \partial c}\dfrac{\partial^2 w}{\partial c \partial a} + \dfrac{\partial^2 w}{\partial b \partial b}\dfrac{\partial^2 w}{\partial c \partial b} + \dfrac{\partial^2 w}{\partial b \partial c}\dfrac{\partial^2 w}{\partial c \partial c} \\[3mm] \left(\dfrac{\partial^2 w}{\partial a \partial c}\right)^2 + \left(\dfrac{\partial^2 w}{\partial b \partial c}\right)^2 + \left(\dfrac{\partial^2 w}{\partial c^2}\right)^2 \end{array} \right) \qquad (7\text{-}131)$$

(b) The moment tensor—

$$\mathscr{M} = \begin{pmatrix} M_{aa} & -M_{ab} & -Q_a c \\[2mm] M_{ba} & M_{bb} & -Q_b c \\[2mm] -Q_a c & -Q_b c & -\dfrac{qc^2}{2} \end{pmatrix} \qquad (7\text{-}132)$$

(c) The equilibrium equations—

$$\operatorname{div} \mathscr{M} = 0$$

or

$$\left(\dfrac{\partial}{\partial a} \quad \dfrac{\partial}{\partial b} \quad \dfrac{\partial}{\partial c} \right) \begin{pmatrix} M_{aa} & -M_{ab} & -Q_a c \\[2mm] M_{ba} & M_{bb} & -Q_b c \\[2mm] -Q_a c & -Q_b c & -\dfrac{qc^2}{2} \end{pmatrix} = 0 \qquad (7\text{-}133)$$

(d) Boundary conditions on moments, torques, and shears—

$$\mathcal{M}\bar{N} = \left(\overline{\frac{M_{tt}}{Q_t c}}\right) \tag{7-134}$$

or

$$\begin{pmatrix} M_{aa} & -M_{ab} & -Q_a c \\ M_{ba} & M_{bb} & -Q_b c \\ -Q_a c & -Q_b c & -\dfrac{qc^2}{2} \end{pmatrix} \begin{pmatrix} l \\ m \\ n \end{pmatrix} = \begin{pmatrix} \overline{M_{aa}} \\ \overline{M_{bb}} \\ -\overline{(Q_t c)} \end{pmatrix} \tag{7-135}$$

(e) Hooke's Law, the *nonlinear* relation between the moment tensor and the curvature tensor (see Ref. (16)) for an incompressible material—

$$\mathcal{M} = \mathcal{J}[(\lambda_1 \mathcal{I}_1 \mathcal{E}_3 + 2\mu_1 \mathcal{R}) + (l_1 \mathcal{I}_1{}^2 - 2m_1 \mathcal{I}_2)\mathcal{E}_3 \\ + 2m_1 \mathcal{I}_1 \mathcal{R} + n_1 \text{co}\,\mathcal{R}]\mathcal{J}^\star \tag{7-136}$$

λ_1, μ_1, l_1, m_1, and n_1 are constants.

$\mathcal{I}_1$ and $\mathcal{I}_2$ are the invariants of the $\mathcal{R}$ tensor.

(f) A set of six nonlinear curvature compatibility conditions of which one, typically, is given by (see Eq. 4-140)

$$\frac{\partial^2\left(-\dfrac{1}{r_{aa}}\right)}{\partial b^2} + \frac{\partial^2\left(-\dfrac{1}{r_{bb}}\right)}{\partial a^2}$$

$$= \frac{2\partial^2\left(\dfrac{1}{r_{ab}}\right)}{\partial a \partial b} + \left[-\frac{\partial\left(-\dfrac{1}{r_{aa}}\right)}{\partial a}\frac{\partial\left(-\dfrac{1}{r_{bb}}\right)}{\partial a} - \frac{\partial\left(-\dfrac{1}{r_{aa}}\right)}{\partial b}\frac{\partial\left(-\dfrac{1}{r_{bb}}\right)}{\partial b} \right.$$

$$- \frac{2\partial\left(-\dfrac{1}{r_{aa}}\right)}{\partial b}\frac{\partial\left(-\dfrac{1}{r_{aa}}\right)}{\partial b} - \frac{2\partial\left(-\dfrac{1}{r_{bb}}\right)}{\partial a}\frac{\partial\left(-\dfrac{1}{r_{bb}}\right)}{\partial a}$$

$$\left. + \frac{2\partial\left(-\dfrac{1}{r_{aa}}\right)}{\partial a}\frac{\partial\left(\dfrac{1}{r_{ab}}\right)}{\partial b} + \frac{2\partial\left(\dfrac{1}{r_{ab}}\right)}{\partial a}\frac{\partial\left(-\dfrac{1}{r_{bb}}\right)}{\partial b} \right] \tag{7-137}$$

In addition to defining a postulated nonlinear plate theory,[6] the above

[6] The inextensional bending theory of thin plates has been studied by various investigators, the analysis generally being restricted to plates that deform to a developable surface. See Ref. (19). See also E. H. Mansfield and P. W. Kleeman, "A large deflection theory for thin plates: a theory based on the assumption of an inextensional middle surface of the plate," *Airc. Engrg.* 27, 314, April 1955; D. G. Ashwell, "The equilibrium equations of the inextensional theory for thin flat plates", *Quart. J. Mech. Appl. Math.* 10, 2, May 1957, and E. H. Mansfield, "The inextensional theory for thin flat plates," *Quart. J. Mech. Appl. Math.* 8, 3, 1955.

equations are analogous to those of the Murnaghan three-dimensional nonlinear theory of elasticity, Ref. (16). Hence, solutions of the non-linear elasticity equations also correspond to *possible* solutions in the postulated nonlinear thin plate theory, and vice versa. It must be emphasized that the theory and relations presented in this article are postulated ones only. No solution to these are given and, in fact, there is no assurance at this point that solutions of physical interest exist.

7-9 Summary. The fundamental equations for small deflection thin plate theory were derived using matrix arguments, and several simple solutions of these were given. Then, an analogy, based on a tensoral generalization, was developed between thin plate theory and elasticity theory. Finally, a brief discussion of Kármán's large deflection theory was presented, and another large deflection theory was derived, based upon the tensoral analogy between the nonlinear elasticity equations and a postulated nonlinear thin plate theory.

Problems

1. One form of the basic beam equation is

$$EI\frac{d^2}{dx^2}\left(\frac{d^2w}{dx^2}\right) = q$$

in which EI = stiffness
w = deflection at any point x
q = transverse loading per unit length

An analogous equation occurs in plate theory. Obtain this equation by substituting the plate tensors for the corresponding linear terms given above, and by noting that q/unit area in the plate equation is an invariant, so that the left-hand side also must contain only invariants.

2. An alternate form of the beam equation is

$$q = \frac{d^2M}{dx^2}$$

in which M is the bending moment at any point x. Obtain the analogous plate equation using the reasoning discussed in Prob. 1.

3. The strain energy stored in a beam is given by

$$\eta = \frac{EI}{2}\int_{E.L.}\left(\frac{d^2w}{dx^2}\right)^2 dx$$

Obtain the corresponding plate energy term by using the tensor generalization and the invariant reasoning described in Prob. 1.

4. The strain energy stored in a beam may also be given by

$$\eta = \frac{1}{2EI}\int_{E.L.} M^2\, dx$$

Obtain the corresponding plate energy using the argument of Prob. 1.

5. The equation for the laterally loaded thin plate can be written in the form

$$\nabla^2(\nabla^2 w) = \frac{q}{D}$$

in which ∇^2 is the Laplacian operator.

(a) Using the polar coordinate form of the Laplacian as given in Chapter 3, obtain the general form of this equation in terms of r, θ.

(b) For circular symmetry (i.e., equation independent of θ) show that this becomes

$$\frac{1}{r}\frac{d}{dr}\left(r\frac{d}{dr}\right)\left[\frac{1}{r}\frac{d}{dr}\left(r\frac{dw}{dr}\right)\right] = \frac{q}{D}$$

(c) For $q = $ constant, show that the general solution of this equation is

$$Dw = \frac{qr^4}{64} + C_1 r^2(\log r - 1) + C_2 r^2 + C_3 \log r + C_4$$

(d) Using the curvilinear coordinate transformations of Chapter 3, obtain

$$M_{rr} = -D\left(\frac{\partial^2 w}{\partial r^2} + \frac{\nu}{r}\frac{\partial w}{\partial r}\right)$$

$$M_{\theta\theta} = -D\left(\frac{1}{r}\frac{\partial w}{\partial r} + \nu\frac{\partial^2 w}{\partial r^2}\right)$$

(e) For a plate of radius a and thickness h without a central hole, show that in the equation above $C_1 = C_3 = 0$. If the plate is simply supported at its outer edge, set up its boundary conditions in term of w and M_{rr}, and prove that

$$Dw = \frac{1}{64}qr^2\left[r^2 - \frac{2(3+\nu)}{1+\nu}a^2\right] + \frac{qa^4}{64}\left(\frac{5+\nu}{1+\nu}\right)$$

(f) Show that the maximum bending stress (see Chapter 5) is therefore given by

$$\sigma_{\max} = \frac{Mc}{I}$$

$$= \frac{3a^2}{8h^2}q(3+\nu)$$

Chapter 8

THE EQUATIONS OF VISCOUS FLOW AND INTRODUCTION TO BOUNDARY LAYER THEORY

8-1 Introduction. A *viscous fluid* is one in which viscous (shearing) forces may appear. This is different from the so-called *perfect fluid* in which shear stresses are assumed to be absent. In this chapter the equations governing the flow of any viscous, non-heat-conducting fluid will be derived. Obviously, this will degenerate to the perfect fluid equations if the terms pertaining to viscous fluids are permitted to approach zero. It will be found that for *incompressible* viscous fluids there will, in general, be four unknown quantities that must be determined at every point in order that the flow field be completely known. These are the three components of the fluid velocity and the pressure. This, in turn, requires that we have four independent equations governing the flow of the fluid. These four equations are

(a) One equation which is a statement of constancy of mass, the so-called continuity equation.

(b) A statement of Newton's Law, which in vector form is

$$\bar{F} = m\bar{a} \tag{8-1}$$

or which, upon equating components of the vectors, becomes three equations—

$$\left. \begin{array}{l} F_x = ma_x \\ F_y = ma_y \\ F_z = ma_z \end{array} \right\} \tag{8-2}$$

These equations of Newton's Law for the specialized viscous flow case are called the Navier–Stokes Equation. It was derived independently by Navier (1827) and Stokes (1845) although Poisson and St. Venant had some influence upon its final derivation.[1]

8-2 The Continuity Equation. To derive the equation of continuity, we consider a cartesian frame as shown in Fig. 8.1.

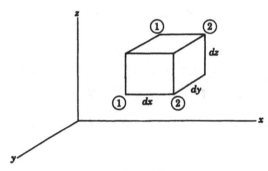

Fig. 8.1

Denoting the velocity components at any point by $V = (u\ v\ w)$, we consider the balance of mass flow which goes into the volume occupied by the differential element dx, dy, dz shown.

Consider the face ①–①. There is a mass of fluid equal to $(\rho u)\ dydz$ which flows into this face in each second. The density of the fluid is denoted by ρ.

Similarly, at face ②–② there is an amount of fluid $[\rho + (\partial \rho / \partial x)\ dx]$ $[u + (\partial u / \partial x)\ dx]\ dydz$ which flows out, and the net gain (for the x direction) in the mass of the volume per second is given by

Amount flowing in less amount flowing out

$$\left.\begin{aligned}
&= \rho u\ dydz - \left(\rho + \frac{\partial \rho}{\partial x}dx\right)\left(u + \frac{\partial u}{\partial x}dx\right)\ dydz \\
&= -\left(u\frac{\partial \rho}{\partial x} + \rho\frac{\partial u}{\partial x}\right)\ dxdydz
\end{aligned}\right\} \tag{8-3}$$

Similarly, for the other two pairs of faces we find a net gain in mass, per second, given by

$$-\left(v\frac{\partial \rho}{\partial y} + \rho\frac{\partial v}{\partial y}\right)\ dxdydz \tag{8-4}$$

---

[1] It must not be inferred that viscosity and heat-conduction are the only modifying effects to otherwise "perfect fluids." Some other typical factors which must be considered in special cases are magnetic effects, chemical reaction effects, nonequilibrium conditions, ionization and disassociation. Generally, the simplest problems occur in perfect fluid theory, and the inclusion of each added factor adds considerably to the complexity of the problem.

and

$$-\left(w\frac{\partial \rho}{\partial z}+\rho\frac{\partial w}{\partial z}\right)dxdydz \tag{8-5}$$

or the total net gain in *density* (mass per unit volume) per second due to the fluid velocity is given by

$$-\left[\frac{\partial(\rho u)}{\partial x}+\frac{\partial(\rho v)}{\partial y}+\frac{\partial(\rho w)}{\partial z}\right] \tag{8-6}$$

This must be balanced by a gain in the density of the fluid in the element. In the interval of time $dt = 1$ sec, the density of the fluid will increase by

$$\frac{\partial \rho}{\partial t}dt = \frac{\partial \rho}{\partial t} \tag{8-7}$$

Equating Eqs. 8-6 and 8-7, we obtain the equation of continuity in unsteady flow:

$$\frac{\partial \rho}{\partial t}+\frac{\partial(\rho u)}{\partial x}+\frac{\partial(\rho v)}{\partial y}+\frac{(\partial \rho w)}{\partial z} = 0 \tag{8-8}$$

In vector notation this becomes

$$\frac{\partial \rho}{\partial t}+\nabla \cdot (\rho \vec{V}) = 0 \tag{8-9}$$

Equations 8-8 and 8-9 hold for all compressible fluids. If the fluid is *incompressible*, then ρ does not vary and the equation of continuity takes the simplified form,

or
$$\left.\begin{array}{l} \nabla \cdot \vec{V} = 0 \\[2mm] \dfrac{\partial u}{\partial x}+\dfrac{\partial v}{\partial y}+\dfrac{\partial w}{\partial z} = 0 \end{array}\right\} \tag{8-10}$$

which is valid even for a time-dependent flow.

8-3 The Navier–Stokes (N.S.) Equations. There are several methods of deriving the Navier–Stokes Equations. Fundamentally, they all assume, in addition to the usual assumptions of homogeneity and isotropy, the following (we list the assumptions first and then discuss them):

1. Laminar flow.

2. A linear relation between stress and time rate of change of deformation of the fluid. That is, it is assumed that Newton's Law for one-dimensional laminar flow,

$$\tau = \mu\frac{\partial u}{\partial y} \tag{8-11}$$

is valid and may be extended to laminar flow for the general shear case,

$$\tau_{xy} = \mu\left(\frac{\partial u}{\partial y} + \frac{\partial v}{\partial x}\right) \tag{8-12}$$

In Eqs. 8-11 and 8-12, τ is the shear stress and μ is the viscosity.

3. A definition of pressure as the mean value of the negative sum of the principal diagonal elements of the stress tensor: i.e., if p is the pressure,

$$p = -\tfrac{1}{3}(\sigma_x + \sigma_y + \sigma_z) \tag{8-13}$$

Discussion of the Assumptions

1. *Laminar flow.* In connection with this assumption, it must be pointed out that in this chapter we are considering *only* laminar flow, as against the more complicated (and, in nature, more common) turbulent flow. We shall not go into a lengthy discussion of the difference between laminar and turbulent flow. It will suffice to state that whether a flow is laminar or turbulent depends upon the magnitude of the Reynolds number—above a certain value for this number the flow is turbulent.[2] The equations governing laminar flows take a much simpler form and the solution of the laminar flow problem is much simpler than those for turbulent flow. The equations which we shall derive and all solutions which we shall give hold only for laminar flows.

2. *Linear relation between stress and time rate of change of deformation.* Newton's Law of Friction was derived by him in an improper manner. However, its theoretical justification may be proved by the kinetic theory of gases and it has been verified experimentally. As given by Newton, it takes the form

$$\tau = \mu\frac{\partial u}{\partial y} \tag{8-14}$$

The extension of this law to the general case, i.e., the assumption that it takes the more general form

$$\tau_{xy} = \mu\left(\frac{\partial u}{\partial y} + \frac{\partial v}{\partial x}\right) \tag{8-15}$$

means in effect that the equations of solid elasticity for linear variation of stress and strain may be extended to the case of fluids with

$$\mu \text{ corresponding to } G$$

and with *time rates* of unit strain corresponding to unit strain.

---

[2] See Ref. (48) for a descriptive discussion of turbulent flows. In this text we use the symbol N_R to represent Reynolds number.

This last is so because (u, v, w) for the fluids are velocities (or time rates of deformation) whereas for solids they are deformations.

Fluids which behave in accordance with the Newton assumption are called "Newtonian fluids." Many fluids do in fact behave in this manner, among them water, mercury, petroleum, some molten resins and other so-called "simple" fluids.

Many more fluids, however, do not behave in accordance with Newton's assumed linear relation between shear stress and rate of deformation. These are called "non-Newtonian" fluids and their behavior is studied in the broad field called *rheology*, the science of flow. Rheologists state, with some justification, that "everything flows," hence their field encompasses the study of everything. However, in the usual scientific manner, we can assume with sufficient accuracy a simplified behavior for many materials under actual conditions and solutions obtained for these cases, in fact, agree well with the observed phenomena. Thus we can consider (using the assumption of a Newtonian fluid) problems in viscous flow as in this chapter and we can consider problems in plasticity theory as in Chapter 9. When these methods do not account for observed phenomena we must then resort to a more accurate (and generally more complicated) rheology theory.

3. *Pressure definition.* In connection with the third assumption we note the following facts: It is known that the pressure at a point in a fluid is the same in all directions at the point.[3] In other words, the pressure is independent of direction (or an invariant). Also, the physical dimension of a pressure is a stress.

Now, in a perfect fluid (shear stresses absent) or in a viscous fluid at rest, the stress tensor becomes, for all axes,

$$\begin{pmatrix} \sigma_x & 0 & 0 \\ 0 & \sigma_y & 0 \\ 0 & 0 & \sigma_z \end{pmatrix} \tag{8-16}$$

In other words, regardless of the axial orientation, it follows that (using the known invariants of this tensor)

$$\left. \begin{array}{l} \sigma_x + \sigma_y + \sigma_z = \sigma'_x + \sigma'_y + \sigma'_z = K_1 \\ \sigma_y\sigma_z + \sigma_z\sigma_x + \sigma_x\sigma_y = \sigma'_y\sigma'_z + \sigma'_z\sigma'_x + \sigma'_x\sigma'_y = K_2 \\ \sigma_x\sigma_y\sigma_z = \sigma'_x\sigma'_y\sigma'_z = K_3 \end{array} \right\} \tag{8-17}$$

and it may be shown (see Prob. 6 at end of this chapter) that these

---

[3] This is essentially a postulate, a fundamental postulate of hydrostatics and hydrodynamics. It was introduced by Euler and, based upon this simple assumption, the entire theory of fluid mechanics as we know it today has been developed. See in this connection Ref. (47).

equations are satisfied only for

$$\sigma_x = \sigma_y = \sigma_z = \sigma'_x = \sigma'_y = \sigma'_z = K \qquad (8\text{-}18)$$

Thus, for the perfect fluid or for the viscous fluid at rest, we have for the stress tensor

$$\begin{pmatrix} \sigma & 0 & 0 \\ 0 & \sigma & 0 \\ 0 & 0 & \sigma \end{pmatrix} \qquad (8\text{-}19)$$

For this case the physical requirement that the dimension of the pressure be that of a stress, when combined with the requirement of invariance can be satisfied by taking

$$p = -\tfrac{1}{3}(\sigma_x + \sigma_y + \sigma_z) = -\sigma_x = -\sigma_y = -\sigma_z = -\sigma \qquad (8\text{-}20)$$

This definition is carried over into the general theory of viscous fluids. That is, pressure is defined as

$$p = -\tfrac{1}{3}(\sigma_x + \sigma_y + \sigma_z) \qquad (8\text{-}21)$$

and this gives for p a quantity that (1) satisfies dimensional requirements, (2) satisfies requirements concerning invariance, and (3) enables us to proceed with the mathematical development of the relations between stresses and strains for a viscous fluid.

We now proceed with the derivation of the N.S. equations.

It will be recalled from the chapter on theory of elasticity that there exists a deformation tensor given by (see footnote p. 91)

$$\begin{pmatrix} \dfrac{\partial u}{\partial x} & \dfrac{\partial u}{\partial y} & \dfrac{\partial u}{\partial z} \\[2mm] \dfrac{\partial v}{\partial x} & \dfrac{\partial v}{\partial y} & \dfrac{\partial v}{\partial z} \\[2mm] \dfrac{\partial w}{\partial x} & \dfrac{\partial w}{\partial y} & \dfrac{\partial w}{\partial z} \end{pmatrix} \qquad (8\text{-}22)$$

in which $(u\ v\ w)$ are components of a deformation.

It may be shown in exactly the same way that a *rate of deformation* tensor occurs in fluid theory, this tensor being given by the same expression as above, except that $(u\ v\ w)$ are now components of a *velocity*. Then, recalling that if A is any tensor (see Eq. 1-31),

$$A = \frac{A + A^\star}{2} + \frac{A - A^\star}{2} \qquad (8\text{-}23)$$

where the first term on the right is symmetric and the second term is skew-symmetric, we find that the rate of deformation tensor can be decomposed into two additional tensors,

$$\Phi = \begin{pmatrix} \dfrac{\partial u}{\partial x} & \dfrac{1}{2}\left(\dfrac{\partial u}{\partial y}+\dfrac{\partial v}{\partial x}\right) & \dfrac{1}{2}\left(\dfrac{\partial u}{\partial z}+\dfrac{\partial w}{\partial x}\right) \\[2ex] \dfrac{1}{2}\left(\dfrac{\partial v}{\partial x}+\dfrac{\partial u}{\partial y}\right) & \dfrac{\partial v}{\partial y} & \dfrac{1}{2}\left(\dfrac{\partial w}{\partial y}+\dfrac{\partial v}{\partial z}\right) \\[2ex] \dfrac{1}{2}\left(\dfrac{\partial w}{\partial x}+\dfrac{\partial u}{\partial z}\right) & \dfrac{1}{2}\left(\dfrac{\partial w}{\partial y}+\dfrac{\partial v}{\partial z}\right) & \dfrac{\partial w}{\partial z} \end{pmatrix} \tag{8-24}$$

which is a symmetric tensor, and

$$\Omega = \begin{pmatrix} 0 & \dfrac{1}{2}\left(\dfrac{\partial u}{\partial y}-\dfrac{\partial v}{\partial x}\right) & \dfrac{1}{2}\left(\dfrac{\partial u}{\partial z}-\dfrac{\partial w}{\partial x}\right) \\[2ex] \dfrac{1}{2}\left(\dfrac{\partial v}{\partial x}-\dfrac{\partial u}{\partial y}\right) & 0 & \dfrac{1}{2}\left(\dfrac{\partial v}{\partial z}-\dfrac{\partial w}{\partial y}\right) \\[2ex] \dfrac{1}{2}\left(\dfrac{\partial w}{\partial x}-\dfrac{\partial u}{\partial z}\right) & \dfrac{1}{2}\left(\dfrac{\partial w}{\partial y}-\dfrac{\partial v}{\partial z}\right) & 0 \end{pmatrix} \tag{8-25}$$

which is a skew-symmetric or antisymmetric tensor. The rate of strain tensor is Φ and the rotation tensor is Ω.

We have for the general viscous fluid

$$\sigma_x + \sigma_y + \sigma_z = -3p \tag{8-26}$$

If the fluid is nonviscous or if the viscous fluid were at rest, then, as we saw,

$$\sigma_x = \sigma_y = \sigma_z = -p \tag{8-27}$$

Therefore, it is logical to assume that it is the *excess* of these stresses over $-p$ which must somehow be connected with the rate of strain properties of the fluid. Also, for small strains, we assume the relation is one which may be approximated by a tangent or, in effect, we assume that this excess is proportional to the rates of pure strain $\partial u/\partial x$, $\partial v/\partial y$, $\partial w/\partial z$, in a linear relation. This assumption is analogous to the one found to hold in solid elasticity for small strains.

We assume, therefore, in view of isotropy and homogeneity,

$$\left. \begin{aligned} \sigma_x-(-p) &= A\frac{\partial u}{\partial x}+B\left(\frac{\partial v}{\partial y}+\frac{\partial w}{\partial z}\right) \\[1.5ex] \sigma_y-(-p) &= A\frac{\partial v}{\partial y}+B\left(\frac{\partial w}{\partial z}+\frac{\partial u}{\partial x}\right) \\[1.5ex] \sigma_z-(-p) &= A\frac{\partial w}{\partial z}+B\left(\frac{\partial u}{\partial x}+\frac{\partial v}{\partial y}\right) \end{aligned} \right\} \tag{8-28}$$

which may be put in the following equivalent form (λ and μ are constants):

$$
\left.
\begin{aligned}
\sigma_x &= -p + \lambda\left(\frac{\partial u}{\partial x} + \frac{\partial v}{\partial y} + \frac{\partial w}{\partial z}\right) + 2\mu\frac{\partial u}{\partial x} \\[4pt]
\sigma_y &= -p + \lambda\left(\frac{\partial u}{\partial x} + \frac{\partial v}{\partial y} + \frac{\partial w}{\partial z}\right) + 2\mu\frac{\partial v}{\partial y} \\[4pt]
\sigma_z &= -p + \lambda\left(\frac{\partial u}{\partial x} + \frac{\partial v}{\partial y} + \frac{\partial w}{\partial z}\right) + 2\mu\frac{\partial w}{\partial z}
\end{aligned}
\right\}
\tag{8-29}
$$

It will be seen later that this assumed form of this equation is also consistent with Newton's Law

$$
\tau_{xy} = \mu\left(\frac{\partial u}{\partial y} + \frac{\partial v}{\partial x}\right)
\tag{8-30}
$$

Add the three equations of Eq. 8-29 to give

$$
-3p = -3p + (3\lambda + 2\mu)\left[\left(\frac{\partial u}{\partial x} + \frac{\partial v}{\partial y} + \frac{\partial w}{\partial z}\right)\right]
\tag{8-31}
$$

and because the term in the brackets is not, in general, equal to zero,

$$
\lambda = -\tfrac{2}{3}\mu
\tag{8-32}
$$

Therefore, from the first of Eq. 8-29,

$$
\sigma_x = -p - \tfrac{2}{3}\mu\left(\frac{\partial u}{\partial x} + \frac{\partial v}{\partial y} + \frac{\partial w}{\partial z}\right) + 2\mu\frac{\partial u}{\partial x}
\tag{8-33}
$$

or, in vector notation, since

$$
\nabla \cdot V = \operatorname{div} V = \frac{\partial u}{\partial x} + \frac{\partial v}{\partial y} + \frac{\partial w}{\partial z}
\tag{8-34}
$$

the expression for σ_x becomes (note that σ_y and σ_z are given by similar equations)

$$
\sigma_x = -p - \tfrac{2}{3}\mu\nabla \cdot V + 2\mu\frac{\partial u}{\partial x}
\tag{8-35}
$$

Now, the stress tensor is a symmetric tensor, and the rate of strain tensor (Eq. 8-24) is a symmetric tensor. Therefore, they may be put in diagonal form, and by inspection of Eq. 8-35 we may assume that these will occur in diagonal form for the same set of axes. The tensor expression for the relation between stress and rate of strain for a viscous fluid, in diagonal form, becomes, therefore,

$$\begin{pmatrix} \sigma_x & 0 & 0 \\ 0 & \sigma_y & 0 \\ 0 & 0 & \sigma_z \end{pmatrix} = -(p + \tfrac{2}{3}\mu\nabla \cdot \vec{V})E_3 + 2\mu \begin{pmatrix} \dfrac{\partial u}{\partial x} & 0 & 0 \\ 0 & \dfrac{\partial v}{\partial y} & 0 \\ 0 & 0 & \dfrac{\partial w}{\partial z} \end{pmatrix} \qquad (8\text{-}36)$$

If Ψ = stress tensor, and Φ = rate of strain tensor, the general tensor equation which holds with respect to *any* set of cartesian axes becomes

$$\Psi = -(p + \tfrac{2}{3}\mu\nabla \cdot \vec{V})E_3 + 2\mu\Phi \qquad (8\text{-}37)$$

which represents nine equations, of which six are independent, three of these being the expressions for the generalized Newton Friction Law:

$$\tau_{xy} = \mu\left(\frac{\partial u}{\partial y} + \frac{\partial v}{\partial x}\right) \qquad (8\text{-}38)$$

plus two other similar expressions.

Let us consider next a differential element of cube, Fig. 8.2, in which, for purposes of clarity, only the x components of the stresses are shown

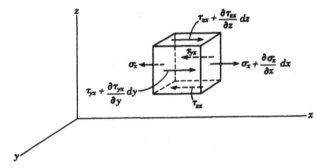

Fig. 8.2

acting. The y and z components are similar and lead to an equation similar to the following and obtained from the following by cyclical interchange of the terms. Summing the forces shown in Fig. 8.2 in the x direction, we have

$$\text{Force} = \frac{\partial \sigma_x}{\partial x}\,dxdydz + \frac{\partial \tau_{yx}}{\partial y}\,dxdydz + \frac{\partial \tau_{zx}}{\partial z}\,dxdydz \qquad (8\text{-}39)$$

or *per unit volume*, the *total* force acting on the element is

$$
\begin{aligned}
\boldsymbol{F} &= (F_x \quad F_y \quad F_z) \\
\boldsymbol{F} &= \left(\frac{\partial}{\partial x} \quad \frac{\partial}{\partial y} \quad \frac{\partial}{\partial z}\right)\begin{pmatrix} \sigma_x & \tau_{xy} & \tau_{xz} \\ \tau_{yx} & \sigma_y & \tau_{yz} \\ \tau_{zx} & \tau_{zy} & \sigma_z \end{pmatrix} \\
&= \operatorname{div} \Psi
\end{aligned}
\right\} \tag{8-40}
$$

Now

$$
\sigma_x = -p - \tfrac{2}{3}\mu \nabla \cdot \boldsymbol{V} + 2\mu\frac{\partial u}{\partial x} \tag{8-41}
$$

so that

$$
\frac{\partial \sigma_x}{\partial x} = -\frac{\partial p}{\partial x} - \tfrac{2}{3}\mu\frac{\partial}{\partial x}(\nabla \cdot \boldsymbol{V}) + 2\mu\frac{\partial^2 u}{\partial x^2} \tag{8-42}
$$

Also

$$
\tau_{xy} = \tau_{yx} = \mu\left(\frac{\partial v}{\partial x} + \frac{\partial u}{\partial y}\right) \tag{8-43}
$$

so that

$$
\frac{\partial \tau_{yx}}{\partial y} = \mu\left(\frac{\partial^2 v}{\partial x \partial y} + \frac{\partial^2 u}{\partial y^2}\right) \tag{8-44}
$$

and

$$
\tau_{zx} = \mu\left(\frac{\partial w}{\partial x} + \frac{\partial u}{\partial z}\right) \tag{8-45}
$$

so that

$$
\frac{\partial \tau_{zx}}{\partial z} = \mu\left(\frac{\partial^2 w}{\partial x \partial z} + \frac{\partial^2 u}{\partial z^2}\right) \tag{8-46}
$$

and therefore

$$
F_x = -\frac{\partial p}{\partial x} - \tfrac{2}{3}\mu\frac{\partial}{\partial x}(\nabla \cdot \boldsymbol{V}) + \mu\nabla^2 u + \mu\frac{\partial}{\partial x}(\nabla \cdot \boldsymbol{V}) \tag{8-47}
$$

Then the general three-dimensional force $\boldsymbol{F}$ per unit volume is given by

$$
\boldsymbol{F} = -\nabla p + \frac{\mu}{3}\nabla(\nabla \cdot \boldsymbol{V}) + \mu\nabla^2 \boldsymbol{V} \tag{8-48}
$$

and if we include a body force per unit volume equal to $\rho\bar{g}$, and apply Newton's Law,

$$\bar{F} = m\bar{a}$$

to the differential fluid element, we find

$$-\nabla p + \frac{\mu}{3}\nabla(\nabla \cdot \bar{V}) + \mu\nabla^2\bar{V} + \rho\bar{g} = \rho\frac{D\bar{V}}{Dt} \qquad (8\text{-}49)$$

in which $D\bar{V}/Dt$ is a total derivative, the so-called Stokes derivative of velocity, i.e., a derivative which includes spatial and time partial derivatives. Noting that $\mu/\rho = \nu$, the kinematic viscosity, the equation takes the final form, the Navier–Stokes Equation,

$$\frac{D\bar{V}}{Dt} = \bar{g} - \frac{1}{\rho}\nabla p + \nu\nabla^2\bar{V} + \frac{\nu}{3}\nabla(\nabla \cdot \bar{V}) \qquad (8\text{-}50)$$

which is three equations, the first of which (x component) in cartesian form becomes

$$\frac{\partial u}{\partial t} + u\frac{\partial u}{\partial x} + v\frac{\partial u}{\partial y} + w\frac{\partial u}{\partial z} = -g_x - \frac{1}{\rho}\frac{\partial p}{\partial x} + \nu\left(\frac{\partial^2 u}{\partial x^2} + \frac{\partial^2 u}{\partial y^2} + \frac{\partial^2 u}{\partial z^2}\right)$$
$$+ \frac{\nu}{3}\frac{\partial}{\partial x}\left(\frac{\partial u}{\partial x} + \frac{\partial v}{\partial y} + \frac{\partial w}{\partial z}\right) \qquad (8\text{-}51)$$

In the indicial subscript tensor notation (see Art. 2-6), the three Navier–Stokes equations become

$$\frac{\partial u_i}{\partial t} + u_k\frac{\partial u_i}{\partial x_k} = g_i - \frac{1}{\rho}\frac{\partial p}{\partial x_i} + \nu\frac{\partial^2 u_i}{\partial x_k \partial x_k} + \frac{\nu}{3}\frac{\partial}{\partial x_i}\left(\frac{\partial u_j}{\partial x_j}\right) \qquad (8\text{-}52)$$

in which u_i represents the velocity components (u v w).

If the fluid is incompressible, div $\bar{V} = 0$, and the above equations are simplified somewhat.

The following is a summary of the chapter to this point:

The equations governing the viscous laminar flow of a liquid are, for the *incompressible* case, four in number. These equations will be shown below in two forms, first the vector form, and secondly the tensor form, using the summation convention.

(a) The equation of continuity, which is a statement of the conservation of mass

or

$$\left.\begin{array}{l} \text{div } \bar{V} = 0 \\[2mm] \dfrac{\partial u_i}{\partial x_i} = 0 \end{array}\right\} \qquad (8\text{-}53)$$

(b) The Navier–Stokes Equation, which is a statement of Newton's Law of motion

or

$$\left.\begin{aligned}
\frac{D\vec{V}}{Dt} &= \vec{g} - \frac{1}{\rho}\nabla p + \nu\nabla^2\vec{V} \\[2ex]
\frac{\partial u_i}{\partial t} + u_k\frac{\partial u_i}{\partial x_k} &= g_i - \frac{1}{\rho}\frac{\partial p}{\partial x_i} + \nu\frac{\partial^2 u_i}{\partial x_k \partial x_k}
\end{aligned}\right\} \tag{8-54}$$

which is three equations, one for each component x, y, and z. Equations 8-53 and 8-54 are four equations in terms of four unknowns, these being u, v, w and p.

The starting point for all problems in incompressible, viscous, non-heat-conducting, laminar flow is with Eq. 8-53 and 8-54. See Ref. (21) for a detailed presentation of the subject of viscous flows.

8-4 Perfect Fluid Theory. We now present the equations which govern the motion of nonviscous, non-heat-conducting fluids—the so-called "perfect fluids." These equations are obtained directly from those derived for the viscous fluid by simply setting the viscous effect equal to zero whenever it appears. Thus for the perfect fluid we have

$$\text{Continuity,} \quad \text{div }\vec{V} = 0 \quad \text{or} \quad \frac{\partial u_i}{\partial x_i} = 0 \tag{8-55}$$

and the statement of Newton's Law, which for this case is frequently called Euler's equation,

or

$$\left.\begin{aligned}
\frac{D\vec{V}}{Dt} &= \vec{g} - \frac{1}{\rho}\nabla p \\[2ex]
\frac{\partial u_i}{\partial t} + u_k\frac{\partial u_i}{\partial x_k} &= g_i - \frac{1}{\rho}\frac{\partial p}{\partial x_i}
\end{aligned}\right\} \tag{8-56}$$

Although there are no perfectly nonviscous (and hence no "perfect") fluids in nature, many fluids do have very small viscosity so that the nonviscous fluid theory may be assumed to apply in a major portion of the field of flow under consideration. See Ref. (22) for a detailed discussion of many of the topics which form "perfect fluid theory." See Arts. 8-7 to 8-12 in this chapter for a brief discussion of boundary layers, a topic which requires an analysis of both perfect fluid and viscous fluid effects.

8-5 A Simple Application of the Navier–Stokes Equation. Consider as a first example, the steady laminar flow in the x direction of an incompressible fluid between two parallel walls, infinite in length, Fig. 8.3. We shall neglect gravity.

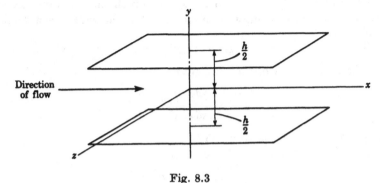

Fig. 8.3

For this case $\partial/\partial t = 0$ and $v = 0$, hence $\partial v/\partial y = 0$, and from Eq. 8-53 it therefore follows that $\partial u/\partial x = 0$, or u is independent of x.

The first N.S. equation, Eq. 8-51, becomes

$$\frac{\partial p}{\partial x} = \mu \frac{d^2 u}{dy^2} \tag{8-57}$$

and from the second one, $\partial p/\partial y = 0$. Since p is independent of y, it follows that $\partial p/\partial x = (dp/dx)$ is independent of y [4] and therefore we can integrate Eq. 8-57 directly twice to obtain

$$u = \frac{1}{2\mu}\left(\frac{dp}{dx}\right)y^2 + Ay + B \tag{8-58}$$

which represents the general solution to the given problem. Now we must introduce the boundary conditions, which will permit the determination of the constants A and B. (The pressure dp/dx is assumed given.) The boundary conditions introduce a most interesting complication. Because there are two constants to be determined, we look for two boundary conditions, and from the form of Eq. 8-58 it is obvious that these must be statements concerning the value of u at the walls. Offhand, one would say that the problem is indeterminate in that we may have any value whatever at the wall. However, the difficulty is resolved if we introduce the so-called "no-slip" condition; that is, the requirement that the fluid particles just adjacent to the wall must, due to viscous effects, move with the velocity of the wall. This no-slip

[4] In fact, since $\partial u/\partial x = 0$, or u is independent of x, it follows that $\partial p/\partial x$ must be independent of x or must be an absolute constant.

condition is invariably assumed in the ordinary theory of laminar viscous flow. Only under very unusual circumstances is this condition relaxed, and then the problem is usually much more complicated.

In the present problem, we assume first that the walls are stationary. Then

$$(u)_{h/2} = (u)_{-(h/2)} = 0 \tag{8-59}$$

and substituting in Eq. 8-58, we find

$$u = \frac{1}{2\mu}\left(\frac{dp}{dx}\right)\left(y^2 - \frac{h^2}{4}\right) \tag{8-60}$$

which is a parabolic distribution of u.

$(u)_{max}$ occurs at $y = 0$ and is given by

$$(u)_{max} = -\frac{1}{2\mu}\left(\frac{dp}{dx}\right)\frac{h^2}{4} \tag{8-61}$$

The shearing stress is given by

$$\tau = \mu\frac{du}{dy} = \left(\frac{dp}{dx}\right)y \tag{8-62}$$

and $(\tau)_{max}$, which occurs at the wall, is given by

$$(\tau)_{max} = 4\mu\frac{(u)_{max}}{h} \tag{8-63}$$

We define a skin friction coefficient C_f by

$$C_f = \frac{|\tau_{max}|}{\frac{1}{2}\rho u^2_{max}} \tag{8-64}$$

Therefore

$$C_f = \frac{8\mu}{h\rho u_{max}} = \frac{8}{N_R} \tag{8-65}$$

where N_R is the Reynolds number based on u_{max} and h, and the important fact results that $C_f \sim 1/N_R$. It will be found in all cases in which the solution requires a balance of viscous and pressure forces only (inertia forces neglected), that $C_f \sim 1/N_R$.

The flow considered in this article is "plane Poiseuille flow."

Problems connected with movement of the walls are given at the end of this chapter.

8-6 Viscous Flow Equations in Cylindrical and Spherical Coordinates.

In many problems, the boundary conditions are such that the differential equations in rectangular form are all but unsolvable.

In general, if the boundaries are straight lines, a rectangular cartesian form is preferred. If, on the other hand, the boundary is circular, spherical, elliptical, or other, curvilinear coordinates will simplify the solution of the problem.

The equations for the viscous flow problem will be put in cylindrical and spherical form using the methods discussed in Chapter 3, Curvilinear Coordinates.

The equations of viscous compressible flow are, in vector form,

Continuity
$$\frac{\partial \rho}{\partial t} + \nabla \cdot (\rho \vec{V}) = 0 \qquad (8\text{-}66)$$

N.S. equations
$$\frac{D\vec{V}}{Dt} = \vec{g} - \frac{1}{\rho}\nabla p + \nu \nabla^2 \vec{V} + \frac{\nu}{3}\nabla(\nabla \cdot \vec{V}) \qquad (8\text{-}67)$$

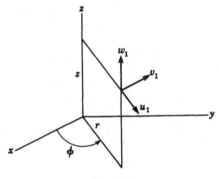

Fig. 8.4

In cylindrical coordinates, (r, ϕ, z), Fig. 8.4, we have (see Art. 3-5) $(h_1\ h_2\ h_3) = (1 \quad 1/r \quad 1)$. Note that we use r instead of ρ.

It was shown that in curvilinear coordinates (see Art. 3-7),

$$\text{div } \vec{V} = h_1 h_2 h_3 \left[\left(\frac{u_1}{h_2 h_3}\right)_r + \left(\frac{v_1}{h_3 h_1}\right)_\phi + \left(\frac{w_1}{h_1 h_2}\right)_z \right]$$

or
$$\text{div}(\rho \vec{V}) = (\rho u_1)_r + \frac{1}{r}(\rho u_1) + \frac{1}{r}(\rho v_1)_\phi + (\rho w_1)_z \qquad (8\text{-}68)$$

where the subscripts represent differentiation.

From this it follows that the continuity equation becomes, in cylindrical coordinates,

$$\frac{\partial \rho}{\partial t} + \frac{\partial(\rho u_1)}{\partial r} + \frac{1}{r}(\rho u_1) + \frac{1}{r}\frac{\partial(\rho v_1)}{\partial \phi} + \frac{\partial(\rho w_1)}{\partial z} = 0 \qquad (8\text{-}69)$$

To obtain the N.S. equation in cylindrical coordinates, we note that the various terms of Eq. 8-67 may be analyzed as follows:

Consider first

$$\frac{D\vec{V}}{Dt} = \frac{\partial \vec{V}}{\partial t} + \vec{V} \cdot \nabla \vec{V} \tag{8-70}$$

The first term on the right becomes, in terms of the cylindrical components,

$$\left(\frac{\partial u_1}{\partial t} \quad \frac{\partial v_1}{\partial t} \quad \frac{\partial w_1}{\partial t} \right) \tag{8-71}$$

The second term is the product (this may easily be verified by expanding the expressions)

$$(u \ v \ w) \begin{pmatrix} \dfrac{\partial u}{\partial x} & \dfrac{\partial v}{\partial x} & \dfrac{\partial w}{\partial x} \\[2mm] \dfrac{\partial u}{\partial y} & \dfrac{\partial v}{\partial y} & \dfrac{\partial w}{\partial y} \\[2mm] \dfrac{\partial u}{\partial z} & \dfrac{\partial v}{\partial z} & \dfrac{\partial w}{\partial z} \end{pmatrix} \tag{8-72}$$

which becomes, in cylindrical coordinates (see Eq. 3-79),

$$(u_1 \ v_1 \ w_1) \begin{pmatrix} \dfrac{\partial u_1}{\partial r} & \dfrac{\partial v_1}{\partial r} & \dfrac{\partial w_1}{\partial r} \\[2mm] \dfrac{1}{r}\dfrac{\partial u_1}{\partial \phi} - \dfrac{v_1}{r} & \dfrac{1}{r}\dfrac{\partial v_1}{\partial \phi} + \dfrac{u_1}{r} & \dfrac{1}{r}\dfrac{\partial w_1}{\partial \phi} \\[2mm] \dfrac{\partial u_1}{\partial z} & \dfrac{\partial v_1}{\partial z} & \dfrac{\partial w_1}{\partial z} \end{pmatrix} \tag{8-73}$$

Therefore, the second term of Eq. 8-70 becomes, in cylindrical coordinates,

$$\left(u_1\frac{\partial u_1}{\partial r} + \frac{v_1}{r}\frac{\partial u_1}{\partial \phi} - \frac{v_1{}^2}{r} + w_1\frac{\partial u_1}{\partial z} \quad u_1\frac{\partial v_1}{\partial r} + \frac{v_1}{r}\frac{\partial v_1}{\partial \phi} + \frac{u_1 v_1}{r} + w_1\frac{\partial v_1}{\partial z} \quad u_1\frac{\partial w_1}{\partial r} + \frac{v_1}{r}\frac{\partial w_1}{\partial \phi} + w_1\frac{\partial w_1}{\partial z} \right) \tag{8-74}$$

Next consider ∇p. In cylindrical coordinates this becomes (see Eq. 3-43)

$$\left(\frac{\partial p}{\partial r} \quad \frac{1}{r}\frac{\partial p}{\partial \phi} \quad \frac{\partial p}{\partial z} \right) \tag{8-75}$$

The term $\nabla^2 \vec{V}$ (see Prob. 12, Chapter 3) is shown as a column matrix:

$$
\begin{pmatrix}
\dfrac{\partial^2 u_1}{\partial r^2}+\dfrac{1}{r}\dfrac{\partial u_1}{\partial r}+\dfrac{1}{r^2}\dfrac{\partial^2 u_1}{\partial \phi^2}+\dfrac{\partial^2 u_1}{\partial z^2}-\dfrac{2}{r^2}\dfrac{\partial v_1}{\partial \phi}-\dfrac{1}{r^2}u_1 \\[2ex]
\dfrac{\partial^2 v_1}{\partial r^2}+\dfrac{1}{r}\dfrac{\partial v_1}{\partial r}+\dfrac{1}{r^2}\dfrac{\partial^2 v_1}{\partial \phi^2}+\dfrac{\partial^2 v_1}{\partial z^2}+\dfrac{2}{r^2}\dfrac{\partial u_1}{\partial \phi}-\dfrac{v_1}{r^2} \\[2ex]
\dfrac{\partial^2 w_1}{\partial r^2}+\dfrac{1}{r}\dfrac{\partial w_1}{\partial r}+\dfrac{1}{r^2}\dfrac{\partial^2 w_1}{\partial \phi^2}+\dfrac{\partial^2 w_1}{\partial z^2}
\end{pmatrix}
\tag{8-76}
$$

Finally, the term $\nabla(\nabla \cdot \vec{V})$ becomes (see Eq. 3-43)

$$
\begin{pmatrix} \dfrac{\partial f}{\partial r} & \dfrac{1}{r}\dfrac{\partial f}{\partial \phi} & \dfrac{\partial f}{\partial z} \end{pmatrix}
\tag{8-77}
$$

with (see Eq. 3-68)

$$
f = \frac{\partial u_1}{\partial r}+\frac{u_1}{r}+\frac{1}{r}\frac{\partial v_1}{\partial \phi}+\frac{\partial w_1}{\partial z}
\tag{8-78}
$$

or, showing $\nabla(\nabla \cdot \vec{V})$ as a column matrix,

$$
\nabla(\nabla \cdot \vec{V}) =
\begin{pmatrix}
\dfrac{\partial^2 u_1}{\partial r^2}-\dfrac{u_1}{r^2}+\dfrac{1}{r}\dfrac{\partial u_1}{\partial r}-\dfrac{1}{r^2}\dfrac{\partial v_1}{\partial \phi}+\dfrac{1}{r}\dfrac{\partial^2 v_1}{\partial \phi \partial r}+\dfrac{\partial^2 w_1}{\partial z \partial r} \\[2ex]
\dfrac{1}{r}\dfrac{\partial^2 u_1}{\partial r \partial \phi}+\dfrac{1}{r^2}\dfrac{\partial u_1}{\partial \phi}+\dfrac{1}{r^2}\dfrac{\partial^2 v_1}{\partial \phi^2}+\dfrac{1}{r}\dfrac{\partial^2 w_1}{\partial \phi \partial z} \\[2ex]
\dfrac{\partial^2 u_1}{\partial r \partial z}+\dfrac{1}{r}\dfrac{\partial u_1}{\partial z}+\dfrac{1}{r}\dfrac{\partial^2 v_1}{\partial \phi \partial z}+\dfrac{\partial^2 w_1}{\partial z^2}
\end{pmatrix}
\tag{8-79}
$$

Combining Eqs. 8-71, 8-74, 8-75, 8-76 and 8-79, we obtain finally the three Navier–Stokes equations in cylindrical coordinates,

$$
\left.
\begin{aligned}
&\frac{\partial u_1}{\partial t}+u_1\frac{\partial u_1}{\partial r}+\frac{v_1}{r}\frac{\partial u_1}{\partial \phi}-\frac{v_1{}^2}{r}+w_1\frac{\partial u_1}{\partial z} \\[1ex]
&\quad = g_r-\frac{1}{\rho}\frac{\partial p}{\partial r}+\nu\left(\frac{\partial^2 u_1}{\partial r^2}+\frac{1}{r}\frac{\partial u_1}{\partial r}+\frac{1}{r^2}\frac{\partial^2 u_1}{\partial \phi^2}+\frac{\partial^2 u_1}{\partial z^2}-\frac{2}{r^2}\frac{\partial v_1}{\partial \phi}-\frac{u_1}{r^2}\right) \\[1ex]
&\qquad +\frac{\nu}{3}\left(\frac{\partial^2 u_1}{\partial r^2}-\frac{u_1}{r^2}+\frac{1}{r}\frac{\partial u_1}{\partial r}-\frac{1}{r^2}\frac{\partial v_1}{\partial \phi}+\frac{1}{r}\frac{\partial^2 v_1}{\partial d\phi \partial r}+\frac{\partial^2 w_1}{\partial z \partial r}\right) \\[2ex]
&\frac{\partial v_1}{\partial t}+u_1\frac{\partial v_1}{\partial r}+\frac{v_1}{r}\frac{\partial v_1}{\partial \phi}+\frac{u_1 v_1}{r}+w_1\frac{\partial v_1}{\partial z} \\[1ex]
&\quad = g_\phi-\frac{1}{\rho r}\frac{\partial p}{\partial \phi}+\nu\left(\frac{\partial^2 v_1}{\partial r^2}+\frac{1}{r}\frac{\partial v_1}{\partial r}+\frac{1}{r^2}\frac{\partial^2 v_1}{\partial \phi^2}+\frac{\partial^2 v_1}{\partial z^2}+\frac{2}{r^2}\frac{\partial u_1}{\partial \phi}-\frac{v_1}{r^2}\right) \\[1ex]
&\qquad +\frac{\nu}{3}\left(\frac{1}{r}\frac{\partial^2 u_1}{\partial r \partial \phi}+\frac{1}{r^2}\frac{\partial u_1}{\partial \phi}+\frac{1}{r^2}\frac{\partial^2 v_1}{\partial \phi^2}+\frac{1}{r}\frac{\partial^2 w_1}{\partial \phi \partial z}\right)
\end{aligned}
\right\}
\tag{8-80}
$$

(continued)

$$\frac{\partial w_1}{\partial t} + u_1 \frac{\partial w_1}{\partial r} + \frac{v_1}{r} \frac{\partial w_1}{\partial \phi} + w_1 \frac{\partial w_1}{\partial z}$$

$$= g_z - \frac{1}{\rho} \frac{\partial p}{\partial z} + \nu \left(\frac{\partial^2 w_1}{\partial r^2} + \frac{1}{r} \frac{\partial w_1}{\partial r} + \frac{1}{r^2} \frac{\partial w_1}{\partial \phi^2} + \frac{\partial^2 w_1}{\partial z^2} \right)$$

$$+ \frac{\nu}{3} \left(\frac{\partial^2 u_1}{\partial r \partial z} + \frac{1}{r} \frac{\partial u_1}{\partial z} + \frac{1}{r} \frac{\partial^2 v_1}{\partial \phi \partial z} + \frac{\partial^2 w_1}{\partial z^2} \right)$$

$$(8\text{-}80)$$
continued

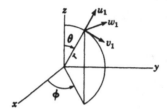

$u_1 =$ velocity parallel to r
$v_1 =$ velocity parallel to $rd\theta$
$w_1 =$ velocity parallel to $r \sin\theta \, d\theta$

Fig. 8.5

For the spherical coordinates, as shown in Fig. 8.5, we proceed as for the cylindrical coordinates. In this case we have

$u_1 = $ velocity parallel to r
$v_1 = $ velocity parallel to $rd\theta$
$w_1 = $ velocity parallel to $r \sin\theta \, d\phi$

Thus, for this case (see Eq. 3-63), and note we are now taking the elements in the order (r, θ, ϕ)

$$(h_1 \ h_2 \ h_3) = (1 \quad 1/r \quad 1/r \sin\theta) \tag{8-81}$$

and (see Eq. 3-69)

$$\operatorname{div}(\rho \vec{V}) = \frac{\partial(\rho u_1)}{\partial r} + \frac{2}{r}(\rho u_1) + \frac{1}{r} \frac{\partial(\rho v_1)}{\partial \theta} + \frac{\cot\theta}{r}(\rho v_1) + \frac{1}{r \sin\theta} \frac{\partial(\rho w_1)}{\partial \phi}$$

$$(8\text{-}82)$$

and therefore the continuity equation becomes

$$\frac{\partial \rho}{\partial t} + \frac{\partial(\rho u_1)}{\partial r} + \frac{2}{r}(\rho u_1) + \frac{1}{r} \frac{\partial(\rho v_1)}{\partial \theta} + \frac{\cot\theta}{r}\rho v_1 + \frac{1}{r \sin\theta} \frac{\partial(\rho w_1)}{\partial \phi} = 0 \tag{8-83}$$

Referring to Eq. 8-67 and considering this equation term by term, we obtain the following expressions in spherical form:

$$\frac{\partial \vec{V}}{\partial t} = \left(\frac{\partial u_1}{\partial t} \quad \frac{\partial v_1}{\partial t} \quad \frac{\partial w_1}{\partial t} \right) \tag{8-84}$$

The product is $\vec{V} \cdot \nabla \vec{V}$:

$$(u_1 \quad v_1 \quad w_1) \begin{pmatrix} \dfrac{\partial u_1}{\partial r} & \dfrac{\partial v_1}{\partial r} & \dfrac{\partial w_1}{\partial r} \\[2mm] \dfrac{1}{r}\dfrac{\partial u_1}{\partial \theta} - \dfrac{v_1}{r} & \dfrac{1}{r}\dfrac{\partial v_1}{\partial \theta} + \dfrac{u_1}{r} & \dfrac{1}{r}\dfrac{\partial w_1}{\partial \theta} \\[2mm] \dfrac{1}{r\sin\theta}\dfrac{\partial u_1}{\partial \phi} - \dfrac{w_1}{r} & \dfrac{1}{r\sin\theta}\dfrac{\partial v_1}{\partial \phi} - \dfrac{\cot\theta}{r}w_1 & \dfrac{1}{r\sin\theta}\dfrac{\partial w_1}{\partial \phi} + \dfrac{u_1}{r} + \dfrac{\cot\theta\, v_1}{r} \end{pmatrix}$$

$$(8\text{-}85)$$

Therefore, $\vec{V} \cdot \nabla \vec{V}$ becomes, in spherical coordinates (showing this as a column matrix),

$$\begin{pmatrix} u_1\dfrac{\partial u_1}{\partial r} + \dfrac{v_1}{r}\dfrac{\partial u_1}{\partial \theta} - \dfrac{v_1{}^2}{r} + \dfrac{w_1}{r\sin\theta}\dfrac{\partial u_1}{\partial \phi} - \dfrac{w_1{}^2}{r} \\[3mm] u_1\dfrac{\partial v_1}{\partial r} + \dfrac{v_1}{r}\dfrac{\partial v_1}{\partial \theta} + \dfrac{u_1 v_1}{r} + \dfrac{w_1}{r\sin\theta}\dfrac{\partial v_1}{\partial \phi} - \dfrac{\cot\theta}{r}w_1{}^2 \\[3mm] u_1\dfrac{\partial w_1}{\partial r} + \dfrac{v_1}{r}\dfrac{\partial w_1}{\partial \theta} + \dfrac{w_1}{r\sin\theta}\dfrac{\partial w_1}{\partial \phi} + \dfrac{w_1 u_1}{r} + \cot\theta\dfrac{v_1 w_1}{r} \end{pmatrix}$$

$$(8\text{-}86)$$

The term ∇p becomes in spherical coordinates

$$\nabla p = \left(\frac{\partial p}{\partial r} \quad \frac{1}{r}\frac{\partial p}{\partial \theta} \quad \frac{1}{r\sin\theta}\frac{\partial p}{\partial \phi} \right) \tag{8-87}$$

The term $\nabla^2 \vec{V}$ (in column matrix form) is

$$\begin{pmatrix} \dfrac{\partial^2 u_1}{\partial r^2} + \dfrac{2}{r}\dfrac{\partial u_1}{\partial r} + \dfrac{1}{r^2}\dfrac{\partial^2 u_1}{\partial \theta^2} + \dfrac{\cot\theta}{r^2}\dfrac{\partial u_1}{\partial \theta} + \dfrac{1}{r^2\sin^2\theta}\dfrac{\partial^2 u_1}{\partial \phi^2} - \dfrac{2u_1}{r^2} - \dfrac{2}{r^2}\dfrac{\partial v_1}{\partial \theta} - \dfrac{2\cot\theta v_1}{r^2} - \dfrac{2}{r^2\sin\theta}\dfrac{\partial w_1}{\partial \phi} \\[3mm] \dfrac{\partial^2 v_1}{\partial r^2} + \dfrac{2}{r}\dfrac{\partial v_1}{\partial r} + \dfrac{1}{r^2}\dfrac{\partial^2 v_1}{\partial \theta^2} + \dfrac{\cot\theta}{r^2}\dfrac{\partial v_1}{\partial \theta} + \dfrac{1}{r^2\sin^2\theta}\dfrac{\partial^2 v_1}{\partial \phi^2} + \dfrac{2}{r^2}\dfrac{\partial u_1}{\partial \theta} - \dfrac{\mathrm{cosec}^2\theta}{r^2}v_1 - \dfrac{2\cos\theta}{r^2\sin^2\theta}\dfrac{\partial w_1}{\partial \phi} \\[3mm] \dfrac{\partial^2 w_1}{\partial r^2} + \dfrac{2}{r}\dfrac{\partial w_1}{\partial r} + \dfrac{1}{r^2}\dfrac{\partial^2 w_1}{\partial \theta^2} + \dfrac{\cot\theta}{r^2}\dfrac{\partial w_1}{\partial \theta} + \dfrac{1}{r^2\sin^2\theta}\dfrac{\partial^2 w_1}{\partial \phi^2} + \dfrac{2}{r^2\sin\theta}\dfrac{\partial u_1}{\partial \phi} + \dfrac{2\cos\theta}{r^2\sin^2\theta}\dfrac{\partial v_1}{\partial \phi} - \dfrac{\mathrm{cosec}^2\theta w_1}{r^2} \end{pmatrix}$$

$$(8\text{-}88)$$

and finally, the term $\nabla(\nabla \cdot \vec{V})$ becomes

$$\left(\frac{\partial f}{\partial r} \quad \frac{1}{r}\frac{\partial f}{\partial \theta} \quad \frac{1}{r\sin\theta}\frac{\partial f}{\partial \phi} \right) \tag{8-89}$$

with

$$f = \frac{\partial u_1}{\partial r} + \frac{2u_1}{r} + \frac{1}{r}\frac{\partial v_1}{\partial \theta} + \frac{\cot\theta\, v_1}{r} + \frac{1}{r\sin\theta}\frac{\partial w_1}{\partial \phi} \tag{8-90}$$

or (as a column matrix)

$$\nabla(\nabla \cdot \bar{V}) = \begin{pmatrix} \dfrac{\partial^2 u_1}{\partial r^2} + \dfrac{2}{r}\dfrac{\partial u_1}{\partial r} - \dfrac{2u_1}{r^2} - \dfrac{1}{r^2}\dfrac{\partial v_1}{\partial \theta} + \dfrac{1}{r}\dfrac{\partial^2 v_1}{\partial \theta \partial r} - \dfrac{\cot\theta}{r^2}v_1 + \dfrac{\cot\theta}{r}\dfrac{\partial v_1}{\partial r} - \dfrac{1}{r^2\sin\theta}\dfrac{\partial w_1}{\partial\phi} + \dfrac{1}{r\sin\theta}\dfrac{\partial^2 w_1}{\partial r \partial\theta} \\[3mm] \dfrac{1}{r}\left(\dfrac{\partial^2 u_1}{\partial r\partial\theta} + \dfrac{2}{r}\dfrac{\partial u_1}{\partial\theta} + \dfrac{1}{r}\dfrac{\partial^2 v_1}{\partial\theta^2} + \dfrac{\cot\theta}{r}\dfrac{\partial v_1}{\partial\theta} - \dfrac{\operatorname{cosec}^2\theta}{r}v_1 + \dfrac{1}{r\sin\theta}\dfrac{\partial^2 w_1}{\partial\phi\partial\theta} - \dfrac{\cot an\,\theta\operatorname{cosec}\theta}{r}\dfrac{\partial w_1}{\partial\phi}\right) \\[3mm] \dfrac{1}{r\sin\theta}\left(\dfrac{\partial^2 u_1}{\partial r\partial\phi} + \dfrac{2}{r}\dfrac{\partial u_1}{\partial\phi} + \dfrac{1}{r}\dfrac{\partial^2 v_1}{\partial\theta\partial\phi} + \dfrac{\cot\theta}{r}\dfrac{\partial v_1}{\partial\phi} + \dfrac{1}{r\sin\theta}\dfrac{\partial^2 w_1}{\partial\phi^2}\right) \end{pmatrix}$$

$$(8\text{-}91)$$

Combining Eqs. 8-84, 8-86, 8-87, 8-88 with Eq. 8-91, the three N.S. equations are obtained in spherical coordinates as shown in equation (8-92), on p. 253.

8-7 The Boundary Layer Problem—Introduction. The number of exact solutions of the equations of laminar viscous flow is very small. One example was given in a previous section, and several others are given as problems at the end of this chapter.

A problem of considerable practical importance occurs in airplane, missile, and spaceship wing or body theory. According to the perfect fluid theory, no "drag" or force opposing the motion should occur when a body is moving through air at a constant velocity. However, there is a very definite drag force present as even the simplest test will indicate. The Prandtl–Blasius solution of the viscous "boundary layer" equations is the starting point for determining these viscous forces. In this section the laminar boundary layer theory will first be described on a purely physical basis. The form which the N.S. equations take based upon the Prandtl theory is then developed, and finally the Blasius solution of these equations is given in some detail. The chapter closes with a short discussion of various "boundary layer" effects in the general field of applied mechanics. See Ref. (23) for a more detailed discussion of boundary layer theory.

8-8 General Discussion of the Boundary Layer Effect. It is known that air has very small viscosity. Also, a theoretical solution for an airfoil based upon perfect fluid theory (nonviscous fluids, see Art. 8-4) can be obtained, and tests indicate that at some distance away from the airfoil (neglecting separation effects), the flow conforms fairly closely to that predicted by perfect fluid theory. Prandtl, in setting up the problem for the wing moving in air, assumed that the fluid (i.e., air) has small viscosity. Blasius, in solving the problem, sought a solution that at a short distance away from the wing was essentially a perfect fluid flow. In other words the problem becomes one in which the effect of viscosity is concentrated in a narrow region close to the wing (the

$$\frac{\partial u_1}{\partial t} + u_1\frac{\partial u_1}{\partial r} + \frac{v_1}{r}\frac{\partial u_1}{\partial \theta} + \frac{w_1}{r\sin\theta}\frac{\partial u_1}{\partial \phi} - \frac{v_1^2}{r} - \frac{w_1^2}{r}$$

$$= -\frac{1}{\rho}\frac{\partial p}{\partial r} + \nu\left(\frac{\partial^2 u_1}{\partial r^2} + \frac{2}{r}\frac{\partial u_1}{\partial r} + \frac{1}{r^2}\frac{\partial^2 u_1}{\partial\theta^2} + \frac{\cot\theta}{r^2}\frac{\partial u_1}{\partial\theta} + \frac{1}{r^2\sin^2\theta}\frac{\partial^2 u_1}{\partial\phi^2} - \frac{2u_1}{r^2} - \frac{2}{r^2}\frac{\partial v_1}{\partial\theta} - \frac{2\cot\theta\, v_1}{r^2} - \frac{2}{r^2\sin\theta}\frac{\partial w_1}{\partial\phi}\right)$$

$$+ \frac{\nu}{3}\left(\frac{\partial^2 u_1}{\partial r^2} + \frac{2}{r}\frac{\partial u_1}{\partial r} + \frac{1}{r^2}\frac{\partial^2 v_1}{\partial\theta\partial r} + \frac{\cot\theta}{r^2}\frac{\partial v_1}{\partial r} + \frac{1}{r\sin\theta}\frac{\partial^2 w_1}{\partial r\partial\phi}\right)$$

$$\frac{\partial v_1}{\partial t} + u_1\frac{\partial v_1}{\partial r} + \frac{v_1}{r}\frac{\partial v_1}{\partial\theta} + \frac{w_1}{r\sin\theta}\frac{\partial v_1}{\partial\phi} + \frac{u_1 v_1}{r} - \frac{\cot\theta}{r}w_1^2$$

$$= -\frac{1}{\rho r}\frac{\partial p}{\partial\theta} + \nu\left(\frac{\partial^2 v_1}{\partial r^2} + \frac{2}{r}\frac{\partial v_1}{\partial r} + \frac{1}{r^2}\frac{\partial^2 v_1}{\partial\theta^2} + \frac{\cot\theta}{r^2}\frac{\partial v_1}{\partial\theta} + \frac{1}{r^2\sin^2\theta}\frac{\partial^2 v_1}{\partial\phi^2} + \frac{2}{r^2}\frac{\partial u_1}{\partial\theta} - \frac{\operatorname{cosec}^2\theta\, v_1}{r^2} - \frac{2\cos\theta}{r^2\sin^2\theta}\frac{\partial w_1}{\partial\phi}\right)$$

$$+ \frac{\nu}{3}\left(\frac{1}{r}\frac{\partial^2 u_1}{\partial r\partial\theta} + \frac{2}{r^2}\frac{\partial u_1}{\partial\theta} + \frac{1}{r^2}\frac{\partial^2 v_1}{\partial\theta^2} + \frac{\cot\theta}{r^2}\frac{\partial v_1}{\partial\theta} - \frac{\operatorname{cosec}^2\theta}{r^2}v_1 + \frac{1}{r^2\sin\theta}\frac{\partial^2 w_1}{\partial\phi\partial\theta}\right)$$

$$\frac{\partial w_1}{\partial t} + u_1\frac{\partial w_1}{\partial r} + \frac{v_1}{r}\frac{\partial w_1}{\partial\theta} + \frac{w_1}{r\sin\theta}\frac{\partial w_1}{\partial\phi} + \frac{u_1 w_1}{r} + \frac{\cot\theta\, v_1 w_1}{r}$$

$$= -\frac{1}{\rho r\sin\theta}\frac{\partial p}{\partial\phi} + \nu\left(\frac{\partial^2 w_1}{\partial r^2} + \frac{2}{r}\frac{\partial w_1}{\partial r} + \frac{1}{r^2}\frac{\partial^2 w_1}{\partial\theta^2} + \frac{\cot\theta}{r^2}\frac{\partial w_1}{\partial\theta} + \frac{1}{r^2\sin^2\theta}\frac{\partial^2 w_1}{\partial\phi^2} + \frac{2}{r^2\sin\theta}\frac{\partial u_1}{\partial\phi} + \frac{2\cos\theta}{r^2\sin^2\theta}\frac{\partial v_1}{\partial\phi} - \frac{\operatorname{cosec}^2\theta}{r^2}w_1\right)$$

$$+ \frac{\nu}{3}\left(\frac{1}{r\sin\theta}\frac{\partial^2 u_1}{\partial r\partial\phi} + \frac{2}{r^2\sin\theta}\frac{\partial u_1}{\partial\phi} + \frac{\cot\theta}{r^2\sin\theta}\frac{\partial v_1}{\partial\theta} + \frac{1}{r^2\sin^2\theta}\frac{\partial^2 w_1}{\partial\phi^2} - \frac{w_1}{r^2}\right)$$

$$(8\text{-}92)$$

boundary layer), and outside this region the flow is, for all practical purposes, nonviscous.

8-9 The Prandtl Laminar-Flow Boundary Layer Equations.
The fluid is assumed incompressible; then the two-dimensional N.S. equations (see Eq. 8-54) become (neglecting gravity),

$$\left. \begin{aligned}
\frac{\partial u}{\partial t}+u\frac{\partial u}{\partial x}+v\frac{\partial u}{\partial y} &= -\frac{1}{\rho}\frac{\partial p}{\partial x}+\nu\left(\frac{\partial^2 u}{\partial x^2}+\frac{\partial^2 u}{\partial y^2}\right) \quad \text{(a)} \\
\frac{\partial v}{\partial t}+u\frac{\partial v}{\partial x}+v\frac{\partial v}{\partial y} &= -\frac{1}{\rho}\frac{\partial p}{\partial y}+\nu\left(\frac{\partial^2 v}{\partial x^2}+\frac{\partial^2 v}{\partial y^2}\right) \quad \text{(b)}
\end{aligned} \right\} \tag{8-93}$$

and continuity becomes (see Eq. 8-53)

$$\frac{\partial u}{\partial x}+\frac{\partial v}{\partial y} = 0 \tag{8-94}$$

Prandtl then assumed ν very small (or the Reynolds number, ul/ν, very large) and, by applying an order-of-magnitude analysis, as indicated in the following discussion, he simplified these equations to the point where they were later solved by Blasius.

For a flat plate, as shown in Fig. 8.6, u and v are zero at $y = 0$. δ, the boundary layer thickness, and all distances y in the boundary layer (as we shall see) are very small. The value of u at the point $y = \delta$ is taken as u_1, the free stream velocity, which we shall assume is uniform at $x = -\infty$. We assume that

$$O\left(\frac{\partial A}{\partial B}\right) = O\left(\frac{A}{B}\right) \tag{8-95}$$

where the symbol $O(A/B)$ reads "order of magnitude of A over B."

We shall obtain nondimensional (i.e., absolute value) forms of the basic equations, Eqs. 8-93 and 8-94, in terms of (a) the free stream velocity u_1 and (b) the distance, l, along the plate. Furthermore, we shall assume

$$\left. \begin{aligned}
O\left(\frac{u}{u_1}\right) &= 1 \\
O\left(\frac{x}{l}\right) &= 1 \\
N_R &= \frac{u_1 l}{\nu} \gg 1
\end{aligned} \right\} \tag{8-96}$$

Then Eq. 8-94, the continuity equation, becomes

$$\frac{\partial(u/u_1)}{\partial(x/l)}+\frac{\partial(v/u_1)}{\partial(y/l)} = 0 \tag{8-97}$$

so that we require

$$O\left(\frac{v}{u_1}\right) = O\left(\frac{y}{l}\right) \qquad (8\text{-}98)$$

The first N.S. equation, Eq. 8-93a, becomes,

$$\frac{\partial(u/u_1)}{\partial(t/l)\,u_1} + \frac{u}{u_1}\frac{\partial(u/u_1)}{\partial(x/l)} + \frac{v}{u_1}\frac{\partial(u/u_1)}{\partial(y/l)} = -\frac{\partial(p/\rho u_1{}^2)}{\partial(x/l)}$$
$$+\frac{v}{u_1 l}\left[\frac{\partial^2(u/u_1)}{\partial(x/l)^2} + \frac{\partial^2(u/u_1)}{\partial(y/l)^2}\right] \qquad (8\text{-}99)$$

Note that the order of magnitude of the second term in Eq. 8-99

$$\frac{u}{u_1}\frac{\partial(u/u_1)}{\partial(x/l)} \qquad (8\text{-}100)$$

is unity. This term will be retained in the equation. Hence, since we want to retain the pressure term, we require that

$$O\left(\frac{p}{\rho u_1{}^2}\right) = O\left(\frac{x}{l}\right) = \text{unity} \qquad (8\text{-}101)$$

Also, by inspection of the last term in Eq. 8-99, we see that, since

$$\frac{v}{u_1 l} \ll 1 \qquad (8\text{-}102)$$

we will have a viscosity contribution (which is certainly necessary if we are to solve a viscous flow problem) only if

$$O\left(\frac{y}{l}\right)^2 = O\left(\frac{1}{N_R}\right) \qquad (8\text{-}103)$$

and we see therefore that y (and all distances in the boundary layer) are very small compared to l. Hence, from Eq. 8-98, v is very small compared to u_1.

The second N.S. equation becomes, in nondimensional form

$$\frac{\partial(v/u_1)}{\partial(t/l)u_1} + \frac{u}{u_1}\frac{\partial(v/u_1)}{\partial(x/l)} + \frac{v}{u_1}\frac{\partial(v/u_1)}{\partial(y/l)} = -\frac{\partial(p/\rho u_1{}^2)}{\partial(y/l)}$$
$$+\frac{v}{u_1 l}\left(\frac{\partial^2(v/u_1)}{\partial(x/l)^2} + \frac{\partial^2(v/u_1)}{\partial(y/l)^2}\right) \qquad (8\text{-}104)$$

or

$$O\left(\frac{y}{l}\right) + O\left(\frac{y}{l}\right) + O\left(\frac{y}{l}\right) = O\left[\frac{p/(\rho u_1{}^2)}{y/l}\right] + O\left[\left(\frac{y}{l}\right)^3 + O\left(\frac{y}{l}\right)\right] \quad (8\text{-}105)$$

from which, at most

$$O\left[\frac{p/(\rho u_1^2)}{y/l}\right] = O\left(\frac{y}{l}\right) \qquad (8\text{-}106)$$

and this is a higher-order infinitesimal, so that $\partial p/\partial y$ may be taken as a lower order of magnitude than $\partial p/\partial x$ and may be neglected. In other words, the pressure is assumed constant along a line normal to the plate.

Thus, Eqs. 8-93 and 8-94 become, in the Prandtl boundary-layer form

$$\begin{aligned} \frac{\partial u}{\partial t}+u\frac{\partial u}{\partial x}+v\frac{\partial u}{\partial y} &= -\frac{1}{\rho}\frac{\partial p}{\partial x}+\nu\frac{\partial^2 u}{\partial y^2} \qquad &(a) \\[2mm] 0 &\rightleftharpoons 0 \qquad &(b) \\[2mm] \frac{\partial u}{\partial x}+\frac{\partial v}{\partial y} &= 0 \qquad &(c) \end{aligned} \qquad\Biggr\} \quad (8\text{-}107)$$

Also, at the junction of the boundary layer and the free stream, the equation corresponding to Eq. 8-56 becomes

$$\frac{\partial u_1}{\partial t}+u_1\frac{\partial u_1}{\partial x} = -\frac{1}{\rho}\frac{dp}{dx} \qquad (8\text{-}108)$$

Now a stream function, ψ, is introduced, defined by

$$u = \frac{\partial \psi}{\partial y} \qquad v = -\frac{\partial \psi}{\partial x} \qquad (8\text{-}109)$$

Thus continuity Eq. 8-107c is satisfied and Eq. 8-107a becomes

$$\frac{\partial^2 \psi}{\partial y\,\partial t}+\frac{\partial \psi}{\partial y}\frac{\partial^2 \psi}{\partial x\,\partial y}-\frac{\partial \psi}{\partial x}\frac{\partial^2 \psi}{\partial y^2} = \nu\frac{\partial^3 \psi}{\partial y^3}+\frac{\partial u_1}{\partial t}+u_1\frac{\partial u_1}{\partial x} \qquad (8\text{-}110)$$

in which Eq. 8-108 was used. The boundary conditions are

$$\begin{aligned} \frac{\partial \psi}{\partial y} &= 0 \quad \text{at} \quad y = 0 \\[2mm] \frac{\partial \psi}{\partial x} &= 0 \quad \text{at} \quad y = 0 \\[2mm] \frac{\partial \psi}{\partial y} &= u_1 \quad \text{at} \quad y \to \infty \end{aligned} \qquad\Biggr\} \quad (8\text{-}111)$$

Equations 8-110 and 8-111, derived by Prandtl, Ref. (30), in 1904 are the general boundary layer equations.

In 1908, Blasius, in his doctoral dissertation, Ref. (31), gave the first solution to this equation for a specific problem. He assumed

1. Steady flow, or $\partial/\partial t = 0$;

2. $dp/dx = 0$ or $\partial u_1/\partial x = 0$ (see Eq. 8-108).

and he considered a flat plate. Then Eqs. 8-107a and 8-109 become

or

$$\left.\begin{array}{c} u\dfrac{\partial u}{\partial x} + v\dfrac{\partial u}{\partial y} = \nu\dfrac{\partial^2 u}{\partial y^2} \\[3mm] \dfrac{\partial\psi}{\partial y}\dfrac{\partial^2\psi}{\partial x\,\partial y} - \dfrac{\partial\psi}{\partial x}\dfrac{\partial^2\psi}{\partial y^2} = \nu\dfrac{\partial^3\psi}{\partial y^3} \end{array}\right\} \qquad (8\text{-}112)$$

with

$$\left.\begin{array}{ll} \dfrac{\partial\psi}{\partial y} = 0 & \text{at} \quad y = 0 \\[3mm] \dfrac{\partial\psi}{\partial x} = 0 & \text{at} \quad y = 0 \\[3mm] \dfrac{\partial\psi}{\partial y} = u_1 & \text{at} \quad y \to \infty \end{array}\right\} \qquad (8\text{-}113)$$

and the problem of the boundary layer flow for a flat plate will be solved if a function ψ can be found which will satisfy Eq. 8-112 and the boundary conditions of Eq. 8-113.

Eq. 8-112 is a partial differential equation, and although some partial differential equations can be solved by more or less formal methods—such as separation of variable, or assuming in Eq. 8-112 that $\psi = X(x)Y(y)$—in general, the solution of a partial differential equation subject to given boundary conditions is extremely difficult. This is particularly so in the above problem. A tremendous simplification would be introduced if there were some method for transforming the above equation into an ordinary differential equation, since the theory of ordinary differential equations is much more simplified and advanced than the theory of partial differential equations.

It was toward this end that Prandtl introduced a decisive step. He saw that if the following change in variable is introduced

$$\eta = \frac{1}{2}\left(\frac{u_1}{\nu x}\right)^{1/2} y \qquad \psi = (\nu u_1 x)^{1/2} f(\eta) \qquad (8\text{-}114)$$

then Eq. 8-112 and the boundary conditions of Eq. 8-113 can be given in terms of f, where f is a function of η only. In other words, the partial differential equation can be transformed into an ordinary differential equation.

We verify this as follows: From Eq. 8-114,

$$\left.\begin{aligned}
u &= \frac{\partial \psi}{\partial y} = (u_1 \nu x)^{1/2} \frac{df}{d\eta} \frac{d\eta}{dy} \\
&= \tfrac{1}{2} u_1 \frac{df}{d\eta} \\
v &= -\frac{\partial \psi}{\partial x} = -(u_1 \nu x)^{1/2} \frac{df}{d\eta} \frac{d\eta}{dx} - \frac{1}{2}\left(\frac{u_1 \nu}{x}\right)^{1/2} f \\
&= \frac{1}{2}\left(\frac{u_1 \nu}{x}\right)^{1/2}\left(\eta \frac{df}{d\eta} - f\right)
\end{aligned}\right\} \tag{8-115}$$

and

$$\left.\begin{aligned}
\frac{\partial u}{\partial y} &= \tfrac{1}{2} u_1 \frac{d^2 f}{d\eta^2} \frac{d\eta}{dy} \\
&= \frac{u_1}{4}\left(\frac{u_1}{\nu x}\right)^{1/2} \frac{d^2 f}{d\eta^2} \\
\frac{\partial u}{\partial x} &= \tfrac{1}{2} u_1 \frac{d^2 f \cdot d\eta}{d\eta^2 \, dx} \\
&= -\frac{1}{4}\frac{u_1}{x}\eta \frac{d^2 f}{d\eta^2} \\
\frac{\partial^2 u}{\partial y^2} &= \frac{u_1}{4}\left(\frac{u_1}{\nu x}\right)^{1/2} \frac{d^3 f}{d\eta^3} \frac{d\eta}{dy} = \frac{u_1}{8}\left(\frac{u_1}{\nu x}\right)\frac{d^3 f}{d\eta^3}
\end{aligned}\right\} \tag{8-116}$$

Substituting the above in Eq. 8-112 and denoting differentiation with respect to η by primes, we obtain

$$-(\tfrac{1}{2} u_1 f')\left(\frac{1}{4}\frac{u_1}{x}\eta f''\right) + \frac{1}{2}\left(\frac{u_1 \nu}{x}\right)^{1/2}(\eta f' - f)\frac{u_1}{4}\left(\frac{u_1}{\nu x}\right)^{1/2} f'' = \nu \frac{u_1}{8}\left(\frac{u_1}{\nu x}\right)f''' \tag{8-117}$$

or

$$f''' + ff'' = 0 \tag{8-118}$$

And the boundary conditions, Eq. 8-113, become

$$\left.\begin{aligned}
f' &= 0 \quad \text{at} \quad \eta = 0 \quad \text{(a)} \\
f &= 0 \quad \text{at} \quad \eta = 0 \quad \text{(b)} \\
f' &= 2 \quad \text{at} \quad \eta \to \infty \quad \text{(c)}
\end{aligned}\right\} \tag{8-119}$$

Blasius solved this problem by obtaining two separate solutions: (a) one solution which holds for small values of η and (b) one solution which holds for large values of η; and he equated these solutions at an arbitrary boundary point to obtain a single solution which held for both small and large values of η. We shall describe another method of solution due to Töpfer, Ref. (32).

In this method we assume that f can be given in series form

$$f = C_0 + \frac{C_1\eta}{1!} + \frac{C_2\alpha\eta^2}{2!} + \frac{C_3\eta^3}{3!} + \frac{C_4\eta^4}{4!} + \frac{C_5\alpha^2\eta^5}{5!} + \cdots \quad (8\text{-}120)$$

The boundary conditions of Eq. 8-119 require that $C_0 = C_1 = 0$. Then substituting Eq. 8-120 in Eq..8-118 and setting the coefficient of each power of η equal to zero, we find

$$\left.\begin{array}{r} C_2{}^2\alpha^2 + C_5\alpha^2 = 0 \\[4pt] C_3 = 0 \\[4pt] C_4 = 0 \\[4pt] C_6 + \dfrac{C_2C_3\alpha}{2!} + \dfrac{C_2C_3\alpha}{3!} = 0 \end{array}\right\} \quad (8\text{-}121)$$

and if we take $f''(0) = \alpha$ (which merely sets the scale or datum for the C values), then $C_2 = 1$, $C_5 = -1$, $C_6 = 0$, $C_7 = 0$, $C_8 = 11$, etc., and Eq. 8-120 becomes

$$f = \frac{\alpha\eta^2}{2!} - \frac{\alpha^2\eta^5}{5!} + \frac{11\alpha^3\eta^8}{8!} - \frac{375\alpha^4\eta^{11}}{11!} \quad (8\text{-}122)$$

and the problem is solved if we determine α.

Now, if $F(\eta)$ is this equation when $\alpha = 1$, then as may be verified by substitution

$$f = \alpha^{1/3}F(\alpha^{1/3}\eta) \quad (8\text{-}123)$$

and

$$\lim_{\eta\to\infty} f' = \lim_{\eta\to\infty} \frac{df}{d\eta} = \alpha^{2/3}\lim_{\eta\to\infty}\frac{dF(\alpha^{1/3}\eta)}{d(\alpha^{1/3}\eta)} = \alpha^{2/3}\lim_{\eta\to\infty}\frac{dF(\eta)}{d\eta} \quad (8\text{-}124)$$

Now, from Eq. 8-119c, Eq. 8-124 becomes

$$\left.\begin{array}{ll} \alpha^{2/3}\lim\limits_{\eta\to\infty}\dfrac{dF(\eta)}{d\eta} = 2 & \text{(a)} \\[10pt] \alpha^{2/3} = \dfrac{2}{\lim\limits_{\eta\to\infty}F'(\eta)} & \text{(b)} \end{array}\right\} \quad (8\text{-}125)$$

or

Equation 8-125b may be solved by numerical methods by evaluating $F'(\eta)$ for various values of η starting at $\eta = 0$ and continuing until

$F'(\eta)$ is constant to a sufficient approximation. If this is done, it is found that

$$\alpha = 1.328 \tag{8-126}$$

Having α, we can determine the drag force on the plate due to the boundary layer. For, from Eq. 8-116,

$$\frac{\partial u}{\partial y} = \frac{u_1}{4}\left(\frac{u_1}{vx}\right)^{1/2} f'' \tag{8-127}$$

and at $y = 0$, see Eq. 8-122

$$\frac{\partial u}{\partial y} = \tfrac{1}{4}\alpha u_1\left(\frac{u_1}{vx}\right)^{1/2} \tag{8-128}$$

so that the shearing stress at the plate is given by

$$\left.\begin{aligned} \tau &= \mu\frac{\partial u}{\partial y} \\[2mm] &= \tfrac{1}{4}\alpha\rho u_1\left(\frac{vu_1}{x}\right)^{1/2} \end{aligned}\right\} \tag{8-129}$$

Note that in this equation, at $x = 0$, $\tau \to \infty$. This is a reflection of the fundamental requirement that the solution holds only for N_R very large—hence it does *not* hold near the leading edge of the plate.

To obtain the drag force on a plate of width b and length l, integrate Eq. 8-129, noting that the shear stress acts on both sides of the plate to obtain

$$\text{Drag} = D = 1.328b\sqrt{\mu\rho l u_1{}^3} \tag{8-130}$$

The dimensionless drag coefficient C_D, defined as

$$C_D = \frac{D}{\tfrac{1}{2}\rho u_1{}^2 A} \tag{8-131}$$

(in which A is the "wetted area" or $2bl$), is then given by

$$C_D = \frac{1.328}{\sqrt{N_R}} \tag{8-132}$$

where N_R is the Reynolds number based on u_1 and l.

8-10 Summary of the Prandtl–Blasius Boundary Layer Solution.

(a) The boundary layer effect was first discussed in purely physical terms.

(b) The Navier–Stokes Equations were then modified in accordance with Prandtl's boundary layer development.

(c) The Blasius form of the Prandtl equations were then obtained.

(d) These equations were solved using Töpfer's method.

(e) The shear stress, drag force, and drag coefficient for boundary layer effects were then obtained for a flat plate of length l and width b.

8-11 The Thickness of the Boundary Layer. In the previous section, the boundary layer equations were solved for a given case, the infinite flat plate. It was pointed out that Prandtl's original hypothesis was that viscous effects were important only in the narrow boundary layer, and that outside of this layer the flow is essentially nonviscous. The question then arises, "How thick is the boundary layer?"

Referring to Fig. 8.6, it is seen that there is no actual boundary line between the viscous and nonviscous portions of the field. In fact, viscosity is theoretically effective throughout the entire region of flow. But, as can also be seen from Fig. 8.6, practically, the effect of viscosity is negligible at a small distance from the plate. The determination of the "small distance" is our next task.

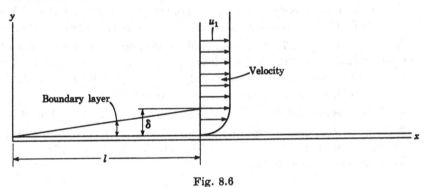

Fig. 8.6

One definition of δ, of course, is the distance from the plate at which the velocity is very nearly equal to the undisturbed free stream velocity u_1; for example,

$$\frac{u}{u_1} = 0.99$$

or other similar value. This means (see Eqs. 8-115 and 8-122) considering only the first term of the series,

$$\frac{u}{u_1} = \frac{1}{2}\frac{df}{d\eta} = \frac{(\alpha\eta)}{2} = 0.99 \qquad (8\text{-}133)$$

and, from Eq. 8-114,

$$\left. \begin{aligned} y = \delta &= 2\eta\left(\frac{\nu x}{u_1}\right)^{1/2} \\ &= \frac{2(0.99)(2)}{1.328}\left(\frac{\nu x}{u_1}\right)^{1/2} \\ &= 2.986\left(\frac{\nu x}{u_1}\right)^{1/2} \end{aligned} \right\} \tag{8-134}$$

(If more terms in the series had been used it would be found that the value of the constant in the above equation would be 5.0 instead of 2.986; see Ref. 23.)

The above definition of boundary layer thickness is a numerical one. In many problems dealing with boundary layers it is more convenient to define a boundary thickness based upon either of the following considerations:

(a) kinematical considerations, i.e., mass flow or streamline distortion considerations. This is called the "displacement thickness" and is usually designated δ^*.

(b) dynamical considerations, i.e., momentum considerations. This is called the "momentum thickness" and is normally designated θ.

As shown in Fig. 8.7, δ^* is essentially a measure of displacement of the streamlines due to boundary layer effects. It is connected with a defect in the mass flow, and we can define it physically in two ways, both of which are kinematical (geometrical) definitions.

1. δ^* represents the deflection of a streamline at infinity, Fig. 8.7a.

2. δ^* is a thickness such that rectangular mass flow through it is equal to the mass defect of the variable velocity flow, Fig. 8.7b.

Then we have, from (1) or (2) above

or

$$\left. \begin{aligned} (\rho u_1 y)_{y\to\infty} &= \rho u_1(y+\delta^*)_{y\to\infty} - \rho\int_0^\infty (u_1 - u)\, dy \\ \delta^* &= \int_0^\infty \left(1 - \frac{u}{u_1}\right) dy \end{aligned} \right\} \tag{8-135}$$

This may be evaluated by numerical means, using Eqs. 8-115 and 8-120, and it gives

$$\delta^* = 1.721\left(\frac{\nu x}{u_1}\right)^{1/2} \tag{8-136}$$

The momentum thickness θ may be defined as the thickness of a layer of fluid of velocity u_1 whose momentum flow is the same as that of the

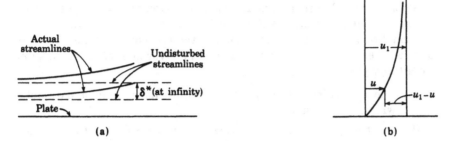

Fig. 8.7

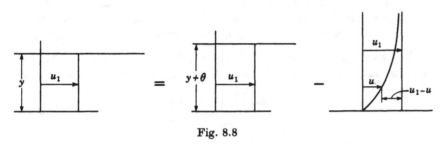

Fig. 8.8

defect in momentum in the actual boundary layer. Or (see Fig. 8.8)

or

$$(\rho u_1 y)u_1 = \rho u_1(y+\theta)u_1 - \int_0^\infty \rho(u_1-u)u\,dy \\ \theta = \int_0^\infty \frac{u}{u_1}\left(1-\frac{u}{u_1}\right)dy \Bigg\} \tag{8-137}$$

and if this expression is integrated numerically, we find

$$\theta = 0.664\left(\frac{\nu x}{u_1}\right)^{1/2} \tag{8-138}$$

Summarizing the definitions of boundary layer thicknesses, we defined three different boundary layer thicknesses:

1. a numerical definition in which the boundary layer thickness is the width to a point at which the velocity is some arbitrarily assigned percentage of the free-stream undisturbed velocity.

2. a kinematical definition, δ^*, this being essentially the deflection of a streamline at infinity.

3. a dynamical definition, θ, this being a thickness based upon momentum considerations.

Numerical values for the three thicknesses were obtained.

8-12 The General Occurrence of Boundary Layer Type. Solutions in Applied Mechanics.[6] It will now be shown that the typical "boundary layer" type of behavior occurs in other fields of applied mechanics. By "typical boundary layer behavior," we assume essentially the following action:

(a) The equations governing the phenomena contain a parameter which is very small.

(b) For this parameter *equal to zero*, a certain solution to the equations is obtained.

(c) For this parameter *very small* a solution is obtained which is the same as the solution for case (b) except in a narrow region within which there is a rapid change of a significant field quantity.

For example, in the case of viscous flow, the Euler equation for nonviscous, non-heat-conducting flow in two dimensions (see Eq. 8-56) is

$$u\frac{\partial u}{\partial x} + v\frac{\partial u}{\partial y} = -\frac{1}{\rho}\frac{\partial p}{\partial x} \tag{8-139}$$

and may be interpreted as a balance of inertia and pressure forces.

The Navier–Stokes equation,

$$u\frac{\partial u}{\partial x} + v\frac{\partial u}{\partial y} = -\frac{1}{\rho}\frac{\partial p}{\partial x} + \nu\nabla^2 u \tag{8-140}$$

is a balance between inertia, pressure, and viscous forces.

Prandtl's boundary layer theory assumes that in Eq. 8-140

$$\frac{\text{viscous effect}}{\text{inertia effect}} \to 0 \tag{8-141}$$

or that the Reynolds number N_R is very large. This leads to the boundary layer solution previously obtained and holds for very small values of viscosity ν, provided the length parameter in the Reynolds number is also not small.

Equations such as Eqs. 8-139 and 8-140 are typical of those which lead to boundary layer effects. It can be shown that if the parameter that multiplies the *highest* term of the differential equation approaches zero (i.e., if $\nu \to 0$ above), then the solution to the equation approaches the solution corresponding to the equations for $\nu = 0$ (Eq. 8-139 above), except for a narrow region within which there is a rapid transition of, in this case, velocity. And this narrow region is the boundary layer.

[6] Most of the material covered in this section appeared in a paper "A Note on Boundary Layer Type Solutions in Applied Mechanics" in the April 1950 issue of the *Journal of Aeronautical Sciences.*

The phenomena are dependent upon the number of boundary conditions required for a unique solution to a partial differential equation. Thus, Eq. 8-140, being one order higher than Eq. 8-139, requires an additional boundary condition, the no-slip condition. For $\nu \to 0$, one is tempted to neglect the last term of Eq. 8-140, but if this were done the order of the equation would be lowered, a boundary condition would be lost, and the solution (at least in the neighborhood of the boundary) would differ from the no-slip solution; or, in effect, we would not obtain the boundary layer in the solution. It would seem, therefore, that a fundamental requirement for the type of boundary layer action considered herein is what may be called a "singularity in the boundary condition"—that is, the requirement of an entirely new boundary condition because of the inclusion of the extremely small term in the field equation. Furthermore, it would seem that a physical description of the boundary layer phenomenon can be given in terms of "ratio of effects" approaching zero as was done in the case of the viscous fluid boundary layer, Eq. 8-141.

Several boundary layer solutions in which the form of the differential equation is similar to Eq. 8-140 have been discussed. Taylor and Maccoll Ref. (33), in their analysis of the shock front of supersonic flows, obtain a boundary-layer equation that leads to a characteristic thin layer (shock thickness) as viscosity and heat conduction approach zero. The laminar flow heat convection problem, Ref. (34), also leads to an equation that is identical in form to the viscous flow equation.

Friedrichs, Ref. (35), discusses a differential equation that occurs in the theory of plates and which physically may be thought of as a balance of

$$plate\ effect\ \text{plus}\ membrane\ effect\ =\ transverse\ load\ effect \qquad (8\text{-}142)$$

and considers the case in which

$$\frac{plate\ effect}{membrane\ effect} \to 0 \qquad (8\text{-}143)$$

It is found in this case that the boundary layer is in the *slope* of the plate at the supports and, in effect, that the structure acts like a membrane everywhere except at the supports, at which place a boundary condition is lost in the pure membrane solution and a rapid transition in slope takes place in a narrow region, this being the boundary layer.

Another similar, although apparently unrelated boundary layer effect occurs in plates Ref. (19). If we consider a long plate, simply supported along its sides, free along the short sides and uniformly loaded transversely over its area, it is found that at the short edges the plate "curls" over in a narrow region—the boundary layer for this case. The phenomenon is undoubtedly due to a boundary layer effect; some distance away from the edges a condition of plane strain exists, whereas at the edges this condition must be given up. The boundary layer described by Friedrichs may be thought of as one in which a condition of plane stress prevails throughout the plate but is given up at the edges.

Fung and Witrick, Ref. (36), consider another boundary layer effect which occurs in large deflection plate theory. This case however, is not what is called a "true boundary layer" in this book, since the behavior appears to depend upon the effect of nonlinearity.

Bromberg and Stoker, Ref. (37), discuss a boundary layer effect that occurs in the theory of shells which is also due to the inclusion on the plate theory of certain nonlinear, higher-order terms of the strain tensor.

Two additional types of boundary layer effects which are due to singularities in the boundary conditions and which lead to characteristic thin layer regions will now be described.

The first occurs in the Lame–Clapeyron theory of axially symmetric deformations of elastic bodies. Consider a rotating circular wheel. It is found, Ref. (11), that the tangential and radial stress distributions for the case in which there is a small central hole in the wheel are the same as the solutions corresponding to no hole, everywhere except for a narrow region adjacent to the hole in which there is a rapid change in the stresses. This is a boundary-layer type effect and is connected with the following singularity in the boundary condition:

For the case of no hole, the boundary condition is as follows: (a) the radial *deformation* is zero at the center of the wheel. For the case of a small hole, the boundary condition is as follows: (b) the radial *stress* at the edge of the hole is zero. The differential equation in deflection is the same for both cases, but in going from small hole to no hole there is a *singularity in the boundary condition*. The solution is similar in many respects to that of the small circular hole in an infinite plate subjected to tension stress. Also, there appears to be some connection with the St. Venant principle, see footnote, page 143.

Another boundary-layer effect occurs in the direct compression-buckling loading of a rod. If we consider the ratio of effects

$$\frac{\text{buckling (bending) stress}}{\text{direct compression stress}} \to 0 \qquad (8\text{-}144)$$

it is found that the governing parameters are the slope θ and the slenderness ratio l/ρ of the bar. The slenderness ratio being fixed in any given problem, it becomes necessary that the slope approach zero.

The problem can be set up as follows: for zero slope, we have direct stress only, or

$$\sigma = \frac{P}{A} \qquad (8\text{-}145)$$

For small slope, we have direct stress combined with bending stress (see Eq. 5-23), or

$$\sigma = \frac{P}{A} \pm \frac{Pyc}{I} \qquad (8\text{-}146)$$

and (see Eq. 5-56)

$$EI\frac{d\theta}{dS} = Py \qquad (8\text{-}147)$$

in which

l = length of bar
ρ = radius of gyration of bar
σ = stress
P = direct centrally applied load
A = area of bar
y = deflection
c = distance from neutral axis of bar
EI = stiffness
dS = differential length along bar

For extremely small θ, the stress Eq. 8-146 is nearly the same as the stress Eq. 8-145 for all values of P less than P_{Euler}. As P approaches and then exceeds P_{Euler}, it is found, Ref. (38), that there is a rapid increase in σ corresponding to a small increase in P. Thus, the boundary-layer "region" in this case is the load P, and the term that corresponds to the velocity of the viscous boundary layer is the stress. For small θ a boundary condition corresponding to the end support is needed. Thus, in going from small slope to zero slope a boundary condition is lost.

A familiar example of some interest,—in that it indicates the importance of *losing* a boundary condition (dropping the highest order term of the differential equation) in the boundary layer type solution is the comparison of Stokes and Oseen flows, Ref. (22). The Stokes flow is a balance of viscous and pressure forces and is described by the equation

$$\nu\nabla^2 u = \frac{1}{\rho}\frac{\partial p}{\partial x} \qquad (8\text{-}148)$$

The Oseen improvement introduces a perturbation inertia term (first order in the differential equation)

$$u\frac{\partial u}{\partial x} + \nu\nabla^2 u = \frac{1}{\rho}\frac{\partial p}{\partial x} \qquad (8\text{-}149)$$

and now it is found that the flows of Eqs. 8-148 and 8-149 are the same *in the boundary layer adjacent to the body* but differ in the large region away from the boundary layer—which is just the reverse of the usual boundary-layer effect.

8-13 Summary. The equations governing the flows of viscous fluids were derived first. These are (in three-dimensional space) (a) one equation of continuity and (b) three Navier–Stokes equations.

It was shown that when the viscous terms are set equal to zero, the above equations govern the flows of "perfect-fluids"—the nonviscous, non-heat-conducting fluid.

A simple exact solution of the viscous flow problem was presented, the plane Poiseuille flow.

The N.S. equations were presented in curvilinear form—in spherical and cylindrical coordinates.

The Prandtl–Blasius boundary layer theory was presented briefly. Finally, the general occurrence of "boundary layers" in applied mechanics was discussed.

Problems

1. Consider the problem of steady flow through a straight pipe of uniform circular section. Let z be the axis of the pipe and assume the velocity is everywhere parallel to z and a function of the distance r from this axis. Let the length of the pipe be l and p_1, p_2 the mean values of the pressure at the two ends.

(a) show that

$$\frac{\partial}{\partial r}\left(r\frac{\partial w}{\partial r}\right) = -\frac{p_1-p_2}{\mu l}r$$

and

$$\frac{\partial p}{\partial r} = 0$$

Interpret physically this last result.

(b) Then show that

$$w = -\frac{p_1-p_2}{4\mu l}r^2 + A\log r + B$$

and if a is the radius of the pipe, and the pipe is flowing full,

$$w = \frac{p_1-p_2}{4\mu l}(a^2-r^2)$$

(c) Hence, the quantity of flow in the pipe is given by

$$Q = \frac{\pi a^4}{8\mu}\frac{p_1-p_2}{l}$$

2. For an *annular* pipe, of inner radius b, outer radius a (see data in Prob. 1), show that

$$w = \frac{p_1 - p_2}{4\mu l}\left[a^2 - r^2 + \frac{b^2 - a^2}{\log(b/a)}\log\frac{r}{a}\right]$$

and the flow is given by

$$Q = \frac{\pi}{8\mu}\frac{p_1 - p_2}{l}\left[b^4 - a^4 - \frac{(b^2 - a^2)^2}{\log(b/a)}\right]$$

3. Consider the three-dimensional equations for a viscous, incompressible fluid. Show that if the inertia terms are neglected the pressure must satisfy

$$\nabla^2 p = 0$$

so that p can be expanded in a series of harmonic functions.

4. Consider the unsteady laminar motion of a viscous incompressible fluid such that $v = 0$, $w = 0$, and u is a function of y only. Then show the following:

(a) $p = $ constant.

(b) $\dfrac{\partial u}{\partial t} = \nu\dfrac{\partial^2 u}{\partial y^2}$.

(c) so that a possible solution is given by

$$u = Ae^{(1+i)\beta y} + Be^{-(1+i)\beta y}$$

provided the motion is a simple harmonic one with a time factor

$$e^{i(\sigma t + \epsilon)} \quad \text{and} \quad \beta = \left(\frac{\sigma}{2\nu}\right)^{1/2}$$

(d) If the fluid lies on the positive side of the x-z plane and if this plane is a rigid surface oscillating in accordance with

$$u = ae^{i(\sigma t + \epsilon)}$$

show that if the fluid extends to infinity in the y direction, we have

$$u = ae^{-(1+i)\beta y + i(\sigma t + \epsilon)}$$

or, taking the real part,

$$u = ae^{-\beta y}\cos(\sigma t - \beta y + \epsilon)$$

which is due to a motion of the boundary given by

$$u = a\cos(\sigma t + \epsilon)$$

(e) Show that the *drag* force on the rigid surface, per unit area, is given by

$$D = \rho\nu^{\frac{1}{2}}\sigma^{\frac{1}{2}}a\cos(\sigma t + \epsilon + \tfrac{1}{4}\pi)$$

and show that this force has its maxima at intervals of one-eighth of a period before the oscillating plane passes through its mean position.

5. If, for the fluid of Prob. 4, the fluid does not extend to infinity but is bounded by a fixed rigid plane at $y = h$, then show
(a) the boundary conditions require

$$A + B = a$$

$$A e^{(1+i)\beta h} + B e^{-(1+i)\beta h} = 0$$

(b) so that

$$u = a \frac{\sinh(1+i)\beta(h-y)}{\sinh(1+i)\beta h} e^{i(\sigma t + \epsilon)}$$

(c) and therefore, the drag per unit area is given by

$$D = \mu(1+i)\beta a \coth(1+i)\beta h e^{i(\sigma t + \epsilon)}$$

the real part of which is given by

$$\sqrt{2}\mu\beta a \frac{\sinh 2\beta h \cos(\sigma t + \epsilon + \tfrac{1}{4}\pi) + \sin 2\beta h \sin(\sigma t + \epsilon + \tfrac{1}{4}\pi)}{\cosh 2\beta h - \cos 2\beta h}$$

6. (a) For the two-dimensional stress tensor, prove that if the tensor is given by

$$\begin{pmatrix} \sigma_x & 0 \\ 0 & \sigma_y \end{pmatrix}$$

for all sets of axes, that $\sigma_x = \sigma_y$. Hint: Use the relations concerning the invariants of the tensors.
(b) Do the same for the three-dimensional stress tensor.

7. Given two parallel infinite walls, a distance h apart, $dp/dx = 0$. Assume the top wall moves with a velocity U, and the bottom wall is stationary. The resulting flow is a "plane Couette" (after the French scientist) flow. Obtain the following:
(a) Velocity distribution, including sketch of same.
(b) Shear stress, τ.
(c) Coefficient of friction, C_f.

8. Mixed flows are superpositions of Poiseuille and Couette flows. Obtain velocity distribution, shear stress and friction coefficient for the following:
(a) $dp/dx < 0$, sometimes called the viscosity turbine.
(b) $dp/dx > 0$, sometimes called the viscosity pump.

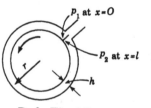

Prob. Fig. 8.9

9. A viscosity pump may be built with a movable inner runner rotating in a stationary outer casing as shown. In the pump shown, $h/r \ll 1$ and centrifugal effects may be neglected. Assuming a unit depth of pump, and letting

$$P = p_1 - p_2 \qquad p_2 > p_1 \qquad H = -\frac{P}{\rho g}$$

(a) Compute (H/H_{max}) as a function of ξ, where $\xi = (Q/Q_{max})$, Q being the discharge.

$$\text{Answer:} \quad \frac{H}{H_{max}} = 1 - \xi$$

(b) Compute the efficiency of the pump, η,

$$\text{Answer:} \quad \eta = \frac{3\xi(1-\xi)}{4-3\xi}$$

(c) Compute the maximum possible efficiency of this pump.

(d) Discuss the operation of the pump and flow condition for $Q = 0$.

10. Obtain the equations of flow for a cylindrical, two-dimensional, axial symmetric motion in concentric circles.

(a) Show that one of these represents a balance of centrifugal and pressure forces and the other an equation of time-rate of change of velocity.

(b) Show that the shear stress is given by

$$\tau = \mu r \frac{d(v_1/r)}{dr}$$

and the rotation (vorticity) by

$$\zeta = \frac{1}{r} \frac{d(v_1 r)}{dr}$$

(c) Discuss and sketch the flow for zero vorticity.

(d) Discuss and sketch the flow for zero shear stress. Is this also zero vorticity flow?

(e) Discuss and sketch the flow for constant vorticity.

11. In the boundary layer substitution of η and ψ, use an exponent n instead of $\frac{1}{2}$ and substitute in the Blasius equation. Show that in order for this equation to be a total differential equation, n must be taken equal to $\frac{1}{2}$.

12. Determine the transverse velocity, v, in the boundary layer (Eq. 8-115) and plot this as a function of x and y (or η). Is this velocity zero at the edge of the boundary layer, as assumed in the theory? Discuss.

Chapter 9

INTRODUCTORY THEORY OF PLASTICITY

9-1 Introduction. In this chapter, the introductory parts of plasticity theory will be developed. As in the other chapters, the presentation will emphasize the fundamental unity (and also the points of difference) between plasticity theory and the other fields of applied mechanics.

The discussion starts with a short review of certain basic ideas in two-dimensional elasticity theory. The place of plasticity theory in the framework of solid mechanics is then brought out by emphasis on the places where elasticity theory *does not hold*. This leads to a discussion of the various directions in which developments in plasticity theory have progressed.

The significance and importance of the postulated plastic stress–strain relations and plasticity conditions are then discussed, and finally a solution is given, using the methods developed, for a simple, well-known problem in plasticity.

9-2 Review of Two-dimensional Elasticity Theory. As an introduction to inelastic theory or plasticity theory, let us review some of the fundamental ideas of two-dimensional elasticity theory. We consider an x–y system of coordinates. Then, for small deformations, the strain tensor is given (see Eq. 4-24) by

$$\eta = \begin{pmatrix} e_x & \tfrac{1}{2}\gamma_{xy} \\ \tfrac{1}{2}\gamma_{yx} & e_y \end{pmatrix} = \begin{pmatrix} \dfrac{\partial u}{\partial x} & \dfrac{1}{2}\left(\dfrac{\partial u}{\partial y}+\dfrac{\partial v}{\partial x}\right) \\ \dfrac{1}{2}\left(\dfrac{\partial v}{\partial x}+\dfrac{\partial u}{\partial y}\right) & \dfrac{\partial v}{\partial y} \end{pmatrix} \tag{9-1}$$

Note: the above form of the strain tensor is independent of any relation between stress and strain. It holds for any small, continuous deformation.

Also, for a body in equilibrium we have, for zero body forces, the equilibrium equations,

$$\text{div } T = 0 \tag{9-2}$$

where T is the stress tensor given by

$$T = \begin{pmatrix} \sigma_x & \tau_{xy} \\ \tau_{yx} & \sigma_y \end{pmatrix} \tag{9-3}$$

and, in expanded form, Eq. 9-2 becomes

$$\left. \begin{array}{ll} \dfrac{\partial \sigma_x}{\partial x} + \dfrac{\partial \tau_{yx}}{\partial y} = 0 & \text{(a)} \\[3mm] \dfrac{\partial \tau_{xy}}{\partial x} + \dfrac{\partial \sigma_y}{\partial y} = 0 & \text{(b)} \end{array} \right\} \tag{9-4}$$

Equations 9-4a and 9-4b are simple equilibrium of force equations, and hold independently of any relation between stresses and strains.

In two-dimensional elasticity theory we generally have either a condition of plane stress or plane strain, see Ref. (13). In plane stress problems we have as an additional unknown e_z, and in plane strain problems we have as an additional unknown σ_z.

Thus in order to solve a problem in two-dimensional elasticity completely we must be able to determine the seven quantities—

$$e_x,\ e_y,\ \gamma_{xy}(=\gamma_{yx}),\ \sigma_x,\ \sigma_y,\ \tau_{xy}(=\tau_{yx}),\quad \text{and} \quad \sigma_z(\text{or } e_z)$$

In other words, there must be seven independent equations in terms of these seven unknowns. As of this point, we have only two such equations, these being Eqs. 9-4a and 9-4b.

For elastic action, that is, for stresses and strains on the straight line portion of the stress–strain curve, we also have Hooke's Law, which in two dimensions is (see Eq. 7-88)

$$\eta = \frac{1+\nu}{E}T - \frac{\nu}{E}\mathscr{I}_1\mathscr{E}_2 \tag{9-5}$$

or, in expanded form, the following four equations (for plane stress or plane strain)

$$\left. \begin{array}{l} e_x = \dfrac{1}{E}(\sigma_x - \nu\sigma_y - \nu\sigma_z) \\[4mm] e_y = \dfrac{1}{E}(\sigma_y - \nu\sigma_x - \nu\sigma_z) \\[4mm] e_z = \dfrac{1}{E}(\sigma_z - \nu\sigma_x - \nu\sigma_y) \\[4mm] \gamma_{xy} = \dfrac{\tau_{xy}}{G} \end{array} \right\} \tag{9-6}$$

As pointed out above, in Eq. 9-6 either σ_z or e_z is zero, depending upon whether we have plane stress or plane strain.

Now there are six independent equations in terms of the seven unknowns, and in order to obtain a unique solution to the problem, it is necessary that we have one more independent equation. In the two-dimensional theory of elasticity, this equation is the compatibility equation (see Eq. 4-131).

$$\frac{\partial^2 e_x}{\partial y^2} + \frac{\partial^2 e_y}{\partial x^2} = \frac{\partial^2 \gamma_{xy}}{\partial x \partial y} \tag{9-7}$$

Equation 9-7 is a consequence of the required continuity, single-valuedness and uniqueness of u and v, and their derivatives, and can be derived entirely from these requirements (see Art. 4-6). Hence, Eq. 9-7 is also an equation that is independent of any relation between stresses and strains.

Summarizing, we have in the two-dimensional theory of elasticity the following seven equations in terms of the seven stress and strain components,

$$\begin{aligned}
\frac{\partial \sigma_x}{\partial x} + \frac{\partial \tau_{yx}}{\partial y} &= 0 & \text{(a)} \\[2mm]
\frac{\partial \tau_{xy}}{\partial x} + \frac{\partial \sigma_y}{\partial y} &= 0 & \text{(b)} \\[2mm]
e_x &= \frac{1}{E}(\sigma_x - \nu\sigma_y - \nu\sigma_z) & \text{(c)} \\[2mm]
e_y &= \frac{1}{E}(\sigma_y - \nu\sigma_x - \nu\sigma_z) & \text{(d)} \\[2mm]
e_z &= \frac{1}{E}(\sigma_z - \nu\sigma_x - \nu\sigma_y) & \text{(e)} \\[2mm]
\gamma_{xy} &= \frac{\tau_{xy}}{G} & \text{(f)} \\[2mm]
\frac{\partial^2 e_x}{\partial y^2} + \frac{\partial^2 e_y}{\partial x^2} &= \frac{\partial^2 \gamma_{xy}}{\partial x \partial y} & \text{(g)}
\end{aligned} \tag{9-8}$$

Of these equations, (a) and (b) and (g) are independent of any stress–strain relation. Equation (g) however, does require uniqueness, finiteness and single-valuedness of u, v, and the derivatives of these.

With this review of two-dimensional elasticity theory, we can now proceed to a discussion of inelastic effects, i.e., plasticity.

9-3   The Equations of Plastic Flow. To explain the form which the plasticity equations take, we must consider a typical metal experimental data, specifically the stress–strain curve. Figure 9.1 is typical of tension stress–strain curves for many metals. It represents the following sequence of events:

1. Zero load and deflection is point O—load is applied slowly to the specimen until point A (the elastic limit) is reached. Between O and A the stress–strain relation is a linear one and Hooke's Law holds. In this region, the equations of elasticity are valid.

2. As the load is increased, the curve AB is traced. This is the region of plastic action, or inelastic action.

3. If at point B the load is gradually released, the curve BC (parallel to OA) is traced. If the load is then increased slowly, it is found that the curve is straight in the region CD, and parallel to OA. Only after D does the inelastic action start again. We have neglected hysteresis or internal friction losses in the portion BCD of the curve, as indicated by the zero distance between the two loops.

4. Steps (2) and (3) repeat at E, F, and G. In EFG also we have neglected hysteresis losses.

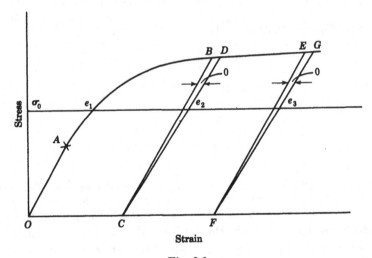

Fig. 9.1

For some metals it is also found that the shape of the stress–strain curve in the plastic region depends upon the rate at which the load is applied to the specimen.

The significance of the above statements is far-reaching, insofar as plasticity theory is concerned. They mean that

(a) The deformations in the plastic region are not single-valued functions of the stress. For example, in Fig. 9.1 the stress σ_0 corresponds to strains e_1 or e_2 or e_3. This effect is sometimes explained by the statement that the state of strain depends upon the prior history of loading of the structure.

(b) The time effect means that deformations are functions not only of x and y, but also of the time, t.

(c) The shape of the stress curve also means that in plastic flow problems it is not always possible to restrict the theory to that of small deformations. In other words,

$$e_x \neq \frac{\partial u}{\partial x}$$

but

$$e_a = \frac{\partial u}{\partial a} + \frac{1}{2}\left[\left(\frac{\partial u}{\partial a}\right)^2 + \left(\frac{\partial v}{\partial a}\right)^2 + \left(\frac{\partial w}{\partial a}\right)^2\right] \tag{9-9}$$

and the higher-order form of the strain tensor may be required to obtain solutions in some large deformation problems of plasticity theory.

In the theory of plasticity, since the equations of equilibrium must also hold for plastic flow, we have, for zero body forces,

$$\text{div } T = 0 \qquad \text{(a)}$$

$$\frac{\partial \sigma_x}{\partial x} + \frac{\partial \tau_{yx}}{\partial y} = 0 \qquad \text{(b)} \tag{9-10}$$

$$\frac{\partial \tau_{xy}}{\partial x} + \frac{\partial \sigma_y}{\partial y} = 0 \qquad \text{(c)}$$

as before.

However, because of (a) and (b) above, the compatibility equations do not hold. Also, because of (b), the relation between stress and strain should contain time effects. And, if the strains are large, the strain components should contain the higher-order terms, as noted in (c).

It follows, therefore, that Eq. 9-10b and c are, at this point, the only exact equations which hold for plastic flow. That is, we have two equations (in two dimensions) among three (or four) stress components (in two dimensions), and no equations as yet among the four (or three) strain components. This is the status of the *exact* mathematical theory of plasticity, in the two-dimensional case. In three dimensions the analogous statement would be: there are three equations of equilibrium among the six stress components and no equations as yet among the six strain components.

Approximate methods for solving inelastic strain problems have developed along two distinct lines. These two methods, along with the exact theory, may be said to describe completely the field of plastic flow and, for our purposes, we list the breakdown as follows:

I—Engineering Plasticity Theory

II—Technical Plasticity Theory

III—Mathematical (Exact) Plasticity Theory

Group I is frequently called "The Theory of Limit Design" and is essentially an engineering extension, to the field of plasticity, of the approximate Bernoulli–Euler theory of bending (see Art. 5-4). It assumes, in most cases, a stress–strain curve which neglects work hardening (i.e., BC is horizontal in Fig. 9.2a), and very frequently neglects elastic effects entirely, and assumes a stress–strain curve as in Fig. 9.2b. This method is described in Refs. (11), (39), (40), and (41) and will not be discussed further here.

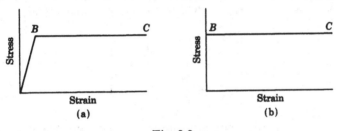

Fig. 9.2

Group II is the theory which attempts to obtain mathematical solutions of plasticity problems by using the available exact equations and by introducing various postulates to permit a solution.[1] The solution then obtained is checked against experimental data for some clue as to its usefulness for the case tested. This is commonly called "Plasticity

---

[1] In many respects, therefore, this theory is very similar to the current status of the general theory of turbulent flow. In the latter case, also, the known exact equations governing the phenomenon are less than the number of unknowns. In order to solve the problem, therefore, various additional equations, in the form of postulates, must be introduced. See Ref. (48) for further discussion of this point.

Theory" and represents the field of greatest activity at this time. We shall devote the remaining sections of the chapter to this field. For more complete treatments see Refs. (42), (43) and (44).

Group III, the exact theory of plasticity, is still awaiting formulation of the exact, necessary and sufficient equations governing plastic action. Hence this portion of the field has not yet gotten underway, and we shall not discuss it further.[2]

9-4   The Technical Theory of Plasticity. For two-dimensional plastic theory, we have again seven unknowns, these being the stress and strain components

$$\sigma_x,\ \sigma_y,\ \tau_{xy}(=\tau_{yx}),\ e_x,\ e_y,\ \gamma_{xy}(=\gamma_{yx})$$
$$\text{and either } \sigma_z \text{ or } e_z$$

The number of available equations (which we list at this time, and then discuss) are

2 equations of equilibrium

4 postulated stress-strain relations

1 plasticity condition

The equilibrium equations of plastic flow are the same as those of elastic action since their mode of derivation is entirely independent of either plasticity or elasticity. These are (for zero body forces)

$$\left.\begin{aligned} \frac{\partial \sigma_x}{\partial x} + \frac{\partial \tau_{yx}}{\partial y} &= 0 \\[2mm] \frac{\partial \tau_{xy}}{\partial x} + \frac{\partial \sigma_y}{\partial y} &= 0 \end{aligned}\right\} \tag{9-11}$$

There are many different postulated stress-strain relations, most of which are similar in form to the equations governing viscous flow (see Chapter 8). One such set of equations, due to St. Venant, is very frequently used, and we shall discuss these in some detail. This discussion shall be given in terms of the three-dimensional case, which will then be specialized to the two-dimensional plane strain condition.

The St. Venant equations for plastic flow can be given in the following

---

[2] There is yet another approach to the plasticity problem, based upon "dislocation theory." The procedures discussed above may be thought of as "macroscopic" or "large scale." They are based essentially on continuum or field theories. In the dislocation theory approach, *microscopic* behavior is considered and the plasticity problem is analyzed by examining the behavior of crystals and the molecular structure of the material. See Ref. (45) for a more complete description of this subject.

form (compare in Chapter 8 the derivation of the equations of viscous flow):

$$\frac{d}{dt}\begin{pmatrix} e_x & \frac{1}{2}\gamma_{xy} & \frac{1}{2}\gamma_{xz} \\ \frac{1}{2}\gamma_{yx} & e_y & \frac{1}{2}\gamma_{yz} \\ \frac{1}{2}\gamma_{zx} & \frac{1}{2}\gamma_{zy} & e_z \end{pmatrix} =$$

$$\frac{1+\nu}{E}\begin{pmatrix} \sigma_x & \tau_{xy} & \tau_{xz} \\ \tau_{yx} & \sigma_y & \tau_{yz} \\ \tau_{zx} & \tau_{zy} & \sigma_z \end{pmatrix} - \frac{\nu}{E}(\sigma_x+\sigma_y+\sigma_z)\begin{pmatrix} 1 & 0 & 0 \\ 0 & 1 & 0 \\ 0 & 0 & 1 \end{pmatrix}$$ (9-12)

in which E is here a *variable*, and t is the time.

It is convenient to give this equation as the sum of two separate equations, one of which is in scalar form, i.e., has all terms multiplying $\mathscr{E}_3$, the unit matrix. To do this, we note first, from the above, that

$$\frac{d}{dt}(e_x+e_y+e_z) = \frac{1-2\nu}{E}(\sigma_x+\sigma_y+\sigma_z)$$ (9-13)

which is in a scalar (invariant) form. In this equation, if $\nu = \frac{1}{2}$ then $e_x+e_y+e_z = 0$. But it may be shown (see Prob. 5, Chapter 4) that for small deformations $e_x+e_y+e_z$ is (neglecting higher-order terms) the change in volume due to deformations of a stressed body. In the St. Venant Theory, ν is taken equal to $\frac{1}{2}$, so that volume change is assumed equal to zero, and we have the so called "incompressible action."

Equation 9-13 is an equation of the required scalar form, since it is equivalent to

$$\frac{d}{dt}(e_x+e_y+e_z)\begin{pmatrix} 1 & 0 & 0 \\ 0 & 1 & 0 \\ 0 & 0 & 1 \end{pmatrix} = \frac{1-2\nu}{E}(\sigma_x+\sigma_y+\sigma_z)\begin{pmatrix} 1 & 0 & 0 \\ 0 & 1 & 0 \\ 0 & 0 & 1 \end{pmatrix}$$ (9-14)

However, in order to make Eq. 9-14 more useful for substitution in Eq. 9-12, we put it in the following equivalent form

$$\frac{d}{dt}\frac{(e_x+e_y+e_z)}{3}\begin{pmatrix} 1 & 0 & 0 \\ 0 & 1 & 0 \\ 0 & 0 & 1 \end{pmatrix} = \frac{1+\nu}{3E}(\sigma_x+\sigma_y+\sigma_z)\begin{pmatrix} 1 & 0 & 0 \\ 0 & 1 & 0 \\ 0 & 0 & 1 \end{pmatrix}$$

$$- \frac{\nu}{E}(\sigma_x+\sigma_y+\sigma_z)\begin{pmatrix} 1 & 0 & 0 \\ 0 & 1 & 0 \\ 0 & 0 & 1 \end{pmatrix}$$ (9-15)

Note that the right-hand side of Eq. 9-15 is just equal to

$$\frac{1-2\nu}{E}p,$$

in which p is the hydrostatic pressure; see Eq. 8-13.

Subtracting Eq. 9-15 from Eq. 9-12 we have, finally,

$$2\lambda \frac{d}{dt} \begin{pmatrix} \dfrac{2e_x - e_y - e_z}{3} & \dfrac{\gamma_{xy}}{2} & \dfrac{\gamma_{xz}}{2} \\[2ex] \dfrac{\gamma_{yx}}{2} & \dfrac{2e_y - e_z - e_x}{3} & \dfrac{\gamma_{yz}}{2} \\[2ex] \dfrac{\gamma_{zx}}{2} & \dfrac{\gamma_{zy}}{2} & \dfrac{2e_z - e_x - e_y}{3} \end{pmatrix}$$

$$= \begin{pmatrix} \dfrac{2\sigma_x - \sigma_y - \sigma_z}{3} & \tau_{xy} & \tau_{xz} \\[2ex] \tau_{yx} & \dfrac{2\sigma_y - \sigma_z - \sigma_x}{3} & \tau_{yz} \\[2ex] \tau_{zx} & \tau_{zy} & \dfrac{2\sigma_z - \sigma_x - \sigma_y}{3} \end{pmatrix} \qquad (9\text{-}16)$$

in which λ is a variable.[3]

Because of assumed volume constancy, $e_x + e_y + e_z = 0$, the left-hand side matrix of Eq. 9-16 is just the left-hand side matrix of Eq. 9-12, and Eq. 9-12 is therefore equivalent to the following *two* equations, which are the St. Venant stress–strain relations,

$$2\lambda \frac{d}{dt} \begin{pmatrix} e_x & \dfrac{\gamma_{xy}}{2} & \dfrac{\gamma_{xz}}{2} \\[2ex] \dfrac{\gamma_{yx}}{2} & e_y & \dfrac{\gamma_{yz}}{2} \\[2ex] \dfrac{\gamma_{zx}}{2} & \dfrac{\gamma_{zy}}{2} & e_z \end{pmatrix}$$

$$= \begin{pmatrix} \dfrac{2\sigma_x - \sigma_y - \sigma_z}{3} & \tau_{xy} & \tau_{xz} \\[2ex] \tau_{yx} & \dfrac{2\sigma_y - \sigma_z - \sigma_x}{3} & \tau_{yz} \\[2ex] \tau_{zx} & \tau_{zy} & \dfrac{2\sigma_z - \sigma_x - \sigma_y}{3} \end{pmatrix} \qquad (9\text{-}17)$$

and

$$e_x + e_y + e_z = 0$$

[3] Experiments indicate that pure hydrostatic pressure does not produce appreciable plastic deformation. In Chapter 8 it was shown that hydrostatic pressure is measured by $p = -\frac{1}{3}(\sigma_x + \sigma_y + \sigma_z)$, and the hydrostatic effect has therefore been essentially removed or subtracted from Equation 9-16.

For two-dimensional plane strain these become, since $e_z = 0$, $\sigma_z \neq 0$,

$$2\lambda\frac{d}{dt}\begin{pmatrix} e_x & \dfrac{\gamma_{xy}}{2} & 0 \\ \dfrac{\gamma_{yx}}{2} & e_y & 0 \\ 0 & 0 & 0 \end{pmatrix}$$

$$= \begin{pmatrix} \dfrac{2\sigma_x - \sigma_y - \sigma_z}{3} & \tau_{xy} & 0 \\ \tau_{yx} & \dfrac{2\sigma_y - \sigma_z - \sigma_x}{3} & 0 \\ 0 & 0 & \dfrac{2\sigma_z - \sigma_x - \sigma_y}{3} \end{pmatrix} \tag{9-18}$$

and

$$e_x + e_y = 0 \tag{9-19}$$

Equations 9-11 and 9-12, for two-dimensional plane strain, and therefore Eqs. 9-11, 9-18 and 9-19 represent six independent equations in terms of the seven stress–strain components. One additional equation is needed in order to permit a solution to the problem. This is the so-called "plasticity condition," which is an equation in terms of stresses and which postulates the condition for the onset of inelastic action.

The "plasticity condition" or "yield condition" is an equation in terms of the principal stresses, σ_1, σ_2 and σ_3. This relation generally requires the determination of a single experimental constant, σ_0, the yield stress in a simple tension text. This value, σ_0, is also assumed to hold for the simple compression test.

In terms of the principal stresses, the St. Venant stress tensor becomes

$$\begin{pmatrix} \dfrac{2\sigma_1 - \sigma_2 - \sigma_3}{3} & 0 & 0 \\ 0 & \dfrac{2\sigma_2 - \sigma_3 - \sigma_1}{3} & 0 \\ 0 & 0 & \dfrac{2\sigma_3 - \sigma_1 - \sigma_2}{3} \end{pmatrix} \tag{9-20}$$

If we assume an isotropic material, then the yield condition,

$$F(\sigma_1, \sigma_2, \sigma_3) = 0 \tag{9-21}$$

must contain the principal stresses in lumped invariant form, i.e.,

$$f(\sigma_0) = K_0 + K_1 I_1 + K_2 I_1^2 + K_{22} I_2 + K_3 I_3 + \cdots \qquad (9\text{-}22)$$

where (using the tensor of Eq. 9-20 and determining the invariants in the usual way)

$$\left.\begin{aligned}
I_1 &= 0 \\[4pt]
I_2 &= -\frac{(\sigma_1 - \sigma_2)^2 + (\sigma_2 - \sigma_3)^2 + (\sigma_3 - \sigma_1)^2}{6} \\[4pt]
I_3 &= \frac{(2\sigma_1 - \sigma_2 - \sigma_3)(2\sigma_2 - \sigma_3 - \sigma_1)(2\sigma_3 - \sigma_1 - \sigma_2)}{27}
\end{aligned}\right\} \qquad (9\text{-}23)$$

Also, the relation should be an even one in the stresses (if we assume that it holds for compression as well as tension stresses). Therefore, a relation in terms of I_2 and I_3 raised to an even power would seem to be called for.

It was shown in Chapter 2, in discussing the inertia tensor, a typical second-order tensor, that the maximum value of I_{xy} is given (in terms of principal moments of inertia) by

$$(I_{xy})_{\max}^2 = \left(\frac{I_{x_p x_p} - I_{y_p y_p}}{2}\right)^2 \qquad (9\text{-}24)$$

In terms of stresses this becomes

$$\tau_{\max}^2 = \left(\frac{\sigma_i - \sigma_j}{2}\right)^2 \qquad (9\text{-}25)$$

where i and j are different and take values 1, 2, and 3.

This is just the "maximum shearing-stress yield condition," as proposed by St. Venant and Tresca. It states that, if $\sigma_1 > \sigma_2 > \sigma_3$ then plasticity occurs when in the given structure,

$$\tau_{\max} = \frac{\sigma_1 - \sigma_3}{2} \qquad (9\text{-}26)$$

reaches the critical value τ_{crit} as determined by a simple tension test.[4]

One other commonly used plasticity condition is the "energy of distortion condition" introduced by von Mises and later by Hencky. It can be shown that the invariant I_2 is proportional to the energy of a

---

[4] This condition, stated mathematically, is that yielding will occur when

$$[(\sigma_2 - \sigma_3)^2 - 4\tau_{\text{crit}}^2][(\sigma_3 - \sigma_1)^2 - 4\tau_{\text{crit}}^2][(\sigma_1 - \sigma_2)^2 - 4\tau_{\text{crit}}^2] = 0$$

and it may be verified that this is given in terms of the invariants I_1, I_2, I_3, Equations 9-23. See Prob. 1 at the end of this chapter.

material subject to the stress–strain relation Eq. 9-16, when given in terms of principal stresses. Because this tensor has had the invariants $e_x + e_y + e_z$ and $\sigma_x + \sigma_y + \sigma_z$ removed from it, and because $\sigma_x + \sigma_y + \sigma_z$ is proportional to the hydrostatic tension or compression (see Chapter 8) it follows that relation Eq. 9-16 is one from which hydrostatic effects have been removed. Also, as pointed out earlier, hydrostatic tensions or compressions cause pure dilatations or compressions. Since these have been removed from the general stress–strain tensors, the energy of relation Eq. 9-16 is, in effect, a "distortion" energy, hence the name given to this plasticity condition. As usually stated, this condition is

$$(\sigma_1 - \sigma_2)^2 + (\sigma_2 - \sigma_3)^2 + (\sigma_3 - \sigma_1)^2 = 2\sigma_0^2 \qquad (9\text{-}27)$$

where σ_0 again is the simple tension test yield stress.

Thus, the solution of the two-dimensional plasticity problem can, in theory, be obtained in terms of

2 equations of equilibrium

4 postulated stress–strain relations

1 postulated plasticity condition

We point out again that five of the above equations are postulates. Hence any solution obtained by the use of the above equations must be compared with experimental results in order to determine the validity of the postulates. To date, no single set of postulates will account for all known plasticity phenomena.

In addition, it may be well to point out that the boundary conditions can be given in terms of stresses or strains or rates of strains (velocities). It may be shown that if the boundary conditions are given in terms of the velocities, then these velocities must satisfy certain compatibility conditions in addition to the equations governing the plastic action. This fact is mentioned in order to emphasize the difficulties which must be overcome to establish an exact theory of plasticity.[5]

However, the theory based upon postulated stress–strain relations and plasticity condition has given some important information concerning plastic action, and in the next section we discuss a simple, well-known solution based upon this set of equations.

9-5 A Solution of a Problem in Plastic Flow. To illustrate a typical application of the foregoing theory, the solution will be given

---

[5] Another difficulty may arise if the stresses and strains vary with time. There is some question as to exactly what requirements must be imposed upon time derivatives of tensors in order that they behave in physically acceptable ways. See Ref. (49).

for a well-known problem in plastic flow, the infinitely long, thick circular tube subjected to internal pressure; see Fig. 9.3. The method used is essentially that of Ref. (43). We assume a long cylinder and consider solutions which will be valid at large distances from the ends. Thus we have a plane strain problem, since at these points far from the ends, $e_z = 0$.

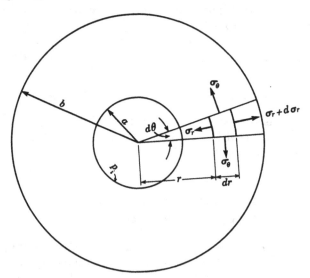

Fig. 9.3

The problem is most easily solved by the use of polar coordinates (r, θ, z) (see Chapter 3). Also, because of the axial symmetry, the problem is independent of the angle θ, and therefore all quantities are functions of r only. This property of axial symmetry, which essentially makes the problem a one-dimensional problem, is a tremendous simplification and leads to a relatively simple solution.

For this case, since the problem is one of plane strain, let

$$e_r = \text{unit strain in radial direction}$$
$$e_\theta = \text{unit strain in tangential direction}$$
$$e_z = \text{unit strain in } z \text{ direction}$$
$$\sigma_r = \text{unit stress in radial direction}$$
$$\sigma_\theta = \text{unit stress in tangential direction}$$
$$\sigma_z = \text{unit stress in } z \text{ direction}$$
$$u_r = \text{deformation in radial direction}$$
$$u_\theta = \text{deformation in tangential direction}$$
$$u_z = \text{deformation in } z \text{ direction}$$

For the plasticity portion of this problem, we have the following:
Assuming plane strain (see Eq. 3-79)

$$e_r = \frac{du_r}{dr}$$
$$e_\theta = \frac{u_r}{r}$$
$$e_z = 0$$
(9-28)

and the incompressibility condition

$$e_r + e_\theta + e_z = 0 \qquad (9\text{-}29)$$

gives at once

$$\frac{du_r}{dr} + \frac{u_r}{r} = 0 \qquad (9\text{-}30)$$

The equation of equilibrium of forces in the radial direction gives
(see Eq. 3-87)

$$\frac{d\sigma_r}{dr} + \frac{(\sigma_r - \sigma_\theta)}{r} = 0 \qquad (9\text{-}31)$$

The equilibrium equations in the θ and z directions are identically satisfied, $0 = 0$.

Let us further assume that the St. Venant plastic stress–strain relations hold and also the Von Mises–Hencky plasticity condition, which for this case become, since σ_r, σ_θ, and σ_z are principal stresses (in view of the fact that the shear stresses are zero on the faces on which these stresses act):

$$2\lambda \frac{d}{dt} \begin{pmatrix} e_r & 0 & 0 \\ 0 & e_\theta & 0 \\ 0 & 0 & e_z \end{pmatrix} = \begin{pmatrix} \dfrac{2\sigma_r - \sigma_\theta - \sigma_z}{3} & 0 & 0 \\ 0 & \dfrac{2\sigma_\theta - \sigma_z - \sigma_r}{3} & 0 \\ 0 & 0 & \dfrac{2\sigma_z - \sigma_r - \sigma_\theta}{3} \end{pmatrix}$$
(9-32)

and

$$(\sigma_r - \sigma_\theta)^2 + (\sigma_\theta - \sigma_z)^2 + (\sigma_z - \sigma_r)^2 = 2\sigma_0^2 \qquad (9\text{-}33)$$

Equations 9-30, 9-31, 9-32 and 9-33 are six equations for the six quantities e_r, e_θ, e_z, σ_r, σ_θ, and σ_z and therefore on purely formal mathematical grounds, a solution to the plasticity portion of the problem may be possible. We shall show that one can actually be obtained.

Physically, the problem to be solved is the following:

The thick circular tube is subjected to internal pressure, p_i. As this pressure increases, the radial and tangential stresses gradually increase as well. For smaller values of p_i these stresses are within the elastic limit, and the problem is one of the theory of elasticity. At a certain value of p_i the elastic limit is reached in accordance with the assumed Von Mises–Hencky condition (we shall show that this occurs at the inner radius, $r = a$), and as p_i is increased further the region of plastic stresses increases, such that the tube is under the condition shown in Fig. 9.4(a), which also shows the assumed stress–strain curve, (b).

We wish to obtain the solution corresponding to Fig. 9.4(a).

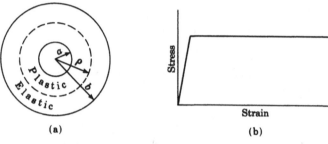

(a)   (b)

Fig. 9.4

The initial steps in the solution require a consideration of the elasticity problem. For elastic action we have equilibrium, as before (see Eq. 9-31),

$$\frac{d\sigma_r}{dr} + \frac{(\sigma_r - \sigma_\theta)}{r} = 0 \tag{9-34}$$

Hooke's Law (see Eq. 9-5),

$$\begin{pmatrix} e_r & 0 & 0 \\ 0 & e_\theta & 0 \\ 0 & 0 & e_z \end{pmatrix} = \frac{1+\nu}{E} \begin{pmatrix} \sigma_r & 0 & 0 \\ 0 & \sigma_\theta & 0 \\ 0 & 0 & \sigma_z \end{pmatrix} - \frac{\nu}{E}(\sigma_r + \sigma_\theta + \sigma_z) \begin{pmatrix} 1 & 0 & 0 \\ 0 & 1 & 0 \\ 0 & 0 & 1 \end{pmatrix} \tag{9-35}$$

and instead of the compatibility condition of elasticity we shall use the incompressibility condition corresponding to Poisson's ratio $\nu = \frac{1}{2}$ or

$$e_r + e_\theta + e_z = 0 \tag{9-36}$$

Using the values of e_r, e_θ and e_z as given in Eq. 9-28 (these values are independent of elasticity or plasticity action), Eq. 9-36 becomes, as before (Eq. 9-30),

$$\frac{du_r}{dr} + \frac{u_r}{r} = 0 \tag{9-37}$$

Equation 9-37 integrates at once to

$$u_r = \frac{C}{r} \tag{9-38}$$

or, in terms of quantities at the inner radius, $r = a$

$$u_r = u_a \frac{a}{r} \tag{9-39}$$

Equations 9-28 and 9-35, with $\nu = \frac{1}{2}$, give

$$\sigma_z = \frac{\sigma_r + \sigma_\theta}{2} \tag{9-40}$$

and therefore, from Eq. 9-35,

$$e_r = \frac{3}{4E}(\sigma_r - \sigma_\theta) \tag{9-41}$$

which, using Eqs. 9-28 and 9-39, is also given by

$$e_r = -\frac{au_a}{r^2} \tag{9-42}$$

or

$$\sigma_r - \sigma_\theta = -\frac{4E}{3}\frac{au_a}{r^2} \tag{9-43}$$

Using this, Eq. 9-34 becomes

$$\frac{d\sigma_r}{dr} - \frac{4E}{3}\frac{au_a}{r^3} = 0 \tag{9-44}$$

which can be integrated at once to give

$$\sigma_r = \frac{2E}{3}\frac{au_a}{r^2} + C_1 \tag{9-45}$$

Equations 9-45, 9-43, 9-40 and conditions at $r = a$ ($\sigma_r = -p_i$ at $r = a$) and at $r = b$($\sigma_r = 0$ at $r = b$) are sufficient to determine the constants C_1, u_a and also σ_r, σ_θ, σ_z. These last are given by

$$\left.\begin{array}{ll} \sigma_r = p_i\dfrac{a^2/b^2 - a^2/r^2}{1 - a^2/b^2} & \text{(a)} \\[3mm] \sigma_\theta = p_i\dfrac{a^2/b^2 + a^2/r^2}{1 - a^2/b^2} & \text{(b)} \\[3mm] \sigma_z = p_i\dfrac{a^2/b^2}{1 - a^2/b^2} & \text{(c)} \end{array}\right\} \tag{9-46}$$

Equation 9-46 shows that

σ_r is a minimum at the inner face

σ_θ is a maximum at the inner face

$\sigma_\theta - \sigma_r$ is a maximum at the inner face

σ_z is a constant.

Hence Eq. 9-46 indicates that plastic action will begin at the inner radius; see Eq. 9-33.

Let us now assume that $p_i > p_{i_{crit}}$ where $p_{i_{crit}}$ is the internal pressure corresponding to the onset of yield. Also, let us assume that p_i is less than the pressure necessary to put the entire tube in a plastic condition. Then the tube will be as shown in Fig. 9.4a, and we now proceed to determine the stresses and strains for this case.

In essence, the procedure used is the following:

1. The solution for the elastic portion has been obtained and is given by Eqs. 9-42 and 9-46.

2. The solution for the inelastic portion will be obtained using Eqs. 9-31, 9-32 and 9-33.

3. The two proceeding sets of solutions will be matched at the junction ρ of the plastic and elastic regions.

According to Eq. 9-33, plasticity will begin when

$$(\sigma_r - \sigma_\theta)^2 + (\sigma_\theta - \sigma_z)^2 + (\sigma_z - \sigma_r)^2 = 2\sigma_0{}^2 \tag{9-47}$$

For the case $p_i = p_{crit}$ and at $r = a$, i.e., the onset of plastic action in the tube, this becomes, upon substituting Eq. 9-46 with $r = a$,

$$(p_i)_{crit} = \frac{1 - a^2/b^2}{\sqrt{3}}\sigma_0 \tag{9-48}$$

Then at the radius $r = \rho$ which is the junction between the elastic and plastic regions, Eq. 9-48 also holds, with $a = \rho$ or

$$(p_i)_{crit} = \frac{1 - \rho^2/b^2}{\sqrt{3}}\sigma_0 \tag{9-49}$$

This value, $(p_i)_{crit}$, is therefore the value corresponding to p_i of the elastic solution, Eq. 9-46, with $a = \rho$ so that, the stresses in the elastic portion of the tube of Fig. 9.4(a) are given by

$$\left.\begin{aligned}
(\sigma_r)_{elast} &= \frac{\sigma_0}{\sqrt{3}}\left(\frac{\rho^2}{b^2} - \frac{\rho^2}{r^2}\right) \\[2mm]
(\sigma_\theta)_{elast} &= \frac{\sigma_0}{\sqrt{3}}\left(\frac{\rho^2}{b^2} + \frac{\rho^2}{r^2}\right) \\[2mm]
(\sigma_z)_{elast} &= \frac{\sigma_0}{\sqrt{3}}\frac{\rho^2}{b^2}
\end{aligned}\right\} \tag{9-50}$$

To obtain the stresses in the plastic portion, we use the equilibrium equations and the St. Venant stress–strain equations. These are, for equilibrium (see Eq. 9-31),

$$\frac{d\sigma_r}{dr} + \frac{(\sigma_r - \sigma_\theta)}{r} = 0 \qquad (9\text{-}51)$$

and for the stress–strain relations, which in expanded form (see Eq. 9-32) are

$$\frac{2\lambda}{dt} de_r = \frac{2\sigma_r - \sigma_\theta - \sigma_z}{3} \qquad (a)$$

$$\frac{2\lambda}{dt} de_\theta = \frac{2\sigma_\theta - \sigma_z - \sigma_r}{3} \qquad (b) \qquad \left.\right\} \qquad (9\text{-}52)$$

$$\frac{2\lambda}{dt} de_z = \frac{2\sigma_z - \sigma_r - \sigma_\theta}{3} \qquad (c)$$

we proceed as follows in order to obtain relations independent of time. Divide Eq. 9-52c by 9-52a and then by 9-52b. Then since $e_z = 0$ it follows that de_z is also equal to zero, and we have, for both relations,

$$2\sigma_z - \sigma_r - \sigma_\theta = 0 \qquad (9\text{-}53)$$

or

$$\sigma_z = \frac{\sigma_r + \sigma_\theta}{2} \qquad (9\text{-}54)$$

Using Eq. 9-33 and the above value of σ_z, we obtain

$$\sigma_\theta - \sigma_r = \frac{2}{\sqrt{3}}\sigma_0 \qquad (9\text{-}55)$$

which, when substituted in Eq. 9-51, gives

$$\frac{d\sigma_r}{dr} - \frac{2}{\sqrt{3}}\frac{\sigma_0}{r} = 0 \qquad (9\text{-}56)$$

This may be integrated to give

$$\sigma_r = \frac{2\sigma_0}{\sqrt{3}} \ln r + C_2 \qquad (9\text{-}57)$$

and C_2 is determined by noting (see Eq. 9-49) that for $r = \rho$, $\sigma_r = -(p_i)_{\text{crit}}$ or

$$C_2 = -\frac{\sigma_0}{\sqrt{3}}\left(2 \ln \rho + 1 - \frac{\rho^2}{b^2}\right) \qquad (9\text{-}58)$$

Now from Eqs. 9-57, 9-55 and 9-53 we get finally

$$
\left.
\begin{aligned}
(\sigma_r)_{\text{plast}} &= \frac{\sigma_0}{\sqrt{3}}\left(2\ln\frac{r}{\rho} - 1 + \frac{\rho^2}{b^2}\right) \qquad \text{(a)} \\[2mm]
(\sigma_\theta)_{\text{plast}} &= \frac{\sigma_0}{\sqrt{3}}\left(2\ln\frac{r}{\rho} + 1 + \frac{\rho^2}{b^2}\right) \qquad \text{(b)} \\[2mm]
(\sigma_z)_{\text{plast}} &= \frac{\sigma_0}{\sqrt{3}}\left(2\ln\frac{r}{\rho} + \frac{\rho^2}{b^2}\right) \qquad \text{(c)}
\end{aligned}
\right\} \qquad (9\text{-}59)
$$

Equation 9-59a permits the determination of the internal pressure corresponding to any radius ρ of the plastic front.

To determine the deformations of the tube, we note that Eqs. 9-29 and 9.30 hold for both the elastic and plastic states. Therefore Eq. 9-38 holds as well, or for both regions of Fig. 9-4(a),

$$
u_r = \frac{C}{r} \qquad (9\text{-}60)
$$

The determination of the constant C, for $r = \rho$, and $a = \rho$ (see Eqs. 9-45 and 9-46a), leads to

$$
u_\rho = \frac{3(p_i)_{\text{crit}}}{2E}\left(\frac{\rho}{1 - \rho^2/b^2}\right) \qquad (9\text{-}61)
$$

or using Eq. 9-48, with $a = \rho$

$$
u_\rho = \frac{3\sigma_0\rho}{2\sqrt{3}E} \qquad (9\text{-}62)
$$

Then, because from Eq. 9-60

$$
u_r = \frac{u_\rho\rho}{r} \qquad (9\text{-}63)
$$

we have, for the plastic region ($r > \rho$), and also for the elastic region ($r < \rho$),

$$
u_r = \frac{3\sigma_0\rho^2}{2\sqrt{3}Er} \qquad (9\text{-}64)
$$

Reference (43) contains additional solutions of the above type.

9-6 Summary. The fundamental difference between elastic and plastic action was discussed from the point of view of obtaining the equations necessary to solve the respective problems. It was pointed out that for the plastic problems, the exact equations have not as yet been obtained. However, by the use of postulated stress–strain relations and plasticity conditions, useful solutions can be obtained in a limited number of cases, and these can then be compared to experimental results for a check on the validity of the postulates. Some of the more commonly used stress–strain relations and plasticity conditions were discussed and finally the solution to a simple elastic–plastic problem was obtained.

Problems

1. Verify that the plasticity condition given in Footnote 4, p. 282 is equal to

$$4I_2{}^3 + 27I_3{}^2 + 36\tau^2_{\text{crit}}I_2{}^2 + 96\tau^4_{\text{crit}}I_2 + 64\tau^6_{\text{crit}} = 0$$

2. The thick-walled spherical shell under internal pressure, in an elastic–plastic condition, can be solved in a manner similar to the solution for the cylinder (see Art. 9-5 and Ref. 43). Obtain the complete solution for this problem.

3. Consider a beam of rectangular cross-section (b = width, h = depth) subjected to bending effects. Assume the beam is stressed in the elastic–plastic range and assume further the beam is to be analyzed using the Engineering Plasticity Theory (see Art. 9-3). Show

(a) The *elastic* moment, M_E, is related to the maximum bending stress, $\sigma_{\max}$ by the equation

$$M_E = \sigma_{\max}\frac{bh^2}{6}$$

(b) The *ultimate* moment, M_{ult}, which is the moment corresponding to *full* yield stress, σ_y, across the cross section is given by

$$M_{\text{ult}} = \sigma_y\frac{bh^2}{4}$$

(c) For *partial* yield stress on the cross section, corresponding to the stress–strain curve of Fig. 9-2a with $2\alpha h$ the depth of *elastic* action, show that

$$M = \sigma_y\frac{bh^2}{4}\left(1 - \frac{4}{3}\alpha^2\right)$$

(d) Recalling that, for a beam in the elastic range (see Eq. 5-56)

$$\frac{d^2v}{dx^2} = \frac{M_z}{EI_z}$$

show that if, at any section, the moment is $M > M_E$, where M

is given by the expression in Part (c) above, then the deformation (i.e., curvature) at this point can be given by

$$\frac{d^2v}{dx^2} = \frac{M}{E_{eff}I_z}$$

in which E_{eff} is an *effective* modulus of elasticity, related to the *linear* modulus of elasticity by the equation

$$E_{eff} = 2E\sqrt{\frac{3}{4}\left(1-\frac{4M}{\sigma_y bh^2}\right)}$$

4. Show that for a simply supported rectangular beam of length, l, with a concentrated load, P, at the center, the ultimate moment is given by

$$M_{ult} = \sigma_y\frac{bh^2}{4} = \frac{Pl}{4}$$

and therefore

$$P_{ult} = \frac{\sigma_y bh^2}{l}$$

5. For the beam of Prob. 4, with the loading a full uniform load of w/unit length, determine the ultimate loading w_{ult}.

6. Given a simply supported beam of length l, with a central load P and a maximum moment M_{max} equal to $M_{E_{max}}$. The cross section is rectangular. Using the results of Prob. 3, show the stress distribution across the cross section, due to bending, at $0.3l$, $0.375l$ and $0.5l$.

7. Given a simply supported beam of length $l = 20$ ft, uniformly loaded. $E = 1,000,000$ psi, $\sigma_y = 2500$ psi, cross section is 3 in. × 12 in. (actual size), $M_{max} = 1.45 M_{E_{max}}$. Using the results of Prob. 3, determine the center deflection. Compare with the value obtained, assuming the elastic deflection formula is valid for this case.

Chapter 10

TENSOR ANALYSIS AS RELATED TO DIMENSIONAL ANALYSIS[1]

10-1 Introduction. In this chapter we describe a type of dimensional reasoning which is based upon matrix-tensor analysis and which has many applications in the various fields of applied mechanics discussed in this book, and other fields as well.

10-2 Outline of the Method. The equations of mathematical physics are given, quantitatively, in the form of partial differential equations. These equations which express laws or relations of nature must, in their fundamental forms, be given in an expression which is independent of the coordinate system used and also independent of the axial orientation of the coordinate system. If this were not so, then different investigators, using different permissible coordinate systems, would arrive at different solutions to the (essentially) same problem. Such a situation is not admissible in our existing physical system.

The method, therefore, depends upon the fact that certain physical quantities must be independent of orientation of axes. For example, when we use the term "energy" in any physical or engineering problem, it is not necessary to specify a system of axes to go along with the word energy. On the other hand if we say that a force vector has a component of 10 pounds in the x direction, it is necessary to state what the x direction is.

Because certain quantities are independent of axial orientation—or invariant—the equations for these quantities must be expressed in invariant form as well. In this discussion we consider the invariants of tensors and we *postulate* a fundamental form of invariant differential equation. Then, knowing the variables that enter into any given problem, and using this postulated form of differential equation, we find

---

[1] The material in this chapter is based upon a paper, "On an application of dimensional analysis", *American Journal of Physics*, Feb. 1951.

that certain equations can be written in *qualitatively* complete form, practically by inspection.

The theory developed herein will be formulated on the basis of a three-dimensional (or two-dimensional) physical world such as we exist in. However, it should be noted that the fundamental theory is independent of the number of dimensions and if, in fact, there exists a four-, five-, or n-dimensional "physical" world then the theory given here would apply as well to these "worlds."

It will be recalled that the 3×3 tensor (see p. 45),

$$A = \begin{pmatrix} a_{11} & a_{12} & a_{13} \\ a_{21} & a_{22} & a_{23} \\ a_{31} & a_{32} & a_{33} \end{pmatrix} \tag{10-1}$$

has three invariants. These are

1. The trace or sum of the diagonal elements,

$$I_1 = a_{11} + a_{22} + a_{33} \tag{10-2}$$

2. The sum of the two-rowed principal minors,

$$I_2 = \begin{vmatrix} a_{22} & a_{23} \\ a_{32} & a_{33} \end{vmatrix} + \begin{vmatrix} a_{11} & a_{13} \\ a_{31} & a_{33} \end{vmatrix} + \begin{vmatrix} a_{11} & a_{12} \\ a_{21} & a_{22} \end{vmatrix} \tag{10-3}$$

3. The determinant of the matrix,

$$I_3 = \begin{vmatrix} a_{11} & a_{12} & a_{13} \\ a_{21} & a_{22} & a_{23} \\ a_{31} & a_{32} & a_{33} \end{vmatrix} \tag{10-4}$$

For the two-dimensional tensor,

$$A = \begin{pmatrix} a_{11} & a_{12} \\ a_{21} & a_{22} \end{pmatrix} \tag{10-5}$$

there are two invariants,

$$I_1 = a_{11} + a_{22} \tag{10-6}$$

and

$$I_2 = \begin{vmatrix} a_{11} & a_{12} \\ a_{21} & a_{22} \end{vmatrix} \tag{10-7}$$

Now, the elements of the tensor (Eq. 10-1) may be thought of as a measure of the variation of a characteristic quantity in a three-dimensional space. As an example, the stress tensor,

$$T = \begin{pmatrix} \sigma_x & \tau_{xy} & \tau_{xz} \\ \tau_{yx} & \sigma_y & \tau_{yz} \\ \tau_{zx} & \tau_{zy} & \sigma_z \end{pmatrix} \tag{10-8}$$

is a measure of the state of stress acting on a differential volume of material.

Also, the differential volume is itself an invariant quantity. There-fore, the following basic hypothesis is made:

Given a physical invariant, η, which is dependent upon the volume V, then if a characteristic tensor for the problem is known, the follow-ing *must* be a form of equation among the variables,

$$\frac{d\eta}{dV} = C_1 K_1 I_1 + K_2(C_{21} I_1{}^2 + C_{22} I_2) + K_3(C_{31} I_1{}^3 + C_{32} I_1 I_2 + C_{33} I_3) + \cdots$$

$$(10\text{-}9)$$

in which

η = the invariant

K's = physical invariants of proper dimensions to give dimensional homogeneity.

C's = constants

I's = tensor invariants

There is not necessarily only one equation of this form for any given problem, since all physical problems have more than one tensor. There may be other secondary forms of the equation which do not use tensor invariants. But if the tensor elements appear in the equation, they must appear in the above form.

The general three-dimensional surface area element dA is a vector quantity. Therefore, its elements are not invariants. An invariant form using area elements may be obtained by taking the scalar product of dA with a vector quantity. This will not be considered in the present discussion. That is, we assume that surface area effects do not enter into the equation for the invariant. However, the specialized two-dimensional form of the three-dimensional equations corresponding to identical conditions in parallel planes does apply and in this case our fundamental equation takes the form

$$\frac{d\eta}{dA} = C_1 K_1 I_1 + K_2(C_{21} I_1{}^2 + C_{22} I_2) + \cdots \qquad (10\text{-}10)$$

In Eqs. 10-9 and 10-10 it will be noted that the expressions assumed assign positive integral values to the exponents of the invariants. The reason for this is as follows:

Each of the tensor invariant quantities may be either positive, nega-tive, or zero at any point. The left-hand sides of the equations are made up of real, finite quantities. In order that the right-hand side may not have imaginary or infinite terms, it is necessary that the exponents be positive integers.

Before proceeding with applications, we list several of the physical invariants which occur in the fields of applied mechanics:

p = pressure
V = volume
t = temperature
η = energy
E, G = isotropic moduli of elasticity
ν = viscosity

There are others, but the ones given above are typical.

In order to illustrate the application of the fundamental hypotheses, Eqs. 10-9 and 10-10, we shall consider various fields in applied mechanics, including some of those considered in other chapters of this text.

10-3 Applications to Theory of Plates and Shells. We consider thin plates with small deflections (see Chapter 7). In this field, we may make use of results obtained for the analogous one-dimensional structure—the beam—with small deflections[2]. We know that the fundamental quantities which enter into beam analysis, Fig. 10.3 (see Ref. (10) and Chapter 5), are (using w to represent the deflection)

$$\frac{d^2w}{dx^2} \quad \text{or} \quad \frac{1}{r}, \text{ the curvature} \qquad (10\text{-}11)$$

$$M, \text{ the bending moment} \qquad (10\text{-}12)$$

$$\left.\begin{array}{c} \dfrac{d^2}{dx^2}\left(\dfrac{d^2w}{dx^2}\right) \\[2ex] \dfrac{d^2M}{dx^2} \end{array}\right\} \qquad (10\text{-}13)$$

and the physical invariant is

$$EI, \text{ the stiffness} \qquad (10\text{-}14)$$

By analogy, the corresponding tensor quantities for the two-dimensional thin plate (see Chapter 7) are given by

$$\begin{pmatrix} \dfrac{\partial^2 w}{\partial x^2} & \dfrac{\partial^2 w}{\partial x\,\partial y} \\[2ex] \dfrac{\partial^2 w}{\partial y\,\partial x} & \dfrac{\partial^2 w}{\partial y^2} \end{pmatrix} \quad \text{or} \quad \begin{pmatrix} -\dfrac{1}{r_{xx}} & \dfrac{1}{r_{xy}} \\[2ex] \dfrac{1}{r_{yx}} & -\dfrac{1}{r_{yy}} \end{pmatrix} \qquad (10\text{-}15)$$

[2] We are, in effect, using in this case a "generalization technique" in which a one-dimensional result is generalized to higher dimensions. The key step in the procedure as used here is the substitution of higher-dimensional tensor quantities for the corresponding one-dimensional term. Generalization techniques in different forms are used

Footnote continued on page 297.

$$\begin{pmatrix} M_{xx} & -M_{xy} \\ M_{yx} & M_{yy} \end{pmatrix} \tag{10-16}$$

$$\begin{pmatrix} \dfrac{\partial^2}{\partial x^2} & \dfrac{\partial^2}{\partial x \partial y} \\[2ex] \dfrac{\partial^2}{\partial y \partial x} & \dfrac{\partial^2}{\partial y^2} \end{pmatrix} \begin{pmatrix} \dfrac{\partial^2 w}{\partial x^2} & \dfrac{\partial^2 w}{\partial x \partial y} \\[2ex] \dfrac{\partial^2 w}{\partial y \partial x} & \dfrac{\partial^2 w}{\partial y^2} \end{pmatrix} = \begin{pmatrix} \dfrac{\partial^4 w}{\partial x^4} + \dfrac{\partial^4 w}{\partial x^2 \partial y^2} & \dfrac{\partial^4 w}{\partial x^3 \partial y} + \dfrac{\partial^4 w}{\partial y^3 \partial x} \\[2ex] \dfrac{\partial^4 w}{\partial x^3 \partial y} + \dfrac{\partial^4 w}{\partial y^3 \partial x} & \dfrac{\partial^4 w}{\partial y^4} + \dfrac{\partial^4 w}{\partial x^2 \partial y^2} \end{pmatrix} \tag{10-17}$$

$$\begin{pmatrix} \dfrac{\partial^2}{\partial x^2} & \dfrac{\partial^2}{\partial x \partial y} \\[2ex] \dfrac{\partial^2}{\partial y \partial x} & \dfrac{\partial^2}{\partial y^2} \end{pmatrix} \begin{pmatrix} M_{xx} & -M_{xy} \\ M_{yx} & M_{yy} \end{pmatrix}$$

$$= \begin{pmatrix} \dfrac{\partial^2 M_{xx}}{\partial x^2} + \dfrac{\partial^2 M_{yx}}{\partial x \partial y} & -\dfrac{\partial^2 M_{xy}}{\partial x^2} + \dfrac{\partial^2 M_{yy}}{\partial x \partial y} \\[2ex] \dfrac{\partial^2 M_{xx}}{\partial x \partial y} + \dfrac{\partial^2 M_{yx}}{\partial y^2} & -\dfrac{\partial^2 M_{xy}}{\partial x \partial y} + \dfrac{\partial^2 M_{yy}}{\partial y^2} \end{pmatrix} \tag{10-18}$$

Corresponding to the stiffness we have the "flexural rigidity" given by

$$D = \frac{Eh^3}{12(1 - \nu^2)} \tag{10-19}$$

Proceeding now to some applications, we know that the strain energy stored in a beam due to bending moments (see Ref. (10)) is

$$\eta = \frac{EI}{2} \int_{\text{E.L.}} \left(\frac{d^2 w}{dx^2} \right)^2 dx \tag{10-20}$$

For the plate, by analogy, and by virtue of Eq. 10-17, we would look for an expression of the form (see Eq. 10-10).

$$\eta = D \iint_A \left[C_{21} \left(\frac{\partial^2 w}{\partial x^2} + \frac{\partial^2 w}{\partial y^2} \right)^2 + C_{22} \left(\frac{\partial^2 w}{\partial x^2} \frac{\partial^2 w}{\partial y^2} - \left(\frac{\partial^2 w}{\partial x \partial y} \right)^2 \right) \right] dA \tag{10-21}$$

and this is the actual form as given in Ref. (19).

Footnote continued from page 296

in different fields. For example, in visco-elastic problems (the exact theory of which is not yet known) several authors have arbitrarily generalized to three dimensions the one-dimensional stress–strain law deducible from models. See also Ref. (50) for other examples.

Other possible forms follow at once as

$$\left.\begin{aligned}\eta &= D \iint_A \left[C_{21}\left(\frac{1}{r_{xx}} + \frac{1}{r_{yy}}\right)^2 + C_{22}\left(\frac{1}{r_{xx}}\frac{1}{r_{yy}} - \left\{\frac{1}{r_{xy}}\right\}^2\right)\right] dA \\ \eta &= \frac{1}{D}\iint_A \left[C'_{21}(M_{xx} + M_{yy})^2 + C'_{22}(M_{xx}M_{yy} - M_{xy}^2)\right] dA\end{aligned}\right\} \quad (10\text{-}22)$$

Now consider a plate subjected to a transverse load of q pounds per unit area.

This loading does not depend upon orientation of the axes and in addition can be thought of as a term of proper fundamental form (see Eq. 10-10)

$$\frac{d(\ )}{dA} \tag{10-23}$$

Referring again to the beam problem, we have for beams, with q equal to the running load per unit length (see Fig. 10.3 and Ref. 10),

$$q = \frac{d^2 M}{dx^2} \tag{10-24}$$

and

$$q = EI\frac{d^2}{dx^2}\left(\frac{d^2 w}{dx^2}\right) \tag{10-25}$$

The extension to plates using Eq. 10-10 and 10-18 would give at once (see Eq. 7-53 and Ref. 19)

$$q = C_1\left(\frac{\partial^2 M_{xx}}{\partial x^2} + \frac{\partial^2 M_{yx}}{\partial x\,\partial y} + \frac{\partial^2 M_{yy}}{\partial y^2} - \frac{\partial^2 M_{xy}}{\partial x\,\partial y}\right) \tag{10-26}$$

and (see Eq. 7-56 and Eq. 10-17)

$$q = DC_1\left(\frac{\partial^4 w}{\partial x^4} + \frac{2\,\partial^4 w}{\partial x^2\,\partial y^2} + \frac{\partial^4 w}{\partial y^4}\right) \tag{10-27}$$

As a further application we consider the thin plate with applied mid-plane edge stresses, and again we make use of beam results. The plate loaded with edge stresses (using the notation of Ref. 19) is shown in Fig. 10.1, from which it is apparent that the N's are identical to the stresses σ and τ and therefore the N's are elements of the tensor

$$N = \begin{pmatrix} N_x & N_{xy} \\ N_{yx} & N_y \end{pmatrix} \tag{10-28}$$

The one-dimensional structure under bending and normal loads is shown in Fig. 10.2, so that, neglecting higher-order terms, the component of F per unit length normal to dS is given by

$$\frac{F}{\rho} = F\frac{\partial^2 w}{\partial x^2} \qquad (10\text{-}29)$$

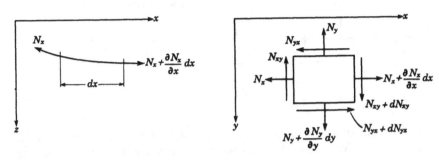

Fig. 10.1

For the plate, therefore, the normal components of the applied edge stresses would be given by

$$
\begin{pmatrix} N_x & N_{xy} \\[2ex] N_{yx} & N_y \end{pmatrix}
\begin{pmatrix} \dfrac{\partial^2 w}{\partial x^2} & \dfrac{\partial^2 w}{\partial x \partial y} \\[2ex] \dfrac{\partial^2 w}{\partial y \partial x} & \dfrac{\partial^2 w}{\partial y^2} \end{pmatrix}
$$

$$\qquad (10\text{-}30)$$

$$
= \begin{pmatrix} N_x\dfrac{\partial^2 w}{\partial x^2} + N_{xy}\dfrac{\partial^2 w}{\partial y \partial x} & N_x\dfrac{\partial^2 w}{\partial x \partial y} + N_{xy}\dfrac{\partial^2 w}{\partial y^2} \\[3ex] N_{yx}\dfrac{\partial^2 w}{\partial x^2} + N_y\dfrac{\partial^2 w}{\partial x \partial y} & N_{yx}\dfrac{\partial^2 w}{\partial x \partial y} + N_y\dfrac{\partial^2 w}{\partial y^2} \end{pmatrix}
$$

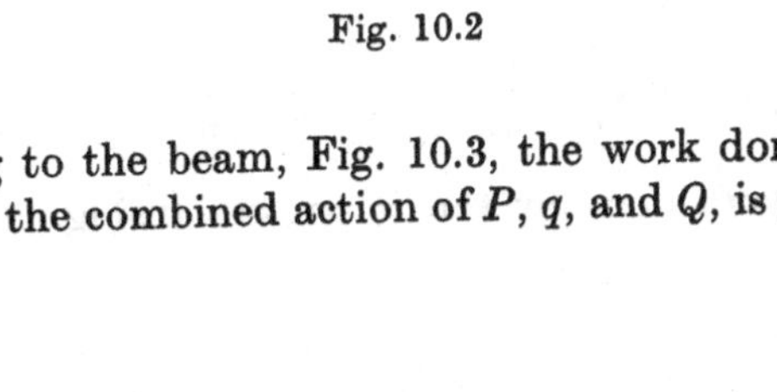

Fig. 10.2

Again referring to the beam, Fig. 10.3, the work done by the known force P, due to the combined action of P, q, and Q, is given (the reader should verify this) by

$$\eta = \int_0^l P\left[\frac{du}{dx} + \frac{1}{2}\left(\frac{dw}{dx}\right)^2\right] dx \tag{10-31}$$

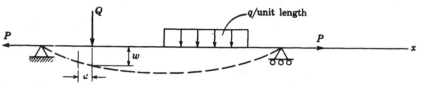

Fig. 10.3

The extension of this result to the thin plate would give, for the work done by the N forces, the corresponding invariant of the following expression:

$$\left[\begin{pmatrix} N_x & N_{xy} \\ \\ N_{yx} & N_y \end{pmatrix} \begin{pmatrix} \dfrac{\partial u}{\partial x} & \dfrac{1}{2}\left(\dfrac{\partial u}{\partial y}+\dfrac{\partial v}{\partial x}\right) \\ \\ \dfrac{1}{2}\left(\dfrac{\partial v}{\partial x}+\dfrac{\partial u}{\partial y}\right) & \dfrac{\partial v}{\partial y} \end{pmatrix} \right. $$

$$\left. + \frac{1}{2}\begin{pmatrix} N_x & N_{xy} \\ \\ N_{yx} & N_y \end{pmatrix} \begin{pmatrix} \dfrac{\partial w}{\partial x} \\ \\ \dfrac{\partial w}{\partial y} \end{pmatrix} \begin{pmatrix} \dfrac{\partial w}{\partial x} & \dfrac{\partial w}{\partial y} \end{pmatrix} \right] \tag{10-32}$$

Equation 10-32, when expanded, becomes

$$\begin{pmatrix} N_x\dfrac{\partial u}{\partial x}+\dfrac{N_{xy}}{2}\left(\dfrac{\partial u}{\partial y}+\dfrac{\partial v}{\partial x}\right) & \dfrac{N_x}{2}\left(\dfrac{\partial u}{\partial y}+\dfrac{\partial v}{\partial x}\right)+N_{xy}\dfrac{\partial v}{\partial y} \\ \\ N_{yx}\dfrac{\partial u}{\partial x}+\dfrac{N_y}{2}\left(\dfrac{\partial v}{\partial x}+\dfrac{\partial u}{\partial y}\right) & \dfrac{N_{yx}}{2}\left(\dfrac{\partial u}{\partial y}+\dfrac{\partial v}{\partial x}\right)+N_y\dfrac{\partial v}{\partial y} \end{pmatrix}$$

$$+ \frac{1}{2}\begin{pmatrix} N_x\left(\dfrac{\partial w}{\partial x}\right)^2+N_{xy}\dfrac{\partial w}{\partial y}\dfrac{\partial w}{\partial x} & N_x\dfrac{\partial w}{\partial x}\dfrac{\partial w}{\partial y}+N_{xy}\left(\dfrac{\partial w}{\partial y}\right)^2 \\ \\ N_{yx}\left(\dfrac{\partial w}{\partial x}\right)^2+N_y\dfrac{\partial w}{\partial y}\dfrac{\partial w}{\partial x} & N_{yx}\dfrac{\partial w}{\partial x}\dfrac{\partial w}{\partial y}+N_y\left(\dfrac{\partial w}{\partial y}\right)^2 \end{pmatrix} \tag{10-33}$$

from which, the work done by the N stresses (using the invariant I_1

of the above) is given by

$$
\eta = C_1 \iint \left[N_x \frac{\partial u}{\partial x} + \frac{N_{xy}}{2}\left(\frac{\partial u}{\partial y} + \frac{\partial v}{\partial x}\right) + N_y \frac{\partial v}{\partial y} + \frac{N_{yx}}{2}\left(\frac{\partial u}{\partial y} + \frac{\partial v}{\partial x}\right) \right.
$$
$$
\left. + \frac{N_x}{2}\left(\frac{\partial w}{\partial x}\right)^2 + \frac{N_{xy}}{2}\frac{\partial w}{\partial y}\frac{\partial w}{\partial x} + \frac{N_y}{2}\left(\frac{\partial w}{\partial y}\right)^2 + \frac{N_{yx}}{2}\frac{\partial w}{\partial x}\frac{\partial w}{\partial y} \right] dA
$$

(10-34)

and therefore the total work done by the plate in bending and by the edge force is given (see Ref. 19) by the sum of the bending work (Eq. 10-21) and the in-plane force work (Eq. 10-34)—

$$
\eta = \iint_A \left\langle C_1 \left[N_x \frac{\partial u}{\partial x} + N_y \frac{\partial v}{\partial y} + N_{xy}\left(\frac{\partial u}{\partial y} + \frac{\partial v}{\partial x}\right) \right. \right.
$$
$$
\left. + \frac{1}{2}\left\{ N_x\left(\frac{\partial w}{\partial x}\right)^2 + N_y\left(\frac{\partial w}{\partial y}\right)^2 + 2N_{xy}\frac{\partial w}{\partial x}\frac{\partial w}{\partial y} \right\} \right]
$$
$$
\left. + C_2 D\left(\frac{\partial^2 w}{\partial x^2} + \frac{\partial^2 w}{\partial y^2}\right)^2 + C_3 D\left[\frac{\partial^2 w}{\partial x^2}\frac{\partial^2 w}{\partial y^2} - \left(\frac{\partial^2 w}{\partial x \partial y}\right)^2\right] \right\rangle dA
$$

(10-35)

The equations corresponding to Eqs. 10-26 and 10-27 follow at once from Eq. 10-30:

$$
q + N_x \frac{\partial^2 w}{\partial x^2} + N_y \frac{\partial^2 w}{\partial y^2} + 2N_{xy}\frac{\partial^2 w}{\partial x \partial y} = C_1\left(\frac{\partial^2 M_{xx}}{\partial x^2} + \frac{\partial^2 M_{yx}}{\partial x \partial y} + \frac{\partial^2 M_{yy}}{\partial y^2} - \frac{\partial^2 M_{xy}}{\partial x \partial y}\right)
$$

(10-36)

and

$$
q + N_x \frac{\partial^2 w}{\partial x^2} + N_y \frac{\partial^2 w}{\partial y^2} + 2N_{xy}\frac{\partial^2 w}{\partial x \partial y} = DC_1\left(\frac{\partial^4 w}{\partial x^4} + 2\frac{\partial^4 w}{\partial x^2 \partial y^2} + \frac{\partial^4 w}{\partial y^4}\right)
$$

(10-37)

Equation 10-37 is just the Kármán large deflection equation for plates (see Eq. 7-121b). Incidentally, the reader should note that although Eq. 7-121a does represent a balance of invariant terms (the reader should prove the terms are invariants), it is not an equation given in the fundamental form postulated in this chapter (Eq. 10-10) and hence the equation can not be obtained using the methods of this chapter.

10-4 Applications to the Theory of Elasticity. The tensors which appear in the theory of isotropic elasticity are the stress and strain tensors (see Chapter 4) given respectively by

$$
T = \begin{pmatrix} \sigma_x & \tau_{xy} & \tau_{xz} \\ \tau_{yx} & \sigma_y & \tau_{yz} \\ \tau_{zx} & \tau_{zy} & \sigma_z \end{pmatrix}
$$

(10-38)

and (in its linear form)

$$
\eta = \begin{pmatrix}
\dfrac{\partial u}{\partial x} & \dfrac{1}{2}\left(\dfrac{\partial u}{\partial y}+\dfrac{\partial v}{\partial x}\right) & \dfrac{1}{2}\left(\dfrac{\partial u}{\partial z}+\dfrac{\partial w}{\partial x}\right) \\[2ex]
\dfrac{1}{2}\left(\dfrac{\partial v}{\partial x}+\dfrac{\partial u}{\partial y}\right) & \dfrac{\partial v}{\partial y} & \dfrac{1}{2}\left(\dfrac{\partial v}{\partial z}+\dfrac{\partial w}{\partial y}\right) \\[2ex]
\dfrac{1}{2}\left(\dfrac{\partial w}{\partial x}+\dfrac{\partial u}{\partial z}\right) & \dfrac{1}{2}\left(\dfrac{\partial w}{\partial y}+\dfrac{\partial v}{\partial z}\right) & \dfrac{\partial w}{\partial z}
\end{pmatrix}
\tag{10-39}
$$

Just as for the plates and shells cases, some information may be obtained by analogy to the simpler types of elastic action. For example, the strain energy stored in an isotropic bar subjected to pure tension (see Ref. 10) is

$$
\eta = \frac{E}{2}\int_0^l \left(\frac{\partial u}{\partial x}\right)^2 dx
\tag{10-40}
$$

This leads at once to the following *general* expression for strain energy:

$$
\eta = E\int_V \left\{ C_{21}\left(\frac{\partial u}{\partial x}+\frac{\partial v}{\partial y}+\frac{\partial w}{\partial z}\right)^2 + C_{22}\left[\left(\frac{\partial v}{\partial y}\frac{\partial w}{\partial z}+\frac{\partial u}{\partial x}\frac{\partial w}{\partial z}+\frac{\partial u}{\partial x}\frac{\partial v}{\partial y}\right)\right.\right.
$$

$$
\left.\left. -\frac{1}{4}\left(\frac{\partial v}{\partial z}+\frac{\partial w}{\partial y}\right)^2 -\frac{1}{4}\left(\frac{\partial u}{\partial z}+\frac{\partial w}{\partial x}\right)^2 -\frac{1}{4}\left(\frac{\partial u}{\partial y}+\frac{\partial v}{\partial x}\right)^2\right]\right\}dV
\tag{10-41}
$$

and although this differs in form from the expression given by Love (Ref. 12) the two equations may easily be shown to be identical.

Now consider the St. Venant torsion problem which was discussed in some detail in Chapter 5. We have, for the strain matrix for this problem,

$$
\eta = \begin{pmatrix}
0 & 0 & \tfrac{1}{2}\gamma_{xz} \\
0 & 0 & \tfrac{1}{2}\gamma_{yz} \\
\tfrac{1}{2}\gamma_{zx} & \tfrac{1}{2}\gamma_{zy} & 0
\end{pmatrix}
\tag{10-42}
$$

(see Eq. 5-101) with

$$
\left.\begin{aligned}
\gamma_{zx} &= \alpha\left(\frac{\partial \phi}{\partial x}-y\right) \\[2ex]
\gamma_{zy} &= \alpha\left(\frac{\partial \phi}{\partial y}+x\right)
\end{aligned}\right\}
\tag{10-43}
$$

The twist per unit length, α, is an invariant. Therefore the invariant for the torsion problem may be taken as

$$I_2 = \alpha^2 \left[\left(\frac{\partial \phi}{\partial x} - y \right)^2 + \left(\frac{\partial \phi}{\partial y} + x \right)^2 \right]$$ (10-44)

with the physical invariants as follows:

α = twist per unit length
G = modulus of elasticity in shear

Then for the energy stored in the bar we have (using the requirements of dimensional homogeneity)

$$\eta = C G \alpha^2 \int_V \left[\left(\frac{\partial \phi}{\partial x} - y \right)^2 + \left(\frac{\partial \phi}{\partial y} + x \right)^2 \right] dV$$ (10-45)

Also, the applied torque M_T is certainly independent of the orientation of the x and y axes, and therefore we look for an equation in the form

$$M_T = C_1 G \alpha \int_A \left[\left(\frac{\partial \phi}{\partial x} - y \right)^2 + \left(\frac{\partial \phi}{\partial y} + x \right)^2 \right] dA$$ (10-46)

and this may easily be put in the form for M_T as given by Eq. 5-130.

10-5 Applications to Capillarity, Membranes, and Vibrations.

The problems in these fields are similar to those in the field of thin plates; hence, in general the same tensors apply.

In capillarity and membrane problems we have a relation between pressure p (i.e., lateral loadings similar to q in the plate theory), edge stresses, and radii of curvature. An elementary dimensional and invariant analysis indicates that the form of the equation is

$$p = SC \left(\frac{1}{r_{xx}} + \frac{1}{r_{yy}} \right)$$ (10-47)

or equivalently, for small deflections w,

$$p = SC \left(\frac{\partial^2 w}{\partial x^2} + \frac{\partial^2 w}{\partial y^2} \right)$$ (10-48)

where S is the edge stress in pounds per unit length, the physical invariant in this problem.

In the vibration of plates, the inertia term $K(\partial^2 w / \partial t^2)$ corresponds to the q loading of the ordinary plate theory. Hence, edges stresses neglected, the equation becomes (see Eq. 10-27)

$$K \frac{\partial^2 w}{\partial t^2} = D C'_1 \left(\frac{\partial^4 w}{\partial x^4} + 2 \frac{\partial^4 w}{\partial x^2 \partial y^2} + \frac{\partial^4 w}{\partial y^4} \right)$$ (10-49)

and edge stresses included (see Eq. 10-37), the equation becomes

$$K\frac{\partial^2 w}{\partial t^2} + N_x \frac{\partial^2 w}{\partial x^2} + N_y \frac{\partial^2 w}{\partial y^2} + 2N_{xy}\frac{\partial^2 w}{\partial x \partial y} = DC'_1\left(\frac{\partial^4 w}{\partial x^4} + 2\frac{\partial^4 w}{\partial x^2 \partial y^2} + \frac{\partial^4 w}{\partial y^4}\right)$$

(10-50)

10-6 Applications to Fluid Mechanics. As a last example we consider the form which the energy equation takes in fluid mechanics. We consider an infinite fluid, steady state, with uniform conditions at infinity. We saw that the rate of strain tensor is a fundamental tensor of fluid mechanics (see Eq. 8-24), this being given by

$$\begin{pmatrix} \dfrac{\partial u}{\partial x} & \dfrac{1}{2}\left(\dfrac{\partial u}{\partial y}+\dfrac{\partial v}{\partial x}\right) & \dfrac{1}{2}\left(\dfrac{\partial u}{\partial z}+\dfrac{\partial w}{\partial x}\right) \\[2mm] \dfrac{1}{2}\left(\dfrac{\partial v}{\partial x}+\dfrac{\partial u}{\partial y}\right) & \dfrac{\partial v}{\partial y} & \dfrac{1}{2}\left(\dfrac{\partial v}{\partial z}+\dfrac{\partial w}{\partial y}\right) \\[2mm] \dfrac{1}{2}\left(\dfrac{\partial w}{\partial x}+\dfrac{\partial u}{\partial z}\right) & \dfrac{1}{2}\left(\dfrac{\partial w}{\partial y}+\dfrac{\partial v}{\partial z}\right) & \dfrac{\partial w}{\partial z} \end{pmatrix}$$

(10-51)

in which u, v, and w are the velocity components.

The energy equation in general invariant form is

$$\frac{d\eta}{dV} =. C_1 K_1 I_1 + K_2[C_{21}I_1{}^2 + C_{22}I_2] + K_3[C_{31}I_1{}^3 + \cdots] + \quad (10\text{-}52)$$

Now, if the fluid is *incompressible*, then the continuity equation requires (see Eq. 8-10) that

$$\frac{\partial u}{\partial x} + \frac{\partial v}{\partial y} + \frac{\partial w}{\partial z} = 0$$

(10-53)

or $I_1 = 0$ and K_2 must have the dimensions of viscosity for dimensional homogeneity. Also, there is no known physical invariant K_3 of the proper dimensions. Therefore, if a fluid is *incompressible* and *non-viscous*, it would seem that the mechanism for dissipating energy is not present, and therefore drag on the body is not possible—which is a statement of D'Alembert's paradox.

It is interesting to speculate on the possible existence of a "world" in which there is a known physical invariant K_3 of the proper dimensions such that the last term in Eq. 10-52 would not be zero (unless $K_3 = 0$). Under these circumstances the D'Alembert paradox would still hold—but not only would we require a non-viscous fluid but we would also require a non-K_3 fluid. If, in this world, K_3 were not equal to zero, then even though the fluid were non-viscous, the D'Alembert paradox would not be true.

Going back to Eq. 10-52, if the fluid is *viscous* and *incompressible* then the constant viscosity permits an expression of the form

$$\frac{d\eta}{dV} = \mu(C_{21}I_1{}^2 + C_{22}I_2)$$ (10-54)

which is equivalent to the energy dissipation equation given in Ref. (22).

If the flow is *compressible* and *nonviscous*, then $I_1 \neq 0$ and a permissible dissipation equation takes the form

$$\frac{d\eta}{dV} = pC\left(\frac{\partial u}{\partial x} + \frac{\partial v}{\partial y} + \frac{\partial w}{\partial z}\right)$$ (10-55)

in which p is the physical invariant, pressure, and which is shown in Ref. (22) to be the rate of dissipation of intrinsic energy of the fluid.

10-7 Summary. A form of dimensional reasoning based upon a tensoral invariant argument has been presented. It is based upon a postulated fundamental form of differential equation, invariant in the tensor terms and also invariant in the physical terms. Using this postulated fundamental form and based upon a knowledge of the tensors which occur in the fields in question, it is shown that many of the differential equations of applied mechanics can be written in qualitative form following a simple dimensional reasoning.[3]

[3] It is possible to derive, using the methods of this chapter, dimensional expressions connecting the various physical constants of macroscopic bodies. Thus, specific heat can be given in terms of the elastic modulus, Poisson's ratio, coefficient of expansion, and density. See S. F. Borg, G. Kopchinski, K. Hoppe, "Extended Applications of Tensoral Dimensional Analysis," Rep. No. 22, Civil Engineering Department, Stevens Institute of Technology, Hoboken, N.J., Oct. 3, 1963. Also S. F. Borg, "Atomic Weights of Solids in Terms of Material Physical Constants," Rep. No. 24, Civil Engineering Department, Stevens Institute of Technology, Hoboken, N.J., Oct. 3, 1963.

References

References numbered in the text are listed below.

(1) Francis D. Murnaghan, *Introduction to Applied Mathematics*, John Wiley and Sons, Inc., New York, 1948.

(2) Henry Margenau and George M. Murphy, *The Mathematics of Physics and Chemistry*, 2nd ed., D. Van Nostrand Co., Inc., Princeton, N.J., 1956.

(3) A. P. Wills, *Vector Analysis With an Introduction to Tensor Analysis*, Dover Publications, Inc., New York, 1958.

(4) J. G. Coffin, *Vector Analysis*, 2nd ed., John Wiley and Sons, New York, 1911.

(5) R. S. Burington and C. C. Torrance, *Higher Mathematics*, McGraw-Hill Book Co., New York, 1939.

(6) C. R. Wylie, Jr., *Advanced Engineering Mathematics*, McGraw-Hill Book Co., New York, 1951.

(7) H. Jeffries, *Cartesian Tensors*, Cambridge Univ. Press, London, 1931.

(8) L. Page, *Introduction to Theoretical Physics*, D. Van Nostrand Co., Inc., Princeton, N.J., 1935.

(9) J. W. Gibbs and E. B. Wilson, *Vector Analysis*, Yale University Press, 1943.

(10) S. F. Borg and J. J. Gennaro, *Advanced Structural Analysis*, D. Van Nostrand Co., Inc., Princeton, N.J., 1959.

(11) S. Timoshenko, *Strength of Materials, Part I and Part II*, D. Van Nostrand Co., Inc., Princeton, N.J., 1956.

(12) A. E. H. Love, *A Treatise on the Mathematical Theory of Elasticity*, 3rd ed., Cambridge University Press, London, 1920.

(13) S. Timoshenko and J. N. Goodier, *Theory of Elasticity*, 2nd ed., McGraw-Hill Book Co., New York, 1951.

(14) R. V. Southwell, *Theory of Elasticity for Engineers*, Oxford University Press, Oxford, 1941.

(15) I. S. Sokolnikoff, *Mathematical Theory of Elasticity*, 2nd ed., McGraw-Hill Book Co., New York, 1956.

(16) F. D. Murnaghan, *Finite Deformations of an Elastic Solid*, John Wiley and Sons, New York, 1951.

(17) C. T. Wang, *Applied Elasticity*, McGraw-Hill Book Co., New York, 1953.

(18) N. I. Muskhelishvili, *Some Basic Problems of the Mathematical Theory of Elasticity*, P. Noordhoff, Groningen, The Netherlands, 1953.

(19) S. Timoshenko and S. Woinowsky-Krieger, *Theory of Plates and Shells*, 2nd ed., McGraw-Hill Book Co., New York, 1959.

(20) J. Prescott, *Applied Elasticity*, Dover Publications, New York, 1946.

(21) S. Goldstein, *Modern Developments in Fluid Mechanics*, Oxford University Press, 1938.

(22) H. Lamb, *Hydrodynamics*, 6th ed., Dover Publications, New York, 1945.

(23) H. Schlichting, *Boundary Layer Theory*, Pergamon Press, New York, 1955.

(24) T. von Kármán, *Festigkeits Problem in Machinenbau*, Vol. 4, Encyk. Math. Wiss., 1910.

(25) Chi-Teh Wang, "Non-linear large deflection boundary value problems of rectangular plates," *NACA*, TN 1425, Mar. 1948.

(26) S. Levy, "Bending of rectangular plates with large deflections," *NACA* TN 846, 1942. Also TN 847, 1942.

(27) S. Way, "Uniformly loaded clamped rectangular plates with large deflections," Proc. Fifth Intern. Cong. App. Mech., Cambridge Mass., 1938.

(28) H. Hencky, "Die Berechnung Dünner rechteckiger Platten mit verschwindender Biegungsteifigkeit," *ZAMM*, Vol. 1, 1921.

(29) R. Kaiser, "Rechnerische und experimentalle Ermittlung der Durchbiegungen und Spannungen von quadratischen Platten ...," *ZAMM*, Vol. 16, 1936.

(30) L. Prandtl, "Über Flüssigkeitsbewegung bei sehr kleiner Reibung," Ver. III, *Intern. Math. Kongress*, Heidelberg, 1904.

(31) H. Blasius, "Grenzschichten in Flüssigkeiten mit kleiner Reibung," *Z. Math. u. Phys.* Bd. 56, 1908.

(32) K. Töpfer, "Bemerkung zu dem Aufsatz von H. Blasius . . .," *Zeitschrift f. Math. u. Phys.* No. 60, p. 397, 398, 1912.

(33) G. I. Taylor and J. W. Maccoll, "The Mechanics of Compressible Fluids," *Aerodynamic Theory*, F. W. Durand, Editor, Vol. 3, Julius Springer, Berlin, 1935.

(34) H. L. Dryden, "Aerodynamics of Cooling"; *Aerodynamic Theory*, F. W. Durand, Editor, Vol. 6, Julius Springer, Berlin, 1935.

(35) K. I. Friedrichs, "The edge effect in bending and buckling with large deflections," *Proc. Symp. in App. Math.*, Vol. 1, Amer. Math. Soc., 1949.

(36) Y. C. Fung and W. H. Witrick, "A boundary layer phenomenon in the large deflexion of thin plates," *Quart. J. Mech. Appl. Math.*, Vol. 8, 1955.

(37) E. Bromberg and S. J. Stoker, "Non-linear theory of curved elastic sheets," *Quart. App. Math.*, Vol. III, 1945.

(38) S. Timoshenko, *Theory of Elastic Stability*, 1st ed., McGraw-Hill Book Co., New York, 1936.

(39) J. Van den Broek, *Theory of Limit Design*, John Wiley and Sons, New York, 1948.

(40) L. S. Beedle, *Plastic Design of Steel Frames*, John Wiley and Sons, New York, 1958.

(41) G. B. Neal, *The Plastic Methods of Structural Analysis*, John Wiley and Sons, New York, 1956.

(42) R. Hill, *Plasticity*, Oxford Univ. Press, New York, 1950.

(43) O. Hoffman and G. Sachs, *Introduction to the Theory of Plasticity for Engineers*, McGraw-Hill Book Co., New York, 1953.

(44) W. Prager and P. G. Hodge, Jr., *Theory of Perfectly Plastic Solids*, John Wiley and Sons, New York, 1951.

(45) L. V. Azaroff, *Introduction to Solids*, McGraw-Hill Book Co., New York, 1960.

(46) P. Moon and D. E. Spencer, *Field Theory for Engineers*, D. Van Nostrand Co., Inc., Princeton, N.J., 1961.

(47) Simon Ince, "Some attempts at theory formation in the history of fluid

mechanics," *Proc. Seminar in Modern Fluid Mechanics*, Michigan State Univ., East Lansing, Mich., 1960.

(48) S. Corrsin, "Turbulent flow," *Am. Scientist*, Vol. 49, No. 3, Sept. 1961.

(49) P. M. Naghdi and W. L. Wainwright, " On the time derivative of tensors in mechanics of continua," *Quart. App. Math.*, Vol. XIX, No. 2, July 1961.

(50) S. F. Borg, *Fundamentals of Engineering Elasticity*, D. Van Nostrand Co., Inc. Princeton, N.J. 1962.

INDEX

APPENDIX

Appendix to Chapters 1 through 5

In this section various topics covered in Chapters 1 through 5 of the text are discussed. In general, additional or different points of view, derivations and explanations are given for the topics considered. It is hoped that these will offer clarifications and additional insights insofar as the subjects discussed are concerned.

The material is identified by referring to the text page and topic that it relates to.

p. 6 — Symmetry and Anti-Symmetry

A simple physical analog of Eq. 1.31 occurs in the static loading of a simply supported beam as shown in Fig. A.1

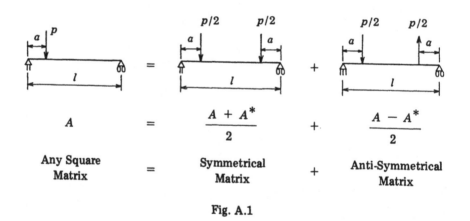

Fig. A.1

p. 76 — Elements of $\dfrac{\partial R^*}{\partial x}$

We may also obtain the elements of Eq. 3.55 more directly as follows (two of the terms will be obtained; the others may be developed similarly).

314

Consider the first row - first column element of $\dfrac{\partial R^*}{\partial x}$.

From Eq. 3.48, this is given by

$$\frac{\partial}{\partial x}\left(h_1\,\frac{\partial x}{\partial \alpha}\right)$$

which $= h_1\left[h_1\,\dfrac{\partial\,\frac{\partial x}{\partial \alpha}}{\partial \alpha} + \dfrac{\partial x}{\partial \alpha}\,\dfrac{\partial h_1}{\partial \alpha}\right]$

$$= h_1\left[h_1\,\frac{\partial\left(\frac{1}{h_1}\right)}{\partial \alpha} + \frac{1}{h_1}\,\frac{\partial h_1}{\partial \alpha}\right]$$

$$= h_1\left[-\frac{1}{h_1} + \frac{1}{h_1}\right]\frac{\partial h_1}{\partial \alpha}$$

$= 0$ as shown, when the axes coincide.

$\hfill$(A.1)

Now consider the first row - second column element. This is given by

$$\frac{\partial}{\partial x}\left(h_2\,\frac{\partial x}{\partial \beta}\right)$$

which $= h_1\,\dfrac{\partial}{\partial \alpha}\left(h_2\,\dfrac{\partial x}{\partial \beta}\right)$

$$= h_1\left[h_2\,\frac{\partial\,\frac{\partial x}{\partial \beta}}{\partial \alpha} + \frac{\partial x}{\partial \beta}\,\frac{\partial h_2}{\partial \alpha}\right]$$

$= h_1 h_2\,\dfrac{\partial\,\frac{\partial x}{\partial \alpha}}{\partial \beta}$ since $\dfrac{\partial x}{\partial \beta} = 0$

$= h_1 h_2\,\dfrac{\partial\left(\frac{1}{h_1}\right)}{\partial \beta}$ as shown, when the axes coincide.

$\hfill$(A.2)

p. 92 — Linear Strain Tensor

The linear strain tensor may be generated as follows:

$$
\text{deformation tensor} = \begin{pmatrix} \dfrac{\partial}{\partial x} \\[2mm] \dfrac{\partial}{\partial y} \\[2mm] \dfrac{\partial}{\partial z} \end{pmatrix} (u \quad v \quad w) = \nabla \delta
$$

$$
= \begin{pmatrix} \dfrac{\partial u}{\partial x} & \dfrac{\partial v}{\partial x} & \dfrac{\partial w}{\partial x} \\[3mm] \dfrac{\partial u}{\partial y} & \dfrac{\partial v}{\partial y} & \dfrac{\partial w}{\partial y} \\[3mm] \dfrac{\partial u}{\partial z} & \dfrac{\partial v}{\partial z} & \dfrac{\partial w}{\partial z} \end{pmatrix} \tag{A.3}
$$

The deformation tensor being a square tensor may be represented by the sum of a symmetric tensor and an anti-symmetric tensor (see p. 6) or

$$
\text{deformation tensor} = \begin{pmatrix} \dfrac{\partial u}{\partial x} & \dfrac{1}{2}\left(\dfrac{\partial u}{\partial y}+\dfrac{\partial v}{\partial x}\right) & \dfrac{1}{2}\left(\dfrac{\partial u}{\partial z}+\dfrac{\partial w}{\partial x}\right) \\[3mm] \dfrac{1}{2}\left(\dfrac{\partial v}{\partial x}+\dfrac{\partial u}{\partial y}\right) & \dfrac{\partial v}{\partial y} & \dfrac{1}{2}\left(\dfrac{\partial v}{\partial z}+\dfrac{\partial w}{\partial y}\right) \\[3mm] \dfrac{1}{2}\left(\dfrac{\partial w}{\partial x}+\dfrac{\partial u}{\partial z}\right) & \dfrac{1}{2}\left(\dfrac{\partial w}{\partial y}+\dfrac{\partial v}{\partial z}\right) & \dfrac{\partial w}{\partial z} \end{pmatrix}
$$

$$
+ \begin{pmatrix} 0 & \dfrac{1}{2}\left(\dfrac{\partial u}{\partial y}-\dfrac{\partial v}{\partial x}\right) & \dfrac{1}{2}\left(\dfrac{\partial u}{\partial z}-\dfrac{\partial w}{\partial x}\right) \\[3mm] \dfrac{1}{2}\left(\dfrac{\partial v}{\partial x}-\dfrac{\partial u}{\partial y}\right) & 0 & \dfrac{1}{2}\left(\dfrac{\partial v}{\partial z}-\dfrac{\partial w}{\partial y}\right) \\[3mm] \dfrac{1}{2}\left(\dfrac{\partial w}{\partial x}-\dfrac{\partial u}{\partial z}\right) & \dfrac{1}{2}\left(\dfrac{\partial w}{\partial y}-\dfrac{\partial v}{\partial z}\right) & 0 \end{pmatrix}
$$

$$= \eta + \frac{1}{2}\Omega \tag{A.4}$$

in which η = the linearized strain tensor
Ω = the rotation tensor which may also be represented by its vector equivalent

$$(\Omega_x, \Omega_y, \Omega_z) = \left(\frac{\partial w}{\partial y} - \frac{\partial v}{\partial z} \quad \frac{\partial u}{\partial z} - \frac{\partial w}{\partial x} \quad \frac{\partial v}{\partial x} - \frac{\partial u}{\partial y} \right)$$

p. 101 Footnote — Sign Convention for Stresses

The sign convention given on p. 101 is a "mathematical" sign convention. There is another sign convention which follows from the equilibrium of force (Newton's) relation which requires that the *net* force on a body in static equilibrium be equal to zero. Thus, when this convention applies, the sign of a *force* is determined by its *direction* — say positive to the right and negative to the left, although both may be tensile forces. We may call this the "equilibrium" sign convention.

p. 111 — The Equilibrium Equations

A simpler, less mathematical derivation of the equilibrium equations follows from a direct application of Newton's Laws to a body in static (i.e., bodies stationary or moving with constant velocity) equilibrium. Note especially that this equation (Eq. 4.85) is given in invariant form, as is also the boundary conditions on stresses, Eq. 4.74. This must be true for the fundamental equations in any discipline. See the later discussion on the compatibility conditions.

See Fig. A.2 which shows an elemental solid with corner closest to the origin at x, y and z and having faces of length dx, dy and dz. The stresses acting on this body are as shown. Note that on the faces closest to the origin the stresses of the tensor, T, act, where

$$T = \begin{pmatrix} \sigma_x & \tau_{xy} & \tau_{xz} \\ \tau_{yz} & \sigma_y & \tau_{yz} \\ \tau_{zx} & \tau_{zy} & \sigma_z \end{pmatrix} \tag{A.5}$$

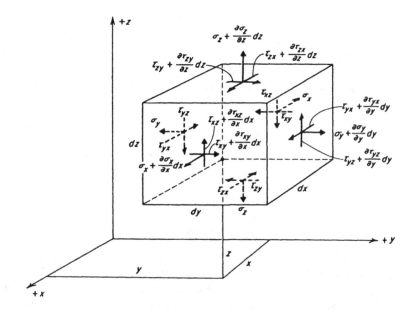

Fig. A.2

whereas on the far faces we have stresses given by

$$
T + dT =
\begin{pmatrix}
\sigma_x + \dfrac{\partial \sigma_x}{\partial x}\,dx & \tau_{xy} + \dfrac{\partial \tau_{xy}}{\partial x}\,dx & \tau_{xz} + \dfrac{\partial \tau_{xz}}{\partial x}\,dx \\[2ex]
\tau_{yx} + \dfrac{\partial \tau_{yx}}{\partial y}\,dy & \sigma_y + \dfrac{\partial \sigma_y}{\partial y}\,dy & \tau_{yz} + \dfrac{\partial \tau_{yz}}{\partial y}\,dy \\[2ex]
\tau_{zx} + \dfrac{\partial \tau_{zx}}{\partial z}\,dz & \tau_{zy} + \dfrac{\partial \tau_{zy}}{\partial z}\,dz & \sigma_z + \dfrac{\partial \sigma_z}{\partial z}\,dz
\end{pmatrix}
$$

$$(A.6)$$

In addition to the stresses shown acting on the body, let us assume that there is acting a *body force* per unit mass. This is, for example, a force due to the weight of the body or a force due to magnetic effects or a similar force. This force per unit mass has components $\mathcal{F}_x$, $\mathcal{F}_y$, and $\mathcal{F}_z$ in the x, y, and z directions respectively. Therefore the *total body force* on the body in, say, the x direction is given (ρ is the density) by

$$\rho \mathfrak{F}_x \, dx \, dy \, dz \ . \tag{A.7}$$

Similar expressions apply in the y and z directions.

If now the sum of the forces (i.e., stresses times area and body force times mass) in the three directions x, y and z are determined for the body of Fig. A.2 we must have, throughout the body

$$\Sigma \, F_x \ = \ 0$$

$$\Sigma \, F_y \ = \ 0 \tag{A.8}$$

$$\Sigma \, F_z \ = \ 0$$

A typical statement of one of the above, say

$$\Sigma \, F_x \ = \ 0$$

is given by the following (noting the positive and negative directions of forces as shown on the figure):

$$- \sigma_x \, (dy \, dz) \ + \ \left(\sigma_x \ + \ \frac{\partial \sigma_x}{\partial x} \ dx \right) dy \, dz \ - \ \tau_{yx} \, (dx \, dz)$$

$$+ \ \left(\tau_{yx} \ + \ \frac{\partial \tau_{yx}}{\partial y} \, dy \right) dx \, dz \ - \ \tau_{zx} \, (dx \, dy) \ + \ \left(\tau_{zx} \ + \ \frac{\partial \tau_{zx}}{\partial z} \ dz \right) dx \, dy$$

$$+ \ \rho \mathfrak{F}_x \, dx \, dy \, dz \ = 0 \ . \tag{A.9}$$

Now, collecting terms, cancelling, and dividing through by the common term $dx \, dy \, dz$, we obtain

$$\frac{\partial \sigma_x}{\partial x} \ + \ \frac{\partial \tau_{yx}}{\partial y} \ + \ \frac{\partial \tau_{zx}}{\partial z} \ + \ \rho \mathfrak{F}_x \ = \ 0 \ . \tag{A.10}$$

The second and third of these equations are obtained in a similar manner, by summing forces in the y and z directions. These give

$$\frac{\partial \tau_{xy}}{\partial x} \ + \ \frac{\partial \sigma_y}{\partial y} \ + \ \frac{\partial \tau_{zy}}{\partial z} \ + \ \rho \mathfrak{F}_y \ = \ 0 \tag{A.11}$$

and

$$\frac{\partial \tau_{xz}}{\partial x} + \frac{\partial \tau_{yz}}{\partial y} + \frac{\partial \sigma_z}{\partial z} + \rho \mathcal{F}_z = 0 . \qquad (A.12)$$

Note that (A.11) and (A.12) are obtained from A.10 by cyclical interchange.

Thus, in the mathematical theory of elasticity, if a body is in static equilibrium, the stresses acting on the body must satisfy the following three equations:

$$\frac{\partial \sigma_x}{\partial x} + \frac{\partial \tau_{yx}}{\partial y} + \frac{\partial \tau_{zx}}{\partial z} + \rho \mathcal{F}_x = 0$$

$$\frac{\partial \tau_{xy}}{\partial x} + \frac{\partial \sigma_y}{\partial y} + \frac{\partial \tau_{zy}}{\partial z} + \rho \mathcal{F}_y = 0 \qquad (A.13)$$

$$\frac{\partial \tau_{xz}}{\partial x} + \frac{\partial \tau_{yz}}{\partial y} + \frac{\partial \sigma_z}{\partial z} + \rho \mathcal{F}_z = 0$$

or, in matrix-tensor notation,

$$\left(\frac{\partial}{\partial x} \ \frac{\partial}{\partial y} \ \frac{\partial}{\partial z} \right) \begin{pmatrix} \sigma_x & \tau_{xy} & \tau_{xz} \\ \tau_{yx} & \sigma_y & \tau_{yz} \\ \tau_{zx} & \tau_{zy} & \sigma_z \end{pmatrix} + \rho (\mathcal{F}_x \ \mathcal{F}_y \ \mathcal{F}_z) = 0$$

or, $\qquad\qquad\qquad\qquad\qquad\qquad\qquad\qquad\qquad\qquad\qquad$ (A.14)

$$\mathrm{div} \ T + \rho \mathcal{F} = 0 .$$

p. 114 — Derivation of Hooke's Law

We may also derive Hooke's Law by referring to the simple tension test of Fig. A.3.

Thus, due to a force in the y direction we have

$$e_y = \frac{\sigma_y}{E}$$

$$e_x = - \frac{\nu \sigma_y}{E} \qquad (A.15)$$

$$e_z = - \frac{\nu \sigma_y}{E}$$

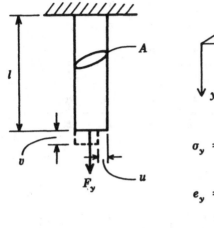

$$\sigma_y = \frac{F_y}{A}$$

$$e_y = \frac{v}{l}$$

$$e_x = e_z = - \nu e_y$$

ν = Poisson Ratio

E = Modulus of Elasticity in Tension-Compression

Fig. A.3

If now all three forces (in the x, y, and z directions) are applied simultaneously then (for small deformations)

$$e_x = \frac{\sigma_x}{E} - \frac{\nu\sigma_y}{E} - \frac{\nu\sigma_z}{E}$$

$$e_y = \frac{\sigma_y}{E} - \frac{\nu\sigma_z}{E} - \frac{\nu\sigma_x}{E} \qquad\qquad (A.16)$$

$$e_z = \frac{\sigma_z}{E} - \frac{\nu\sigma_x}{E} - \frac{\nu\sigma_y}{E}$$

(Note that the second and third of these equations are obtained from the first by cyclical interchange.)
    Then, clearly

$$e_x = \frac{\sigma_x}{E} - \frac{\nu\sigma_y}{E} - \frac{\nu\sigma_z}{E} + \frac{\nu\sigma_x}{E} - \frac{\nu\sigma_x}{E}$$

$$e_y = \frac{\sigma_y}{E} - \frac{\nu\sigma_z}{E} - \frac{\nu\sigma_x}{E} + \frac{\nu\sigma_y}{E} - \frac{\nu\sigma_y}{E} \tag{A.17}$$

$$e_z = \frac{\sigma_z}{E} - \frac{\nu\sigma_x}{E} - \frac{\nu\sigma_y}{E} + \frac{\nu\sigma_z}{E} - \frac{\nu\sigma_z}{E}$$

which, in *matrix* form becomes

$$\begin{pmatrix} e_x & 0 & 0 \\ 0 & e_y & 0 \\ 0 & 0 & e_z \end{pmatrix} = \frac{1+\nu}{E} \begin{pmatrix} \sigma_x & 0 & 0 \\ 0 & \sigma_y & 0 \\ 0 & 0 & \sigma_z \end{pmatrix}$$

$$- \frac{\nu}{E}(\sigma_x + \sigma_y + \sigma_z) \begin{pmatrix} 1 & 0 & 0 \\ 0 & 1 & 0 \\ 0 & 0 & 1 \end{pmatrix} \tag{A.18}$$

or, in *tensor* notation

$$\eta = \frac{1+\nu}{E} T - \frac{\nu}{E} \mathscr{I}_1 \mathscr{E}_3 \tag{A.19}$$

in which,

$$1, \nu, E, \mathscr{I}_1 = \text{invariants}$$

$$\eta = \text{linearized strain tensor}$$

$$T = \text{stress tensor}$$

$$\mathscr{I}_1 = \text{trace invariant of } T$$

$$\mathscr{E}_3 = \text{unit tensor}$$

Note particularly that η and T, because they are symmetrical tensors may be given in diagonal form (see Eq. 2.85).

Also, because the above equation is a *tensor* equation, it holds with respect to *all* axial orientations, even those for which η and T are *not* diagonal. Also the invariant (i.e., scalar) terms do not change when the axes are rotated. Thus, if the axes are rotated the equation becomes, in unexpanded matrix form

$$
\begin{pmatrix}
e_x & \dfrac{1}{2}\gamma_{xy} & \dfrac{1}{2}\gamma_{xz} \\[2mm]
\dfrac{1}{2}\gamma_{yx} & e_y & \dfrac{1}{2}\gamma_{yz} \\[2mm]
\dfrac{1}{2}\gamma_{zx} & \dfrac{1}{2}\gamma_{zy} & e_z
\end{pmatrix}
=
\frac{1+\nu}{E}
\begin{pmatrix}
\sigma_x & \tau_{xy} & \tau_{xz} \\[2mm]
\tau_{yx} & \sigma_y & \tau_{yz} \\[2mm]
\tau_{zx} & \tau_{zy} & \sigma_z
\end{pmatrix}
$$

$$
-\frac{\nu}{E}(\sigma_x + \sigma_y + \sigma_z)
\begin{pmatrix}
1 & 0 & 0 \\
0 & 1 & 0 \\
0 & 0 & 1
\end{pmatrix}
\tag{A.20}
$$

and this equation now *predicts* a result *independent* of the original test, namely

$$
\frac{1}{2}\gamma_{xy} = \frac{1+\nu}{E}\,\tau_{xy}
$$

or

$$
\tau_{xy} = \frac{E}{2(1+\nu)}\,\gamma_{xy}
\tag{A.21}
$$

the pure *shear* behavior of a stressed block. All the terms in the expression are measurable and experiments verify the accuracy of the prediction, thus verifying the validity of the linearized form of Hooke's Law and hence of the derivation.

p. 119 — The Compatibility Equations

The fundamental equations in any discipline should be given in invariant form (i.e., independent of the coordinate system and also of the axial orientation). Thus, Eq. 4.85 stated the three equations of static equilibrium in terms of tensors and a scalar, Eq. 4.75, stated the boundary condition on the stresses in invariant tensor form and Eq. 4.116 stated Hooke's Law in terms of invariant tensor and isotropic scalar quantities.

One should expect therefore that the six equations of compatibility also can be stated in invariant form and it will now be shown that this is, in fact, the case.

We begin by pointing out that the following two facts have been established insofar as the strain compatibility equations of linearized elasticity theory are concerned. These are

1. There are six equations required.

2. The six equations are given in two entirely different sets of three each, with the equations in each set obtainable from each other by cyclical interchange.

In the following analyses it will be shown

1. Why there are six equations required.

2. Why they are given in two sets of three, markedly different from each other.

3. They will be derived in a concise invariant form.

Fundamental to each of the above will be the requirement that "rotations must be considered as of equal importance to the strains" in the compatibility analyses.

Insofar as the term "invariance" is concerned, as used in this discussion, invariance of an expression or equation can be represented in either of two forms:

1. The equation is given entirely in terms of *tensors*. Thus, η is the tensor representation of the complete nine-term strain expression. η by itself is entirely independent of any coordinate system and is thus an "invariant". Hooke's Law, for example is given in terms of η and also T, the stress tensor and is therefore, in invariant form.

2. The equation is given in terms of the "invariants" of the tensors. These terms are invariant under a rotation of axes and may be transformed into any orthogonal curvilinear coordinate system using well-known techniques. In Hooke's Law, for example, in addition to the stress and strain tensors, one term includes an invariant of the stress (or strain) tensor.

The strain tensor in linearized elasticity theory may be generated in the following manner. We begin by considering deformations, represented by the first order tensor, i.e., vector, as either

$$\begin{pmatrix} u \\ v \\ w \end{pmatrix} \quad \text{or} \quad (u \quad v \quad w) \tag{A.22}$$

To introduce the small deformation theory, we have the differential operator, also a first order tensor

$$\Delta = \begin{pmatrix} \dfrac{\partial}{\partial x} \\[2ex] \dfrac{\partial}{\partial y} \\[2ex] \dfrac{\partial}{\partial z} \end{pmatrix} \quad \text{or} \quad \begin{pmatrix} \dfrac{\partial}{\partial x} & \dfrac{\partial}{\partial y} & \dfrac{\partial}{\partial z} \end{pmatrix} \tag{A.23}$$

We may obtain a second order tensor from A.22 and A.23 by performing the operation

$$\begin{pmatrix} \dfrac{\partial}{\partial x} \\[2ex] \dfrac{\partial}{\partial y} \\[2ex] \dfrac{\partial}{\partial z} \end{pmatrix} (u \quad v \quad w) = \begin{pmatrix} \dfrac{\partial u}{\partial x} & \dfrac{\partial v}{\partial x} & \dfrac{\partial w}{\partial x} \\[2ex] \dfrac{\partial u}{\partial y} & \dfrac{\partial v}{\partial y} & \dfrac{\partial w}{\partial y} \\[2ex] \dfrac{\partial u}{\partial z} & \dfrac{\partial v}{\partial z} & \dfrac{\partial w}{\partial z} \end{pmatrix} \tag{A.24}$$

This tensor may be further decomposed into a symmetrical and skew-symmetrical term, each of which is also a second order tensor as follows:

$$\begin{pmatrix} \dfrac{\partial u}{\partial x} & \dfrac{\partial v}{\partial x} & \dfrac{\partial w}{\partial x} \\[2ex] \dfrac{\partial u}{\partial y} & \dfrac{\partial v}{\partial y} & \dfrac{\partial w}{\partial y} \\[2ex] \dfrac{\partial u}{\partial z} & \dfrac{\partial v}{\partial z} & \dfrac{\partial w}{\partial z} \end{pmatrix} \qquad \text{[continued on next page]}$$

$$
= \begin{pmatrix}
\dfrac{\partial u}{\partial x} & \dfrac{1}{2}\left(\dfrac{\partial v}{\partial x} + \dfrac{\partial u}{\partial y}\right) & \dfrac{1}{2}\left(\dfrac{\partial w}{\partial x} + \dfrac{\partial u}{\partial z}\right) \\[3mm]
\dfrac{1}{2}\left(\dfrac{\partial u}{\partial y} + \dfrac{\partial v}{\partial x}\right) & \dfrac{\partial v}{\partial y} & \dfrac{1}{2}\left(\dfrac{\partial v}{\partial z} + \dfrac{\partial w}{\partial y}\right) \\[3mm]
\dfrac{1}{2}\left(\dfrac{\partial u}{\partial z} + \dfrac{\partial w}{\partial x}\right) & \dfrac{1}{2}\left(\dfrac{\partial v}{\partial z} + \dfrac{\partial w}{\partial y}\right) & \dfrac{\partial w}{\partial z}
\end{pmatrix}
\tag{A.25a}
$$

$$
+ \begin{pmatrix}
0 & \dfrac{1}{2}\left(\dfrac{\partial v}{\partial x} - \dfrac{\partial u}{\partial y}\right) & \dfrac{1}{2}\left(\dfrac{\partial w}{\partial x} - \dfrac{\partial u}{\partial z}\right) \\[3mm]
\dfrac{1}{2}\left(\dfrac{\partial u}{\partial y} - \dfrac{\partial v}{\partial x}\right) & 0 & \dfrac{1}{2}\left(\dfrac{\partial w}{\partial y} - \dfrac{\partial v}{\partial z}\right) \\[3mm]
\dfrac{1}{2}\left(\dfrac{\partial u}{\partial z} - \dfrac{\partial w}{\partial x}\right) & \dfrac{1}{2}\left(\dfrac{\partial v}{\partial z} - \dfrac{\partial w}{\partial y}\right) & 0
\end{pmatrix}
$$

$$
= \eta + \frac{1}{2}\,\Omega
\tag{A.25b}
$$

in which η is just the symmetrical strain tensor as shown and Ω is the curl or rotation tensor which may be represented more compactly and usefully for our purposes as

$$
\Omega = (\Omega_x \ \Omega_y \ \Omega_z) = \left(\frac{\partial v}{\partial z} - \frac{\partial w}{\partial y} \quad \frac{\partial w}{\partial x} - \frac{\partial u}{\partial z} \quad \frac{\partial u}{\partial y} - \frac{\partial v}{\partial x}\right)
\tag{A.26}
$$

Two significant facts emerge from an examination of Eqs. (A.25a) and (A.25b):

1. It is clear that the rotation components are as important as the strain components in the separation of the fundamental differential deformation tensor, Eq. (A.24) into its symmetrical and skew-symmetrical parts.
2. It is obvious that there are *nine* independent strain and rotation terms (*six* strains and *three* rotations) and these are given in terms of *three* deformations. Therefore, there are *six* additional independent equations necessary among the strain and rotation elements.

The introduction of rotation is essential. It explains why *six* additional compatibility equations or conditions are required. And in addition, as will be shown, it is an essential element in the derivation of one invariant set of these compatibility conditions. The rotations were also used in Sec. 4.6 of this text in the earlier derivation of the compatibility conditions.

A third fundamental tensor, in elasticity theory, is the ∇^2 tensor, obtained as follows:

$$\nabla^2 = \begin{pmatrix} \dfrac{\partial}{\partial x} \\[2mm] \dfrac{\partial}{\partial y} \\[2mm] \dfrac{\partial}{\partial z} \end{pmatrix} \begin{pmatrix} \dfrac{\partial}{\partial x} & \dfrac{\partial}{\partial y} & \dfrac{\partial}{\partial z} \end{pmatrix}$$

(A.27)

$$= \begin{pmatrix} \dfrac{\partial^2}{\partial x^2} & \dfrac{\partial^2}{\partial x\,\partial y} & \dfrac{\partial^2}{\partial x\,\partial z} \\[3mm] \dfrac{\partial^2}{\partial y\,\partial x} & \dfrac{\partial^2}{\partial y^2} & \dfrac{\partial^2}{\partial y\,\partial z} \\[3mm] \dfrac{\partial^2}{\partial z\,\partial x} & \dfrac{\partial^2}{\partial z\,\partial y} & \dfrac{\partial^2}{\partial z^2} \end{pmatrix}$$

Note, this also is a symmetrical second order tensor, if the order of differentiation is immaterial.

The First Set of Compatibility Conditions

It will now be shown that there are two separate invariant sets (three equations in each) of compatibility equations. The first set will be given in terms of the invariants of the two-dimensional forms of the strain tensor and the ∇^2 tensor, these being, for example, in terms of x, y.

$$\eta = \begin{pmatrix} e_x & \dfrac{1}{2}\,\gamma_{xy} \\[3mm] \dfrac{1}{2}\,\gamma_{yx} & e_y \end{pmatrix} \quad , \quad \nabla^2 = \begin{pmatrix} \dfrac{\partial^2}{\partial x^2} & \dfrac{\partial^2}{\partial x\,\partial y} \\[3mm] \dfrac{\partial^2}{\partial y\,\partial x} & \dfrac{\partial^2}{\partial y^2} \end{pmatrix} \quad \text{(A.28)}$$

Obviously, the two tensors can be given, in two-dimensional form, entirely independent of the third dimension, z in this case.

The invariants, trace, of ∇^2 and η are

$$I_{\nabla^2} = \left(\frac{\partial^2}{\partial x^2} + \frac{\partial^2}{\partial y^2} \right) \quad , \quad I_\eta = (e_x + e_y) \tag{A.29}$$

also

$$\nabla^2\eta = \begin{pmatrix} \dfrac{\partial^2}{\partial x^2} & \dfrac{\partial^2}{\partial x\, \partial y} \\[2ex] \dfrac{\partial^2}{\partial y\, \partial x} & \dfrac{\partial^2}{\partial y^2} \end{pmatrix} \begin{pmatrix} e_x & \dfrac{1}{2}\,\gamma_{xy} \\[2ex] \dfrac{1}{2}\,\gamma_{yx} & e_y \end{pmatrix} \tag{A.30a}$$

$$= \begin{pmatrix} \dfrac{\partial^2 e_x}{\partial x^2} + \dfrac{1}{2}\dfrac{\partial^2 \gamma_{yx}}{\partial x\, \partial y} & \dfrac{1}{2}\dfrac{\partial^2 \gamma_{xy}}{\partial x^2} + \dfrac{\partial^2 e_y}{\partial x\, \partial y} \\[3ex] \dfrac{\partial^2 e_x}{\partial y\, \partial x} + \dfrac{1}{2}\dfrac{\partial^2 \gamma_{yx}}{\partial y^2} & \dfrac{1}{2}\dfrac{\partial^2 \gamma_{xy}}{\partial y\, \partial x} + \dfrac{\partial^2 e_y}{\partial y^2} \end{pmatrix} \tag{A.30b}$$

is a second order tensor, and its trace invariant is

$$I_{\nabla^2\eta} = \frac{\partial^2 e_x}{\partial x^2} + \frac{1}{2}\frac{\partial^2 \gamma_{yx}}{\partial x\, \partial y} + \frac{1}{2}\frac{\partial^2 \gamma_{xy}}{\partial y\, \partial x} + \frac{\partial^2 e_y}{\partial y^2} \tag{A.31}$$

If, now it is assumed that the order of differentiation is immaterial (as it is for finite, single valued continuous functions) and noting, that for these tensors,

$$I_{\nabla^2}I_\eta = I_{\nabla^2\eta} \tag{A.32}$$

we obtain at once from Eqs. A.29, A.31 and A.32,

$$\frac{\partial^2 e_x}{\partial y^2} + \frac{\partial^2 e_y}{\partial x^2} = \frac{\partial^2 \gamma_{xy}}{\partial x\, \partial y} \tag{A.33}$$

which is the first equation of the first set of three (the other two are obtained by cyclical interchange).

Equation A.32 is the invariant form of the first three equations of compatibility.

If the three-dimensional forms of η and ∇^2 are used, we have, again, the invariant identity,

$$I_{\nabla^2} I_\eta = I_{\nabla^2 \eta} \tag{A.34}$$

and we obtain a single equation which is just the sum of the three equations A.33 above. Obviously, this single equation, while correct, is a much weaker statement than the three separate equations A.33.

The Second Set of Compatibility Conditions

The second set of three required equations is of an entirely different form from that of Eq. A.33. For one thing, each equation includes the three-dimensional terms (i.e., elements in x, y and z). It is not surprising that this is so, since this second set is, fundamentally, a statement involving the rotation elements.

We begin by repeating Eq. A.26

$$\Omega = (\Omega_x \ \Omega_y \ \Omega_z) = \left(\frac{\partial v}{\partial z} - \frac{\partial w}{\partial y} \quad \frac{\partial w}{\partial x} - \frac{\partial u}{\partial z} \quad \frac{\partial u}{\partial y} - \frac{\partial v}{\partial x} \right) \tag{A.35}$$

and point out the well-known fact that

$$\nabla \cdot \Omega = 0 \tag{A.36}$$

Statement A.36 is a single equation. We are looking for *three* equations among the Ω components. These are obtained by taking the gradient of A.36, i.e.,

$$\nabla \nabla \cdot \Omega = \nabla^2 \Omega = 0 \tag{A.37}$$

Equation A.37 is the invariant form of the second set of three compatibility conditions. In matrix-tensor notation, it is given by

$$\begin{pmatrix} \dfrac{\partial^2}{\partial x^2} & \dfrac{\partial^2}{\partial x \, \partial y} & \dfrac{\partial^2}{\partial x \, \partial z} \\[2ex] \dfrac{\partial^2}{\partial y \, \partial x} & \dfrac{\partial^2}{\partial y^2} & \dfrac{\partial^2}{\partial y \, \partial z} \\[2ex] \dfrac{\partial^2}{\partial z \, \partial x} & \dfrac{\partial^2}{\partial z \, \partial y} & \dfrac{\partial^2}{\partial z^2} \end{pmatrix} \begin{pmatrix} \Omega_x \\[2ex] \Omega_y \\[2ex] \Omega_z \end{pmatrix} = 0 \tag{A.38a}$$

In expanded form, Eq. A.38a is *three* equations, each of which equals zero, or

$$\frac{\partial}{\partial x} \left[\frac{\partial \Omega_x}{\partial x} + \frac{\partial \Omega_y}{\partial y} + \frac{\partial \Omega_z}{\partial z} \right] = 0$$

$$\frac{\partial}{\partial y} \left[\frac{\partial \Omega_x}{\partial x} + \frac{\partial \Omega_y}{\partial y} + \frac{\partial \Omega_z}{\partial z} \right] = 0 \qquad \text{(A.39)}$$

$$\frac{\partial}{\partial z} \left[\frac{\partial \Omega_x}{\partial x} + \frac{\partial \Omega_y}{\partial y} + \frac{\partial \Omega_z}{\partial z} \right] = 0$$

These are just the second set of three compatibility conditions, in terms of rotations.

If we want these equations in terms of the strains, we introduce the connections between the γ and Ω terms, namely:

$$\frac{1}{2}\Omega_x = \frac{1}{2}\gamma_{yz} - \frac{\partial w}{\partial y}$$

$$\frac{1}{2}\Omega_y = \frac{1}{2}\gamma_{zx} - \frac{\partial u}{\partial z} \qquad \text{(A.40)}$$

$$\frac{1}{2}\Omega_z = \frac{1}{2}\gamma_{xy} - \frac{\partial v}{\partial x}$$

Substituting these in, for example, the first of (A.39) we obtain

$$\frac{\partial}{\partial x} \left[\frac{1}{2} \frac{\partial \gamma_{yz}}{\partial x} - \frac{\partial^2 w}{\partial x\, \partial y} + \frac{1}{2} \frac{\partial \gamma_{zx}}{\partial y} - \frac{\partial^2 u}{\partial y\, \partial z} + \frac{1}{2} \frac{\partial \gamma_{xy}}{\partial z} - \frac{\partial^2 v}{\partial x\, \partial z} \right] = 0$$

$$\text{(A.41)}$$

Collecting terms assuming the order of differentiation is immaterial, this becomes,

$$2 \frac{\partial^2 e_x}{\partial y\, \partial z} = \frac{\partial}{\partial x} \left[\frac{\partial \gamma_{xy}}{\partial z} + \frac{\partial \gamma_{zx}}{\partial y} - \frac{\partial \gamma_{yz}}{\partial x} \right] \qquad \text{(A.42)}$$

with the other two of Eq. (A.39) obtained by cyclical interchange. Equation (A.42) will be recognized as the familiar form in which the compatibility equation is given, Eq. 4.132.

Conclusion

It was first demonstrated that the strain tensor, the rotation tensor, and the ∇^2 tensor are fundamental in the analysis of the compatibility conditions in linearized elasticity.

From a study of these, it became clear that *six* compatibility equations are needed, in order that the *nine* strain and rotation elements be expressible in terms of *three* finite, continuous, single-valued displacements.

These six compatibility conditions in concise, invariant form are

$$I_{\nabla^2} I_{\eta} = I_{\nabla^2 \eta} \qquad (A.32)$$

which is three equations, and

$$\nabla^2 \Omega = 0 \qquad (A.37)$$

which is three equations.

p. 167 — Tabular Summary of the Torsion Problem

We may summarize, in tabular form as follows, the solution to the torsion problem as given in terms of four of the commonly used torsion functions. See p. 332.

p. 171 — Torsion of the Rectangular Cross-Section

One cross-section of practical interest for which an exact solution exists is the rectangular bar, Fig. A.4.

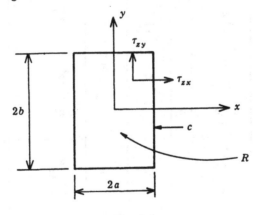

Fig. A.4

SUMMARY — THE TORSION PROBLEM

Torsion Function	ϕ	ψ	Ψ	λ
Everywhere in R	$\nabla^2\phi = 0$	$\nabla^2\psi = 0$	$\nabla^2\Psi = -2$	$\nabla^2\lambda = -2G\alpha$
Everywhere on C	$\left(\dfrac{\partial\phi}{\partial x} - y\right)l + \left(\dfrac{\partial\phi}{\partial y} + x\right)m = 0$ $\dfrac{\partial\phi}{\partial n} = ly - mx = \dfrac{\partial}{\partial s}\dfrac{1}{2}(x^2+y^2)$	$\psi - \dfrac{1}{2}(x^2+y^2) = 0$ $= K$	$\Psi = K$ (constant) $= 0$	$\lambda = K$ $= 0$
τ_{zx}	$G\alpha\left(\dfrac{\partial\phi}{\partial x} - y\right)$	$G\alpha\left(\dfrac{\partial\psi}{\partial y} - y\right)$	$G\alpha\dfrac{\partial\Psi}{\partial y}$	$\dfrac{\partial\lambda}{\partial y}$
τ_{zy}	$G\alpha\left(\dfrac{\partial\phi}{\partial y} + x\right)$	$G\alpha\left(-\dfrac{\partial\psi}{\partial x} + x\right)$	$-G\alpha\dfrac{\partial\Psi}{\partial x}$	$-\dfrac{\partial\lambda}{\partial x}$
Torsional Rigidity D	$G\iint_R\left(x^2+y^2+x\dfrac{\partial\phi}{\partial y} - y\dfrac{\partial\phi}{\partial x}\right)dA$	$G\iint_R\left(x^2+y^2-x\dfrac{\partial\psi}{\partial x} - y\dfrac{\partial\psi}{\partial y}\right)dA$	$-G\iint_R\left(x\dfrac{\partial\Psi}{\partial x} + y\dfrac{\partial\Psi}{\partial y}\right)dA$ $= 2G\iint_R\Psi\,dx\,dy$	$\dfrac{2}{\alpha}\iint_R\lambda\,dA$
Notes	$\dfrac{\partial\phi}{\partial x} = \dfrac{\partial\psi}{\partial y}$ $\dfrac{\partial\phi}{\partial y} = -\dfrac{\partial\psi}{\partial x}$	$\phi + i\psi = f(z)$ $z = x + iy$	$\Psi = \psi - \dfrac{1}{2}(x^2+y^2)$	$\lambda = G\alpha\Psi$

For solution we use the ϕ torsion function. Then ϕ must satisfy

$$\nabla^2 \phi = 0 \quad \text{in} \quad R \tag{A.38}$$

and from the boundary condition equations we find

$$\left(-y + \frac{\partial \phi}{\partial x} \right) l + \left(x + \frac{\partial \phi}{\partial y} m \right) = 0 \quad \text{on} \quad C \tag{A.39}$$

or on $x = \pm a, m = 0, l = \pm 1$ so that

$$\frac{\partial \phi}{\partial x} - y = 0 = \frac{\tau_{zx}}{G\alpha} \tag{A.40}$$

and on $y = \pm b, l = 0, m = \pm 1$ so that

$$x + \frac{\partial \phi}{\partial y} = 0 = \frac{\tau_{zy}}{G\alpha}. \tag{A.41}$$

Introduce a function ϕ' defined by

$$\phi' = \phi - xy \tag{A.42}$$

then the above equations become

$$\nabla^2 \phi' = 0 \quad \text{in} \quad R \tag{A.43}$$

$$\frac{\partial \phi'}{\partial x} = 0 \qquad \text{when} \quad x = \pm a$$

$$\frac{\partial \phi'}{\partial y} + 2x = 0 \quad \text{when} \quad y = \pm b \tag{A.44}$$

also

$$\frac{\tau_{zx}}{G\alpha} = \frac{\tau_{zy}}{G\alpha} = 0 \quad \text{as before.} \tag{A.45}$$

To solve the differential equation and boundary condition we assume

$$\phi' = \Phi(y) \sin Kx \tag{A.46}$$

where Φ is a function of y only and K is a constant. Then

$$\frac{d^2\phi'}{dx^2} = -K^2\Phi(y)\sin Kx \qquad (A.47)$$

and Eq. A.43 becomes

$$\frac{d^2\Phi}{dy^2} = K^2\Phi \qquad (A.48)$$

which has the solution

$$\Phi = Ae^{Ky} + Be^{-Ky} \qquad (A.49)$$

or

$$\phi' = (Ae^{Ky} + Be^{-Ky})\sin Kx . \qquad (A.50)$$

Now, since τ_{zx} is certainly anti-symmetrical with respect to the x axes, $\frac{\partial\phi'}{\partial x}$ must change its sign but not its magnitude when y is changed to $-y$. Thus, $B = -A$ and we have

$$\phi' = A(e^{Ky} - e^{-Ky})\operatorname{Sin} Kx \qquad (A.51)$$

and since $\frac{\partial\phi'}{\partial x} = 0$ when $x = \pm a$ (Eq. A.44), we have

$$\cos Ka = 0$$

since

$$(e^{Ky} - e^{-Ky}) \neq 0 \qquad (A.52)$$

or

$$K = \frac{2n+1}{a}\frac{\pi}{2} \qquad (A.53)$$

where n is any integer, $n = 0, 1, 2, ...$

Therefore Eq. A.43 is satisfied by

$$\phi' = \sum_{n=0}^{\infty} A_n (e^{Ky} - e^{-Ky})\operatorname{Sin} Kx \qquad (A.54)$$

now

$$\frac{\partial\phi'}{\partial y} = \sum_{n=0}^{\infty} KA_n (e^{Ky} + e^{-Ky})\operatorname{Sin} Kx \qquad (A.55)$$

and for this to satisfy Eq. A.44, we must have when $y = \pm b$

$$- 2x = \sum_{n=0}^{\infty} KA_n \, (e^{Kb} + e^{-Kb}) \, \text{Sin } Kx$$

$$= \sum_{n=0}^{\infty} KB_n \, \text{Sin } Kx \text{ , say} \tag{A.56}$$

where

$$B_n = A_n \, (e^{Kb} + e^{-Kb}) \, . \tag{A.57}$$

Multiply both sides of Eq. A.56 by $\text{Sin } K'x \, dx$, where $K' = \dfrac{2n'+1}{a} \dfrac{\pi}{2}$ and integrate between the limits $\displaystyle\int_{-a}^{+a}$. Then, since $K = \dfrac{2n+1}{a} \dfrac{\pi}{2}$, we have

$$\int_{-a}^{+a} \text{Sin } Kx \, \text{Sin } K'x \, dx = 0 \quad \text{when } K \neq K' \tag{A.58}$$

and

$$- 2 \int_{-a}^{+a} x \, \text{Sin } Kx \, dx = KB_n \int_{-a}^{+a} \text{Sin}^2 Kx \, dx, \quad \text{when } K = K'. \tag{A.59}$$

Note

$$\int_{-a}^{+a} x \, \text{Sin } Kx \, dx = \frac{2(-1)^n}{K^2} \tag{A.60}$$

and

$$\int_{-a}^{+a} \text{Sin}^2 Kx \, dx = a \tag{A.61}$$

from which

$$KB_n a = \frac{-4\,(-1)^n}{K^2} \tag{A.62}$$

and

$$B_n = \frac{32 a^2 (-1)^{n+1}}{\pi^3 (2n+1)^3}$$
(A.63)

and therefore

$$\phi = xy - \frac{32 a^2}{\pi^3} \sum_{n=0}^{\infty} \frac{(-1)^n}{(2n+1)^3} \frac{e^{Ky} - e^{-Ky}}{e^{Kb} + e^{-Kb}} \, \text{Sin} \, Kx.$$
(A.64)

Having ϕ, the stresses and torsional rigidity D may be computed (the series converges rapidly) and we find the following values for the two aspect ratios shown

b/a	$\dfrac{D}{G\,(2a)^3\,(2b)}$	$\dfrac{\tau_{max}}{2\,G\alpha\,a}$
1	0.1406	0.675
10	0.312	1.000

p. 173 — Eq. 5.172

The boundary condition $\dfrac{\partial \beta}{\partial n} = 0$ may be proven by applying Green's Theorem to the relation

$$\iint_R \left[\frac{\partial}{\partial x}\left(\frac{\partial \beta}{\partial x}\right) + \frac{\partial}{\partial y}\left(\frac{\partial \beta}{\partial y}\right) \right] dx \, dy = 0 .$$
(A.65)